E
SOURCE
The Prentice Hall
ENGINEERING SOURCE

Introduction to Java™

Stephen J. Chapman
British Aerospace Australia

Prentice Hall
Upper Saddle River, NJ 07458

Library of Congress Cataloging-in-Publication Data

Chapman, Stephen J.
Introduction to Java / Stephen J. Chapman
p. cm. -- (Esource--the Prentice Hall Engineering Source)
ISBN 0-13-919416-9
1. Java (Computer program language) I. Title. II. Series.
QA76.73.J38C476 1999
005.7'2--dc21 99-17739
CIP

Editor-in-chief: **MARCIA HORTON**
Acquisitions editor: **ERIC SVENDSEN**
Director of production and manufacturing: **DAVID W. RICCARDI**
Managing editor: **EILEEN CLARK**
Editorial supervision/Page composition: **SCOTT DISANNO**
Cover director: **JAYNE CONTE**
Creative director: **AMY ROSEN**
Marketing manager: **DANNY HOYT**
Manufacturing buyer: **PAT BROWN**
Editorial assistant: **GRIFFIN CABLE**

Printed in the United States of America

10 9 8 7 6 5 4 3 2 1

ISBN 0-13-919416-9

Prentice-Hall International (UK) Limited, *London*
Prentice Hall of Australia Pty. Limited, *Sydney*
Prentice-Hall Canada, Inc., *Toronto*
Prentice-Hall Hispanoamericana, S.A., *Mexico*
Prentice-Hall of India Private Limited, *New Delhi*
Prentice-Hall of Japan, Inc., *Tokyo*
Prentice-Hall (Singapore) Pte., Ltd., *Singapore*
Editora Prentice-Hall do Brazil, Ltda., *Rio de Janeiro*

About ESource

The Challenge

Professors who teach the Introductory/First-Year Engineering course popular at most engineering schools have a unique challenge—teaching a course defined by a changing curriculum. The first-year engineering course is different from any other engineering course in that there is no real cannon that defines the course content. It is not like Engineering Mechanics or Circuit Theory where a consistent set of topics define the course. Instead, the introductory engineering course is most often defined by the creativity of professors and students, and the specific needs of a college or university each semester. Faculty involved in this course typically put extra effort into it, and it shows in the uniqueness of each course at each school.

Choosing a textbook can be a challenge for unique courses. Most freshmen require some sort of reference material to help them through their first semesters as a college student. But because faculty put such a strong mark on their course, they often have a difficult time finding the right mix of materials for their course and often have to go without a text, or with one that does not really fit. Conventional textbooks are far too static for the typical specialization of the first-year course. How do you find the perfect text for your course that will support your students educational needs, but give you the flexibility to maximize the potential of your course?

ESource—The Prentice Hall Engineering Source http://emissary.prenhall.com/esource

Prentice Hall created ESource—The Prentice Hall Engineering Source—to give professors the power to harness the full potential of their text and their freshman/first year engineering course. In today's technologically advanced world, why settle for a book that isn't perfect for your course? Why not have a book that has the exact blend of topics that you want to cover with your students?

More then just a collection of books, ESource is a unique publishing system revolving around the ESource website—http://emissary.prenhall.com/esource/. ESource enables you to put your stamp on your book just as you do your course. It lets you:

Control You choose exactly what chapters or sections are in your book and in what order they appear. Of course, you can choose the entire book if you'd like and stay with the author's original order.

Optimize Get the most from your book and your course. ESource lets you produce the optimal text for your student's needs.

Customize You can add your own material anywhere in your text's presentation, and your final product will arrive at your bookstore as a professionally formatted text.

ESource Content

All the content in ESource was written by educators specifically for freshman/first-year students. Authors tried to strike a balanced level of presentation, one that was not either too formulaic and trivial, but not focusing heavily on advanced topics that most introductory students will not encounter until later classes. A developmental editor reviewed the books and made sure that every text was written at the appropriate level, and that the books featured a balanced presentation. Because many professors do not have extensive time to cover these topics in the classroom, authors prepared each text with the idea that many students would use it for self-instruction and independent study. Students should be able to use this content to learn the software tool or subject on their own.

While authors had the freedom to write texts in a style appropriate to their particular subject, all followed certain guidelines created to promote the consistency a text needs. Namely, every chapter opens with a clear set of objectives to lead students into the chapter. Each chapter also contains practice problems that tests a student's skill at performing the tasks they have just learned. Chapters close with extra practice questions and a list of key terms for reference. Authors tried to focus on motivating applications that demonstrate how engineers work in the real world, and included these applications throughout the text in various chapter openers, examples, and problem material. Specific Engineering and Science **Application Boxes** are also located throughout the texts, and focus on a specific application and demonstrating its solution.

Because students often have an adjustment from high school to college, each book contains several **Professional Success Boxes** specifically designed to provide advice on college study skills. Each author has worked to provide students with tips and techniques that help a student better understand the material, and avoid common pitfalls or problems first-year students often have. In addition, this series contains an entire book titled ***Engineering Success*** by Peter Schiavone of the University of Alberta intended to expose students quickly to what it takes to be an engineering student.

Creating Your Book

Using ESource is simple. You preview the content either on-line or through examination copies of the books you can request on-line, from your PH sales rep, or by calling(1-800-526-0485). Create an on-line outline of the content you want in the order you want using ESource's simple interface. Either type or cut and paste your own material and insert it into the text flow. You can preview the overall organization of the text you've created at anytime (please note, since this preview is immediate, it comes unformatted.), then press another button and receive an order number for your own custom book . If you are not ready to order, do nothing—ESource will save your work. You can come back at any time and change, re-arrange, or add more material to your creation. You are in control. Once you're finished and you have an ISBN, give it to your bookstore and your book will arrive on their shelves six weeks after the order. Your custom desk copies with their instructor supplements will arrive at your address at the same time.

To learn more about this new system for creating the perfect textbook, go to **http://emissary.prenhall.com/esource/**. You can either go through the on-line walkthrough of how to create a book, or experiment yourself.

Community

ESource has two other areas designed to promote the exchange of information among the introductory engineering community, the Faculty and the Student Centers. Created and maintained with the help of Dale Calkins, an Associate Professor at the University of Washington, these areas contain a wealth of useful information and tools. You can preview outlines created by other schools and can see how others organize their courses. Read a monthly article discussing important topics in the curriculum. You can post your own material and share it with others, as well as use what others have posted in your own documents. Communicate with our authors about their books and make suggestions for improvement. Comment about your course and ask for information from others professors. Create an on-line syllabus using our custom syllabus builder. Browse Prentice Hall's catalog and order titles from your sales rep. Tell us new features that we need to add to the site to make it more useful.

Supplements

Adopters of ESource receive an instructor's CD that includes solutions as well as professor and student code for all the books in the series. This CD also contains approximately **350 Powerpoint Transparencies** created by Jack Leifer—of University South Carolina—Aiken. Professors can either follow these transparencies as pre-prepared lectures or use them as the basis for their own custom presentations. In addition, look to the web site to find materials from other schools that you can download and use in your own course.

Titles in the ESource Series

Introduction to Unix
0-13-095135-8
David I. Schwartz

Introduction to Maple
0-13-095133-1
David I. Schwartz

Introduction to Word
0-13-254764-3
David C. Kuncicky

Introduction to Excel
0-13-254749-X
David C. Kuncicky

Introduction to MathCAD
0-13-937493-0
Ronald W. Larsen

Introduction to AutoCAD, R. 14
0-13-011001-9
Mark Dix and Paul Riley

Introduction to the Internet, 2/e
0-13-011037-X
Scott D. James

Design Concepts for Engineers
0-13-081369-9
Mark N. Horenstein

Engineering Design—A Day in the Life of Four Engineers
0-13-085089-6
Mark N. Horenstein

Engineering Ethics
0-13-784224-4
Charles B. Fleddermann

Engineering Success
0-13-080859-8
Peter Schiavone

Mathematics Review
0-13-011501-0
Peter Schiavone

Introduction to C
0-13-011854-0
Delores M. Etter

Introduction to C++
0-13-011855-9
Delores M. Etter

Introduction to MATLAB
0-13-013149-0
Delores M. Etter with David C. Kuncicky

Introduction to FORTRAN 90
0-13-013146-6
Larry Nyhoff and Sanford Leestma

Introduction to Java
0-13-919416-9
Stephen J. Chapman

http://emissary.prenhall.com/esource/

About the Authors

No project could ever come to pass without a group of authors who have the vision and the courage to turn a stack of blank paper into a book. The authors in this series worked diligently to produce their books, provide the building blocks of the series.

Delores M. Etter is a Professor of Electrical and Computer Engineering at the University of Colorado. Dr. Etter was a faculty member at the University of New Mexico and also a Visiting Professor at Stanford University. Dr. Etter was responsible for the Freshman Engineering Program at the University of New Mexico and is active in the Integrated Teaching Laboratory at the University of Colorado. She was elected a Fellow of the Institute of Electrical and Electronic Engineers for her contributions to education and for her technical leadership in digital signal processing. IN addition to writing best-selling textbooks for engineering computing, Dr. Etter has also published research in the area of adaptive signal processing.

Sanford Leestma is a Professor of Mathematics and Computer Science at Calvin College, and received his Ph.D from New Mexico State University. He has been the long time co-author of successful textbooks on Fortran, Pascal, and data structures in Pascal. His current research interests are in the areas of algorithms and numerical computation.

Larry Nyhoff is a Professor of Mathematics and Computer Science at Calvin College. After doing bachelors work at Calvin, and Masters work at Michigan, he received a Ph.D. from Michigan State and also did graduate work in computer science at Western Michigan. Dr. Nyhoff has taught at Calvin for the past 34 years—mathematics at first and computer science for the past several years. He has co-authored several computer science textbooks since 1981 including titles on Fortran and C++, as well as a brand new title on Data Structures in C++.

Acknowledgments: We express our sincere appreciation to all who helped in the preparation of this module, especially our acquisitions editor Alan Apt, managing editor Laura Steele, development editor Sandra Chavez, and production editor Judy Winthrop. We also thank Larry Genalo for several examples and exercises and Erin Fulp for the Internet address application in Chapter 10. We appreciate the insightful review provided by Bart Childs. We thank our families—Shar, Jeff, Dawn, Rebecca, Megan, Sara, Greg, Julie, Joshua, Derek, Tom, Joan; Marge, Michelle, Sandy, Lori, Michael—for being patient and understanding. We thank God for allowing us to write this text.

Mark Dix began working with AutoCAD in 1985 as a programmer for CAD Support Associates, Inc. He helped design a system for creating estimates and bills of material directly from AutoCAD drawing databases for use in the automated conveyor industry. This system became the basis for systems still widely in use today. In 1986 he began collaborating with Paul Riley to create AutoCAD training materials, combining Riley's background in industrial design and training with Dix' s background in writing, curriculum development, and programming. Dix and Riley have created tutorial and teaching methods for every AutoCAD release since Version 2.5. Mr. Dix has a Master of Arts in Teaching from Cornell University and a Masters of Education from the University of Massachusetts. He is currently the Director of Dearborn Academy High School in Arlington, Massachusetts.

Paul Riley is an author, instructor, and designer specializing in graphics and design for multimedia. He is a founding partner of CAD Support Associates, a contract service and professional training organization for computer-aided design. His 15 years of business experience and 20 years of teaching experience are supported by degrees in education and computer science. Paul has taught AutoCAD at the University of Massachusetts at Lowell and is presently teaching AutoCAD at Mt. Ida College in

Newton, Massachusetts. He has developed a program, Computer-Aided Design for Professionals that is highly regarded by corporate clients and has been an ongoing success since 1982.

David I. Schwartz is a Lecturer at SUNY-Buffalo who teaches freshman and first-year engineering, and has a Ph.D from SUNY-Buffalo in Civil Engineering. Schwartz originally became interested in Civil engineering out of an interest in building grand structures, but has also pursued other academic interests including artificial intelligence and applied mathematics. He became interested in Unix and Maple through their application to his research, and eventually jumped at the chance to teach these subjects to students. He tries to teach his students to become incremental learners and encourages frequent practice to master a subject, and gain the maturity and confidence to tackle other subjects independently. In his spare time, Schwartz is an avid musician and plays drums in a variety of bands.

Acknowledgments: I would like to thank the entire School of Engineering and Applied Science at the State University of New York at Buffalo for the opportunity to teach not only my students, but myself as well; all my EAS140 students, without whom this book would not be possible—thanks for slugging through my lab packets; Andrea Au, Eric Svendsen, and Elizabeth Wood at Prentice Hall for advising and encouraging me as well as wading through my blizzard of e-mail; Linda and Tony for starting the whole thing in the first place; Rogil Camama, Linda Chattin, Stuart Chen, Jeffrey Chottiner, Roger Christian, Anthony Dalessio, Eugene DeMaitro, Dawn Halvorsen, Thomas Hill, Michael Lamanna, Nate "X" Patwardhan, Durvejai Sheobaran, "Able" Alan Sumlo, Ben Stein, Craig Sutton, Barbara Umiker, and Chester "JC" Zeshonski for making this book a reality; Ewa Arrasjid, "Corky" Brunskill, Bob Meyer, and Dave Yearke at "the Department Formerly Known as ECS" for all their friendship, advice, and respect; Jeff, Tony, Forrest, and Mike for the interviews; and, Michael Ryan and Warren Thomas for believing in me.

Ronald W. Larsen is an Associate Professor in Chemical Engineering at Montana State University, and received his Ph.D from the Pennsylvania State University. Larsen was initially attracted to engineering because he felt it was a serving profession, and because engineers are often called on to eliminate dull and routine tasks. He also enjoys the fact that engineering rewards creativity and presents constant challenges. Larsen feels that teaching large sections of students is one of the most challenging tasks he has ever encountered because it enhances the importance of effective communication. He has drawn on a two year experince teaching courses in Mongolia through an interpreter to improve his skills in the classroom. Larsen sees software as one of the changes that has the potential to radically alter the way engineers work, and his book *Introduction to MATHCAD* was written to help young engineers prepare to be productive in an ever-changing workplace.

Acknowledgments: To my students at Montana State University who have endured the rough drafts and typos, and who still allow me to experiment with their classes—my sincere thanks.

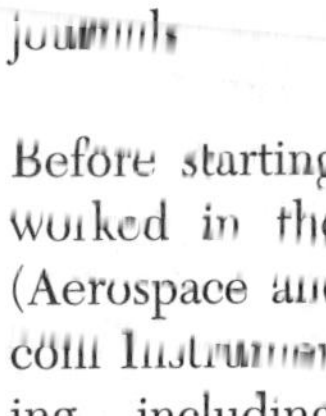

Peter Schiavone is a professor and student advisor in the Department of Mechanical Engineering at the University of Alberta. He received his Ph.D. from the University of Strathclyde, U.K. in 1988. He has authored several books in the area of study skills and academic success as well as numerous papers in scientific research journals.

Before starting his career in academia, Dr. Schiavone worked in the private sector for Smith's Industries (Aerospace and Defence Systems Company) and Marconi Instruments in several different areas of engineering including aerospace, systems and software engineering. During that time he developed an interest in engineering research and the applications of mathematics and the physical sciences to solving real-world engineering problems.

http://emissary.prenhall.com/esource/

His love for teaching brought him to the academic world. He founded the first Mathematics Resource Center at the University of Alberta: a unit designed specifically to teach high school students the necessary survival skills in mathematics and the physical sciences required for first-year engineering. This led to the Students' Union Gold Key award for outstanding contributions to the University and to the community at large.

Dr. Schiavone lectures regularly to freshman engineering students, high school teachers, and new professors on all aspects of engineering success, in particular, maximizing students' academic performance. He wrote the book *Engineering Success* in order to share with you the *secrets of success in engineering study*: the most effective, tried and tested methods used by the most successful engineering students.

Acknowledgments: I'd like to acknowledge the contributions of: Eric Svendsen, for his encouragement and support; Richard Felder for being such an inspiration; the many students who shared their experiences of first-year engineering—both good and bad; and finally, my wife Linda for her continued support and for giving me Conan.

Scott D. James is a staff lecturer at Kettering University (formerly GMI Engineering & Management Institute) in Flint, Michigan. He is currently pursuing a Ph.D. in Systems Engineering with an emphasis on software engineering and computer-integrated manufacturing. Scott decided on writing textbooks after he found a void in the books that were available. "I really wanted a book that showed how to do things in good detail but in a clear and concise way. Many of the books on the market are full of fluff and force you to dig out the really important facts." Scott decided on teaching as a profession after several years in the computer industry. "I thought that it was really important to know what it was like outside of academia. I wanted to provide students with classes that were up to date and provide the information that is really used and needed."

Acknowledgments: Scott would like to acknowledge his family for the time to work on the text and his students and peers at Kettering who offered helpful critique of the materials that eventually became the book.

David C. Kuncicky is a native Floridian. He earned his Baccalaureate in psychology, Master's in computer science, and Ph.D. in computer science from Florida State University. Dr. Kuncicky is the Director of Computing and Multimedia Services for the FAMU-FSU College of Engineering. He also serves as a faculty member in the Department of Electrical Engineering. He has taught computer science and computer engineering courses for the past 15 years. He has published research in the areas of intelligent hybrid systems and neural networks. He is actively involved in the education of computer and network system administrators and is a leader in the area of technology-based curriculum delivery.

Acknowledgments: Thanks to Steffie and Helen for putting up with my late nights and long weekends at the computer. Thanks also to the helpful and insightful technical reviews by the following people: Jerry Ralya, Kathy Kitto of Western Washington University, Avi Singhal of Arizona State University, and Thomas Hill of the State University of New York at Buffalo. I appreciate the patience of Eric Svendsen and Rose Kernan of Prentice Hall for gently guiding me through this project. Finally, thanks to Dean C.J. Chen for providing continued tutelage and support.

Mark Horenstein is an Associate Professor in the Electrical and Computer Engineering Department at Boston University. He received his Bachelors in Electrical Engineering in 1973 from Massachusetts Institute of Technology, his Masters in Electrical Engineering in 1975 from University of California at Berkeley, and his Ph.D. in Electrical Engineering in 1978 from Massachusetts Institute of Technology. Professor Horenstein's research interests are in applied electrostatics and electromagnetics as well as microelectronics, including sensors, instrumentation, and measurement. His research deals with the simulation, test, and measurement of electromagnetic fields. Some topics include electrostatics in manufacturing processes, electrostatic instrumentation, EOS/ESD control, and electromagnetic wave propagation.

Professor Horenstein designed and developed a class at Boston University, which he now teaches entitled Senior Design Project (ENG SC 466). In this course, the student gets real engineering design experience by working for a virtual company, created by Professor Horenstein, that does real projects for outside companies—almost like an apprenticeship. Once in "the company" (Xebec Technologies), the student is assigned to an engineering team of 3-4 persons. A series of potential customers are recruited, from which the team must accept an engineering project. The team must develop a working prototype deliverable engineering system that serves the need of the customer. More than one team may be assigned to the same project, in which case there is competition for the customer's business.

Acknowledgements: Several individuals contributed to the ideas and concepts presented in Design Principles for Engineers. The concept of the Peak Performance design competition, which forms a cornerstone of the book, originated with Professor James Bethune of Boston University. Professor Bethune has been instrumental in conceiving of and running Peak Performance each year and has been the inspiration behind many of the design concepts associated with it. He also provided helpful information on dimensions and tolerance. Several of the ideas presented in the book, particularly the topics on brainstorming and teamwork, were gleaned from a workshop on engineering design help bi-annually by Professor Charles Lovas of Southern Methodist University. The principles of estimation were derived in part from a freshman engineering problem posed by Professor Thomas Kincaid of Boston University.

I would like to thank my family, Roxanne, Rachel, and Arielle, for giving me the time and space to think about and write this book. I also appreciate Roxanne's inspiration and help in identifying examples of human/machine interfaces.

Dedicated to Roxanne, Rachel, and Arielle

Charles B. Fleddermann is a professor in the Department of Electrical and Computer Engineering at the University of New Mexico in Albuquerque, New Mexico. He is a third generation engineer—his grandfather was a civil engineer and father an aeronautical engineer—so "engineering was in my genetic makeup." The genesis of a book on engineering ethics was in the ABET requirement to incorporate ethics topics into the undergraduate engineering curriculum. "Our department decided to have a one-hour seminar course on engineering ethics, but there was no book suitable for such a course." Other texts were tried the first few times the course was offered, but none of them presented ethical theory, analysis, and problem solving in a readily accessible way. "I wanted to have a text which would be concise, yet would give the student the tools required to solve the ethical problems that they might encounter in their professional lives."

Stephen J. Chapman received a BS in Electrical Engineering from Louisiana State University (1975), an MSE in Electrical Engineering from the University of Central Florida (1979), and pursued further graduate studies at Rice University. Mr. Chapman is currently Manager of Technical Systems for British Aerospace Australia, in Melbourne, Australia. In this position, he provides technical direction and design authority for the work of younger engineers within the company. He is also continuing to teach at local universities on a part-time basis.

Mr. Chapman is a Senior Member of the Institute of Electrical and Electronic Engineers (and several of its component societies). He is also a member of the Association for Computing Machinery and the Institution of Engineers (Australia).

Reviewers

ESource benefited from a wealth of reviewers who on the series from its initial idea stage to its completion. Reviewers read manuscripts and contributed insightful comments that helped the authors write great books. We would like to thank everyone who helped us with this project.

Concept Document

Naeem Abdurrahman- University of Texas, Austin
Grant Baker- University of Alaska, Anchorage
Betty Barr- University of Houston
William Beckwith- Clemson University
Ramzi Bualuan- University of Notre Dame
Dale Calkins- University of Washington
Arthur Clausing- University of Illinois at Urbana-Champaign
John Glover- University of Houston
A.S. Hodel- Auburn University
Denise Jackson- University of Tennessee, Knoxville
Kathleen Kitto- Western Washington University
Terry Kohutek- Texas A&M University
Larry Richards- University of Virginia
Avi Singhal- Arizona State University
Joseph Wujek- University of California, Berkeley
Mandochehr Zoghi- University of Dayton

Books

Stephen Allan- Utah State University
Naeem Abdurrahman - University of Texas Austin
Anil Bajaj- Purdue University
Grant Baker - University of Alaska - Anchorage
Betty Barr - University of Houston
William Beckwith - Clemson University
Haym Benaroya- Rutgers University
Tom Bledsaw- ITT Technical Institute
Tom Bryson- University of Missouri, Rolla
Ramzi Bualuan - University of Notre Dame
Dan Budny- Purdue University
Dale Calkins - University of Washington
Arthur Clausing - University of Illinois
James Devine- University of South Florida
Patrick Fitzhorn - Colorado State University
Dale Elifrits- University of Missouri, Rolla
Frank Gerlitz - Washtenaw College
John Glover - University of Houston
John Graham - University of North Carolina-Charlotte
Malcom Heimer - Florida International University
A.S. Hodel - Auburn University
Vern Johnson- University of Arizona
Kathleen Kitto - Western Washington University
Robert Montgomery- Purdue University
Mark Nagurka- Marquette University
Ramarathnam Narasimhan- University of Miami
Larry Richards - University of Virginia
Marc H. Richman - Brown University
Avi Singhal-Arizona State University
Tim Sykes- Houston Community College
Thomas Hill- SUNY at Buffalo
Michael S. Wells - Tennessee Tech University
Joseph Wujek - University of California - Berkeley
Edward Young- University of South Carolina
Mandochehr Zoghi - University of Dayton

Contents

1

Introduction to Java

Java is a relatively new but powerful programming language that it is taking the world by storm. The language has enormous appeal for many reasons. One major reason is that it is largely *platform independent*, meaning that an application written for one computer is very likely to run unchanged on another computer.[1] Thus, an engineer can write a single application that will execute across all of a company's computers, whether they are PCs, Macs, or Unix workstations. This "write once, run anywhere" philosophy means that an organization is not locked into a single type of computer hardware.

A second advantage of Java is that it is *object oriented.* As we will see, object oriented programming languages make the design and maintenance of large programs easier, by encapsulating data and the methods for modifying that data into discrete units, called **objects**. Because objects interact with each other only through well-defined interfaces, unintended side effects can be minimized, and the objects can be re-used more easily in different programs.

Another advantage of Java is that the basic language is relatively *simple.* The Java language itself has a simpler syntax than C and C++ (on which it is based), making it easier

[1]There have been teething pains associated with achieving true platform independence, and the goal of complete platform independence has not been achieved yet. However, significant progress has been made, and if Microsoft's offerings are ignored, platform independence seems to be an achievable goal.

SECTIONS

OBJECTIVES

After reading this chapter, you should be able to:

- Identify the elements of Java
- Understand objects
- Know the difference between Applets and Applications
- Read a Java program
- Compile and execute a Java program

to master. Many of the trickiest and most error-prone portions of the C language (such as pointer manipulation) simply do not exist in Java, and other features are either greatly simplified or handled automatically. For example, memory allocation and deallocation is a major source of errors in C programs. In Java, memory allocation and deallocation happens automatically. The difficult part of Java is not the language itself—it is learning how to use the bewildering variety of classes that come in the Application Programming Interface (API) bundled with the language.

In addition, the standard Java language includes *device-independent graphics.* While graphical output can be created in other languages such as C and Fortran, the code required is not standard, differing from computer to computer and even from device to device within the same computer. For example, the code to print the graphics on a screen will be different from that to print the same graphics on a printer. In contrast, *Java has device independent graphics built directly into the language.* A program that generates a graph on one computer will also generate the same graph when executed on another computer, even if it is a different type and has different operating system. Finally, the Java language is *free.* The Java Development Kit may be downloaded for free from `http://java.sun.com`. This kit includes a Java compiler (`javac`), a Java run-time interpreter (`java`), a debugger (`jdb`), and all of the standard Java libraries. Fancier development environments may be purchased from IBM, Symantec, Microsoft, and many others vendors, but the basic language is free.

One significant disadvantage of Java is that it is a new language that is evolving rapidly. While the basic language has been pretty much fixed, the Java Application Programming Interface (API) has changed rapidly between Java Development Kit (JDK) versions 1.0, 1.1, and 1.2 (now known as Java 2). Features of programs written with older versions of the JDK are now considered obsolescent only months after they have been created. Hopefully, the core portion of the Java API will soon mature and stabilize so that programmers can work with a consistent environment. This book teaches the Java API as it appears in the JDK for Java 2.

1.1 ELEMENTS OF JAVA

Java is composed of three distinct elements:

1. The Java Programming Language
2. The Java Virtual Machine
3. The Java Application Programming Interface (API)

Java differs from other computer languages in that all Java programs are compiled to execute on a special computer known as the *Java Virtual Machine (Java VM).* The machine language of the Java VM is known as *bytecode.* All Java compilers produce bytecodes, which can be executed directly on a Java Virtual Machine. Since the processors in real computers such as PCs and Unix workstations are not Java VMs, they cannot execute bytecode directly. Instead, each type of computer has an interpreter (or a just-in-time compiler) that converts Java VM bytecodes into the machine language of the particular computer on-the-fly as a Java program is executed. The process of compiling and executing a Java program is shown in Figure 1.1. The Java program may be created using any text editor, and is stored in a file with the special file extension `.java`. The Java compiler compiles this program into bytecode for execution on the Java VM, and

Figure 1.1. A Java program is created using an editor and stored in a disk file with the file extent `.java`. Each program is compiled once using the java compiler, and producing a file of Java bytecodes with the file extent `.class`. This file is interpreted by the Java interpreter each time that the program is executed.

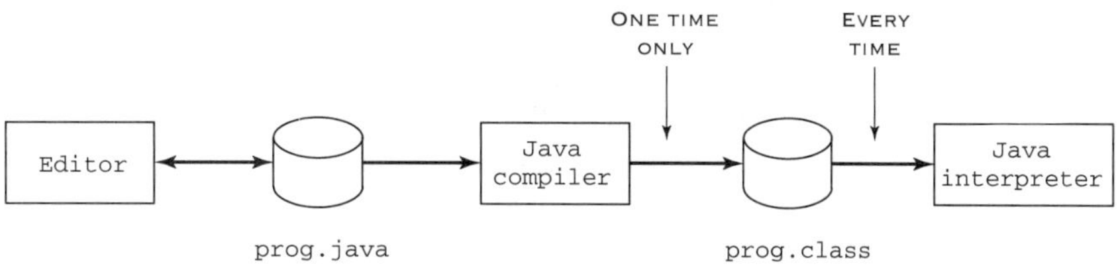

stores the bytecode in a file with the special file extension `.class`. This compilation only occurs once. When the program is executed on a computer, the Java interpreter translates the Java VM bytecode on-the-fly into instructions for the actual computer executing the program. This interpretation process happens every time that the Java program is executed. Note that the Java bytecode is independent of any particular computer hardware, so any computer with a Java interpreter can execute the compiled Java program, no matter what type of computer the program was compiled on.

The Java API is a large collection of ready-made software components that provide many useful capabilities. These components provide standard ways to read and write files, manipulate strings, build Graphical User Interfaces, and perform many other essential functions. The components of the Java API are grouped into libraries (called *packages*) of related components. A programmer can save an enormous amount of time by using the objects in these standard packages to perform tasks instead of trying to "reinvent the wheel" each time that he or she writes a program. The components in these packages are standard across all implementations of Java, so a program that uses them to implement some function will run properly on any computer system that implements Java. In addition, the components are already debugged, so using them reduces the total effort required to write and debug a program.

We will concentrate on the study of the Java language itself in the first seven chapters of this book. The remainder of the book will concentrate on how to use selected contents of the Java API.

1.2 AN INTRODUCTION TO OBJECT-ORIENTED PROGRAMMING

This section provides an introduction to the basic concepts of object-oriented programming. It is intended for individuals who have had prior experience with procedural programming languages such as C, Fortran, or Pascal. *Novice programmers may skip this material with no loss of continuity,* and refer back to it once Chapter 7 is reached.

Object-oriented programming (OOP) is the process of programming by modeling objects in software. The principal features of OOP are described in the following sections.

Objects

The physical world is full of objects: cars, pencils, trees, and so on. *Object-oriented programming* is the process of modeling the properties and behavior of real objects in software.

Any real object can be characterized by its *properties* and its *behavior*. For example, a car can be modeled as an object. A car has certain properties (color, speed, direction, fuel consumption) and certain behaviors (starting, stopping, turning and so on).

In the software world, an **object** is a software component whose structure is like that of objects in the real world. Each object consists of a combination of data (called **properties**) and behaviors (called **methods**). The properties are variables describing the essential characteristics of the object, while the methods describe how the object behaves and how the properties of the object can be modified. Thus, an object is a software bundle of variables and related methods.

A software object is often represented as shown in Figure 1.2. The object can be thought of as a cell, with a central nucleus of variables and an outer layer of methods that form an interface between the object's variables and the outside world. The nucleus of data is hidden from the outside world by the outer layer of methods. The object's variables are said to be *encapsulated* within the object, meaning that no code outside of the object can see or directly manipulate them. Any access to the object's data must be through calls to the object's methods.

The variables and methods in a Java object are formally known as **instance variables** and **instance methods** to distinguish them from class variables and class methods (described later in the Class Variables and Methods section).

Typically, encapsulation is used to hide the implementation details of an object from other objects in the program. If the other objects in the program cannot see the internal state of an object, they cannot introduce bugs by accidentally modifying the object's state. In addition, changes to the internal operation of the object will not affect the operation of the other objects in a program. As long as the interface to the outer world is unchanged, the implementation details of an object can change at any time without affecting other parts of the program.

Encapsulation provides two primary benefits to software developers:

- **Modularity:** An object can be written and maintained independently of the source code for other objects. Therefore, the object can be easily re-used and passed around in the system.
- **Information Hiding:** An object has a public interface that other objects can use to communicate with it. However, the object's instance variables are not directly accessible to other objects. Therefore, if the public interface is not changed, an object's variables and methods can be changed at any time without introducing side-effects in the other objects that depend on it.

Figure 1.2. An object may be represented as a nucleus of data (instance variables) surrounded and protected by methods, which implement the object's behavior and form an interface between the variables and the outside world.

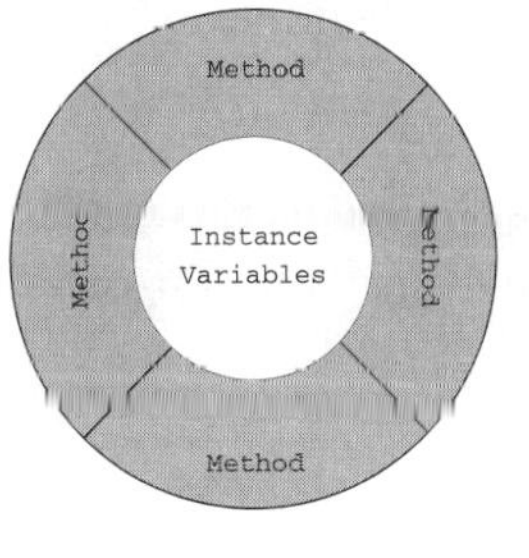

Sometimes, an object will make some of its instance variables **`public`** so that they can be accessed directly by other objects. This is occasionally done for reasons of efficiency if a variable has to be used very often, because invoking a method for each access will make the program be unacceptably slow. This is strictly speaking a violation of the object-oriented methodology, but it is a compromise that is sometimes made in the real world. Normally, an instance variable should *never* be made `public`.

GOOD PROGRAMMING PRACTICE

Always make instance variables private, so that they are hidden within an object. Such encapsulation makes your programs more modular and easier to modify.

Messages

Objects communicate by passing messages back and forth among themselves. If Object A wants Object B to perform some action for it, it sends a message to Object B requesting the object to execute one of its methods (see Figure 1.3). The message causes Object B to execute the specified method.

Each message has three components, which provide all the information necessary for the receiving object to perform the desired method:

1. The object to whom the message is addressed.
2. The name of the method to perform on that object.
3. Any parameters needed by the method.

An object's behavior is expressed through its methods, so message passing supports all possible interactions between objects.

Figure 1.3. If object `ObjA` wants object `ObjB` to do some work for it, it sends a message to that object. The message contains three parts: the name of the object to which it is addressed, the name of the method within the object that will do the work, and required parameters. Note that the names of the object and method are separated by a period.

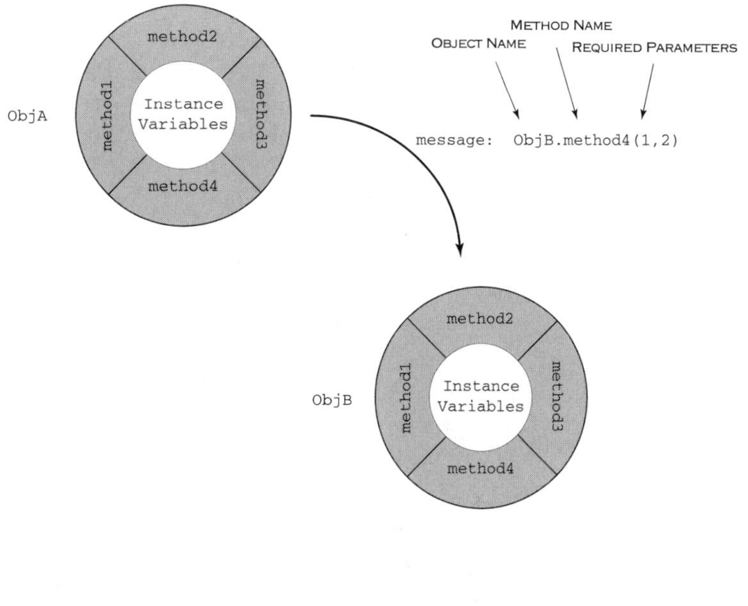

Figure 1.4. Many objects can be instantiated from a single class. In this example, three objects `a`, `b`, and `c` have been instantiated from class `Complex`.

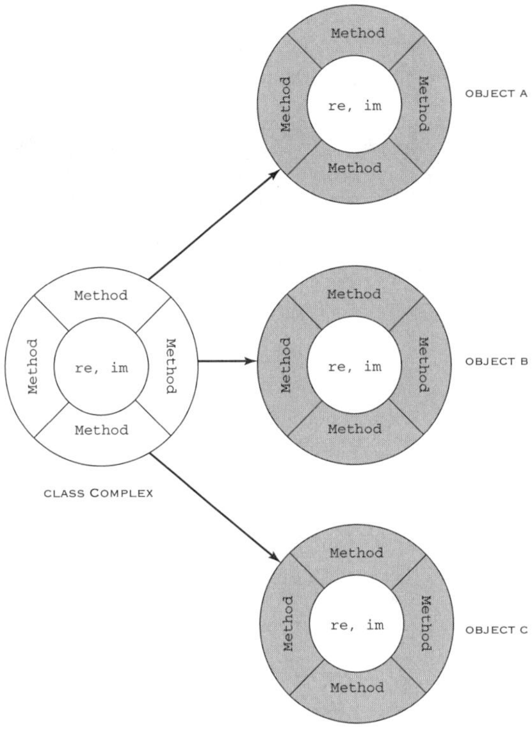

Note that objects don't need to be in the same process or even on the same computer to send and receive messages to each other. As long as a path to transmit messages exists, the objects can interact. This characteristic makes object-oriented programs highly suited to client-server applications, in which the object sending the message resides on a different computer than the object performing the action.

Classes

Classes are the software blueprints from which objects are made. A class is a software construct that specifies the number and type of instance variables to be included in an object, and the instance methods that will be applied to the object. Each component of a class is known as a **member**. The two types of members are **fields**, which specify the data types defined by the class, and **methods**, which specify the operations on these fields. For example, suppose that we wish to create an object to represent a complex number. Such an object would have two instance variables, one for the real part of the number (`re`) and one for the imaginary part of the number (`im`). In addition, it would have methods describing how to add, subtract, multiply, divide, etc., with complex numbers. To create such objects, we would write a class `Complex` that defines the required fields `re` and `im`, together with their associated methods.

Note that a class is a *blueprint* for an object, not an object itself. The class describes what an object will look and behave like once it is created. Each object is created or *instantiated* in memory from the blueprint provided by a class, and many different objects can be instantiated from the same class. For example, Figure 1.4 shows a class `Complex`, and three objects `a`, `b`, and `c` created from that class. Each of the three

Figure 1.5. An example class containing both instance variables and class variables. The instance variables `re` and `im` are different in objects `a` and `b`, while the class variable `count` is are common to both objects.

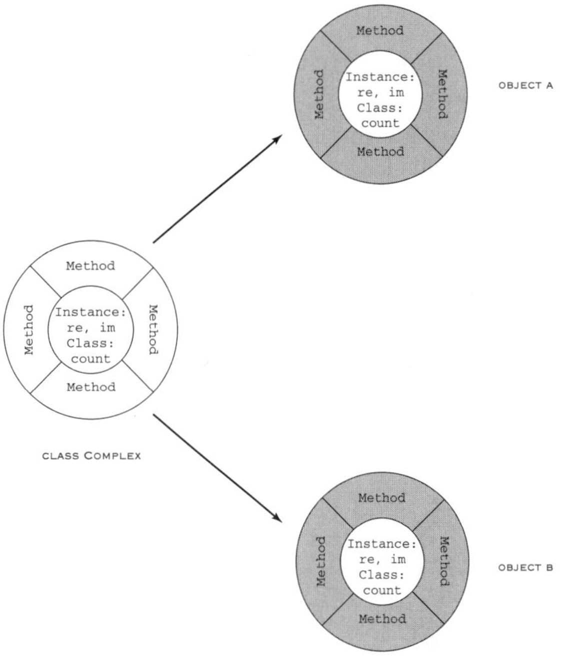

objects has its own copies of the instance variables `re` and `im`, while sharing a single set of methods to modify them.

Class Variables and Methods

As we described above, each object created from a class receives its own copies of all the instance variables defined in the class. The instance variables in each object are independent of the instance variables in all other objects.

In addition to instance variables, it is possible to define **class variables**. Class variables differ from instance variables in that *there is only one variable for all objects created from the class, and every object has access to it.* Class variables are effectively "common" to all of the objects created from the class in which they are defined. They are created when an object is first instantiated from a class, and remain in existence until the program finishes executing. This idea is illustrated in Figure 1.5, which shows a new version of the `Complex` class containing two instance variables (`re` and `im`) and one class variable (`count`). The instance variables `re` and `im` will contain the real and imaginary part of the complex number stored in the object, while the class variable `count` might contain the number of objects instantiated from this class. Every object instantiated from this class will contain a unique copy of the variables `re` and `im`, but all the objects will use a single copy of the variable `count`.

Class variables are typically used to keep track of data that is common across all instances of a class. For example, `count` could be used to count the number of objects created using class `Complex`. Each time that a new `Complex` object is created, the value stored in `count` would be increased by one, and that new value will be available to every `Complex` object.

It is also possible to define **class methods** (also known as static methods). Class methods are methods that exist independently of any objects defined from the class. These methods can access and modify class variables, but they cannot access instance variables or invoke instance methods.

Java API Packages

Groups of related Java classes are usually collected together into special libraries called *packages.* The Java API includes many packages implementing important features of the language. A few of the more important packages in the standard Java environment are summarized in Table 1-1. Learning to use the classes and methods implemented in these packages is the largest part of learning to program in Java.

TABLE 1-1 Some Important Java API Packages

JAVA API PACKAGE	DESCRIPTION
`java.awt`	**The Abstract Windowing Toolkit Package** This package contains many of the classes and interfaces required to support old-style Graphical User Interfaces. Portions of this package are also used with the new "Swing" Graphical User Interfaces.
`java.beans`	**The Java Beans Package** This package contains classes that enable programmers to create reusable software components.
`java.io`	**The Java Input / Output Package** This package contains classes that allow a program to input and output data.
`java.lang`	**The Java Language Package** This package contains the basic classes and interfaces required by most Java programs.
`java.net`	**The Java Networking Package** This package contains classes that allow a program to communicate via networks.
`javax.swing`	**The Swing Package** This package contains many of the classes and interfaces required to support the new "Swing" Graphical User Interfaces.
`java.text`	**The Java Text Package** This package contains classes and interfaces that allow a program to manipulate numbers, dates, characters, and strings.
`java.util`	**The Java Utility Package** This package contains utility classes and interfaces: data and time manipulations, random number generation, etc.

1.3 APPLETS VERSUS APPLICATIONS

Java supports two different types of programs, *applets* and *applications.* An applet is a special type of program that runs within a World Wide Web browser when an HTML document containing the applet is loaded into the browser. Applets have a graphical user interface that must follow strict rules to ensure proper integration with the browser. Applets tend to be small, so that they can be downloaded over the Internet in a small amount of time.

In contrast, applications are complete stand-alone programs designed to be loaded and executed independently within your computer. They can have either command line interfaces or graphical user interfaces, depending on the application's design. Applications are the sort of programs that are traditionally used for engineering calculations, so all of the examples in the first portion of this book are applications. We will introduce applets in Chapter 9, and also show how to design a single program that can run either as an application or as an applet.

1.4 A FIRST JAVA PROGRAM

We will begin our study of Java with an old programming tradition: a program that does nothing more than print out "Hello, World!". A program of this sort is very simple, but it illustrates many important features that we will see in more complex programs. We will analyze this program, and show how to compile and execute it in the Java development environment.

Every Java program must contain at least one class, and that class must contain at least one method. When the program is executed, the class will be used to create an object that executes the application. The execution of a Java application always begins with a method named `main` in the principal class of the program.

The "Hello, World" application is shown in Figure 1.6. Note that it contains one class (`HelloWorld`) and one method within the class (`main`). This is the simplest possible Java program.

```
/*
  This program prints out "Hello, World!" on the
  standard output and quits. It defines a class
  "HelloWorld", and a "main" method within that
  class.
*/
public class HelloWorld {
  // Define the main method
  public static void main(String[] args) {
    System.out.println("Hello World!"); //Print line
  }
}
```

Figure 1.6. The "Hello, world!" program

The first six lines of this program are *comments.* A comment is a note written by the programmer to explain what a portion of a program is doing. Comments are extremely important for understanding the purpose of a program. Every program should begin with comments describing the purpose of the program, and should include comments explaining how the various portions of the program function. The Java compiler completely ignores comments-they are for the benefit of humans looking at the code.

GOOD PROGRAMMING PRACTICE

Always begin every program with comments describing the purpose of the program. Use comments liberally throughout the program to explain how each portion of the code works.

This program illustrates two of the three types of Java comments. Multi-line comments may be created by beginning the comment with the symbol `/*` and ending the comment with the symbol `*/`. All of the text between the two symbols is a comment.

Single-line comments begin with the symbol // and continue to the end of the line. Thus the entire block of text

```
/*
  This program prints out "Hello, World!" on the
  standard output and quits. It defines a class
  "HelloWorld", and a "main" method within that
  class.
*/
```

is a comment, and the words `// Define the main method` and `//Print line` are additional comments. The third form of Java comment, a special version used with the Java documentation system, will not be covered in this text.

Line 7 is the beginning of the definition of class `HelloWorld`. At the end of line 7, the **left brace** (`{`) begins the **body** of the class definition. The corresponding **right brace** (`}`) on line 12 ends the class definition. By convention, the left brace opening the body is always included at the end of the line declaring the class's name, and the right brace is placed on a line by itself indented at the same level as the `class` statement.

As we mentioned previously, a class can contain data (variables) and methods. This particular class does not define any variables, but it does define the method `main`. The line `public static void main(String[] args)` declares the start of the `main` method. The keyword `public` means that the method can be invoked by any caller. It must always be present in the `main` method. The keyword `static` means that this is method is a class method instead of an instance method. The keyword `void` means that the method does not return a result when it finishes executing. The method's parameter list `(String[] args)` contains any command-line arguments passed to the program when it starts to execute. All of these features will be discussed in detail in later chapters.

The body of the method begins with the left brace (`{`) at the end of line 9 and ends with the corresponding right brace (`}`) on line 11. By convention, the left brace opening the method body is always included at the end of the line declaring the method's name, and the right brace is placed on a line by itself indented at the same level as the `method's` declaration.

The only executable statement in this method is on line 10. The statement `System.out.println("Hello World!")` invokes method `println` on variable `out` in the `System` class. The class variable `System.out` represents the standard output device for the computer on which the program is executing, so invoking the `println` method prints the words `"Hello World!"` on the computer's standard output device.

Notice that the Java statement on line 10 ends in a semicolon. Every Java statement must end with a semicolon. A statement can occupy as many lines as desired, or several statements can fit on a single line. In either case, the compiler knows that the statement is complete when it sees the semicolon.

1.5 COMPILING AND EXECUTING A JAVA PROGRAM

To compile the `HelloWorld` application, it must be placed in a file called `HelloWorld.java`. Note that *the name of the class being defined must be the same as the name of the file containing the class*, with the file extension `.java` added. The name of the file must be *exactly* the same as the name of the class, including any capitalization, or the Java compiler will report an error.

PROGRAMMING PITFALL

Be sure that the name of file containing a class is exactly the same as the name of the class being defined, with the addition of the file extension `.java`. It is an error for the name of the file not to agree with the name of the class.

This program can be compiled with the Java compiler `javac` by typing the command "`javac HelloWorld.java`" at the command prompt:[2]

```
D:\book\java\chap1>javac HelloWorld.java
```

Note that the command `javac` is followed by the name of the file, including the file extension. If there were an error in `HelloWorld`, the compiler would list the errors after this command is entered. If there are no errors, then the compiler will compile the class into bytecode, and place the bytecode into a file called `HelloWorld.class`.

Once the program has been compiled, it may be executed using the Java interpreter or another Java Virtual Machine. This is done by invoking the Java interpreter together with the *name of the class* to execute:

```
D:\book\java\chap1>java HelloWorld
Hello World!
```

As you can see, the program prints out the words "Hello, World!".

Note that the Java interpreter expects the name of the class to execute, *not* the name of the file containing that class. If the filename is used in the command, an error results.

```
D:\book\java\chap1>java HelloWorld.class
Can't find class HelloWorld\class
```

PROGRAMMING PITFALLS

Always specify the *file name* (not the class name) when using the Java compiler, and specify the *class name* (not the file name) when executing the compiled program.

When a program is executed, the Java runtime system first invokes a *class loader,* which loads the bytecodes for all the required classes from disk. Once the bytecodes are loaded, a *bytecode verifier* confirms that all bytecodes are valid, and that they do not violate Java's security restrictions. After the bytecodes are verified, they are passed to the Java interpreter or just-in-time compiler for execution. All three of these steps occur when the user types the `java` command.

[2]Note that the command prompt may vary from computer to computer. Usually, the command prompt will begin with the drive letter followed by a colon and a blackslash and then the directory location and a greater-than sign. Possible PC command prompts include `C:\java>` and `D:\compilers\java>`. Unix command prompts depend on the type of shell that a user is running.

SUMMARY

- Java is ideally platform independent, meaning that programs written on one type of computer will run unchanged on another type of computer. This ideal has not yet been achieved, but progress is being made towards it.
- Java is object oriented.
- An object is a self-contained software component that consists of properties (variables) and methods.
- Objects communicate with each other via messages. An object uses a message to request another object to perform a task for it.
- Classes are the software blueprints from which objects are made. When an object is instantiated from a class, a separate copy of each instance variable is created for the object. All objects derived from a given class share a single copy of each class variable.
- Groups of related Java classes are usually collected together into special libraries called packages.
- An applet is a special type of program that runs within a World Wide Web browser when an HTML document containing the applet is loaded into the browser.
- An application is a complete stand-alone program designed to be loaded and executed independently within a computer.
- Every Java application must contain a `main` method within its principal class. Program execution always starts in the `main` method.
- A comment that begins with `//` is a single-line comment.
- A comment that begins with `/*` and ends with `*/` may span multiple lines.
- The body of a class is enclosed in braces (`{}`).
- A application's class name is used as a part of its file name (with the file extension `.java`).
- A Java statement is always terminated by a semicolon. It may stretch over multiple lines, if necessary.
- The Java compiler expects a file name as an argument, not a class name.
- The Java runtime expects a class name as an argument, not a file name.

KEY TERMS

applet
application
bytecode
class method
class variable
comments
device independent graphics
fields
information hiding
instance method
instance variable
instantiated
Java Virtual Machine (Java VM)
members of a class
method
modularity
object
object oriented programming
package
platform independent
properties
`System.out.println` method

Problems

1. Compile and execute the `HelloWorld` application on your computer. Are the results the same as shown in this chapter?
2. Delete the final } in the file `HelloWorld.java` and attempt to compile the program. What happens?

2

Basic Elements of Java

2.1 INTRODUCTION

The core of the Java language is relatively simple, but the Java API is extremely large and complex. In the next three chapters, we will be concentrating on the fundamental core of the Java language, while postponing the complications of the Java API until later chapters. By the end of Chapter 4, you will be able to write Java programs that perform complex calculations, including branches, loops, and disk input and output.

This chapter introduces the very basic elements of the Java language, such as Java names, the types of variables in the language, and some types of operations. By the end of the chapter, you will be able to write simple, but functional, Java programs.

2.2 JAVA NAMES

As we saw in the previous chapter, Java classes, methods, and variables all have names. A name in Java may consist of any combination of letters, numbers, and underscore characters (_), but the first character of the name must be a letter. A Java name may be as short as one character or as long as desired (there is no maximum length). The following names are legal in Java:

SECTIONS

OBJECTIVES

After reading this chapter, you should be able to:

- Learn how to create Java constants and variables
- Understand how to use assignment statements
- Learn the types of operations supported in Java, and the order in which they are executed
- Learn about type conversion, including promotion of operands and casting
- Larn how to use standard Java mathematical methods
- Learn how to read from the standard input stream, and write to the standard output stream

TABLE 2-1 List of Reserved Java Keywords

RESERVED KEYWORDS				
abstract	boolean	break	byte	case
catch	char	class	continue	default
do	double	else	extends	false
final	finally	float	for	if
implements	import	instanceof	int	interface
long	native	new	null	package
private	protected	public	return	short
static	super	switch	synchronized	this
throw	throws	transient	true	try
void	volatile	while		
Reserved but Not Used by Java				
const	goto			

```
ThisIsATest
Hello
ABC
A1B2
a_12
```

and the following names are illegal:

```
1Day                    // Begins with a number
_toupper                // Begins with an underscore
```

By convention, Java class names always begin with a capital letter, and Java instance methods and variables begin with a lowercase letter. For example, we capitalized the first letter of the class name `HelloWorld` in Chapter 1, and did not capitalize the first letter of method `main`. If a method or variable name consists of more than one word, such as `toUpper`, the words are joined together and each word after the first begins with an uppercase letter.

The Java language includes a number of **keywords** that have special meanings, such as `if`, `else`, `while`, and so forth. These words are **reserved**, and no Java name can be the same as one of these keywords. A complete list of these illegal names is given in Table 2-1.

GOOD PROGRAMMING PRACTICES

1. Always capitalize the first letter of a class name, and use a lowercase first letter for method and variable names.
2. If a name consists of more than one word, the words are joined together and each succeeding word should begin with an uppercase letter.

2.3 CONSTANTS AND VARIABLES

A **constant** is a data item whose value does not change during program execution, and a **variable** is a data item whose value can change during program execution. There are four basic types of data in Java (known as **primitive data types**): integer, real, bool-

TABLE 2-2 Java Primitive Data Types

TYPE	BITS	RANGE	COMMENT
`boolean`	1	`true` or `false`	
`char`	16	`'\u0000'` to `'\uFFFF'`	ISO Unicode Character set
`byte`	8	−128 to +127	
`short`	16	−32,768 to +32,767	
`int`	32	−2,147,483,648 to +2,147,483,648	
`long`	64	−9,223,372,036,854,775,808 to 9,223,372,036,854,775,807	
`float`	32	−3.40292347E+38 to +3.40292347E+38	IEEE 754 single-precision floating point. Numbers are represented with about 6–7 decimal digits of precision.
`double`	64	−1.79769313486231570E+308 to +1.79768313486231570E+308	IEEE 754 double-precision floating point. Numbers are represented with about 15–16 decimal digits of precision.

ean, and character. Integers are data types that can represent integers, such as 0, 23, and −1000. Reals are data types that can represent numbers with decimal points, such as 3.14159. Booleans are logical values that are either `true` or `false`. Characters hold a single Unicode[1] character.

There are four versions of integer data types (`byte`, `short`, `int`, and `long`) and two versions of real data types (`float` and `double`), with differing ranges and precisions. They are summarized in Table 2-2. Note that unlike other languages, *the size and range of values supported by each data type is the same on any computer running Java.* This feature helps to guarantee that a Java program written on one computer will run properly on any other computer.

Java constants are written directly into a program. For example, in the line

```
x = y + 12;
```

the characters 12 represent an integer constant.

A variable is a data item of a primitive data type that can change value during the execution of a program. Java is a *strongly typed language*. This means that every variable must be declared with an explicit type before it is used. (We will learn how to declare variables in the next few sections.) When a Java compiler encounters a variable declaration, it reserves a location in memory for the variable and then references that memory location whenever the variable is used in the program.

It is a good idea to give your variables names that describe their contents. This mnemonic aid will help you or anyone else who may be working with your program to understand what it is doing. For example, if a variable in a program contains a currency exchange rate, it could be given the name `exchangeRate`.

GOOD PROGRAMMING PRACTICE

Use meaningful variable names whenever possible to make your programs clearer.

[1]Unicode is an international standard character set that uses 16 bits to code each character, allowing for more than 65,000 possible characters. The Unicode character set supports the alphabets of essentially every modern world language, including Arabic, English, Chinese, Hebrew, Japanese, and Russian.

It is also important to include a **data dictionary** in the body of any classes or methods that you write. A data dictionary is a set of comments that lists the definition *each variable* used in a program. The definition should include both descriptions of the contents of the item and the units in which it is measured. A data dictionary may seem unnecessary while the program is being written, but it is invaluable when you or another person have to go back and modify the program at a later time.

GOOD PROGRAMMING PRACTICE

Create a data dictionary for each program to make program maintenance easier.

2.3.1 Integer Constants and Variables

An integer constant is an integer value written directly into a Java program. It must be written without embedded commas, and may it be preceded by a + or – sign. By default, an integer constant is of type `int`, so it is restricted to be in the range –2,147,483,648 to +2,147,483,648. If a constant is to be of type `long`, it must be concluded with a letter `L`. The following examples show legal literal constants:

```
12
0
-123456
9999999999L                  // Type long
```

The following constants are illegal, and will produced compile-time errors:

```
1,024                       // Embedded comma
9999999999                  // Value too large for int
```

When a Java compiler encounters a constant, it places the value of the constant in a specific location in memory and then references that memory location whenever the constant is used in the program. If a program uses the same constant value in more than one location, each of these constants refers to the same location in memory. This optimization helps to reduce the size of Java programs.

An integer variable is declared in a **declaration statement**. The form of a declaration statement is the name of a primitive data type followed by one or more variable names. For example, the statements

```
int var1, var2;
short var3;
```

declare two integers of type `int` and one integer of type `short`. When a declaration statement is encountered, Java automatically creates a variable of the specified type and refers to it by the specified name.

When an integer is created, its value is undefined. An initial value can be assigned to the integer by including it in the declaration:

```
int var1 = 100;      // Creates var1 and initializes it to 100
```

2.3.2 Real Constants and Variables

Real or **floating-point** numbers are values stored in the computer in a kind of scientific notation. The bits used to store a real number are divided into two separate portions, a **mantissa** and an **exponent**. A single-precision real number (type `float`) occupies 32 bits of memory, divided into a 24-bit mantissa and an 8-bit exponent, as

Figure 2.1. Representation of a single-precision real number. The number is divided into two fields, a mantissa and an exponent.

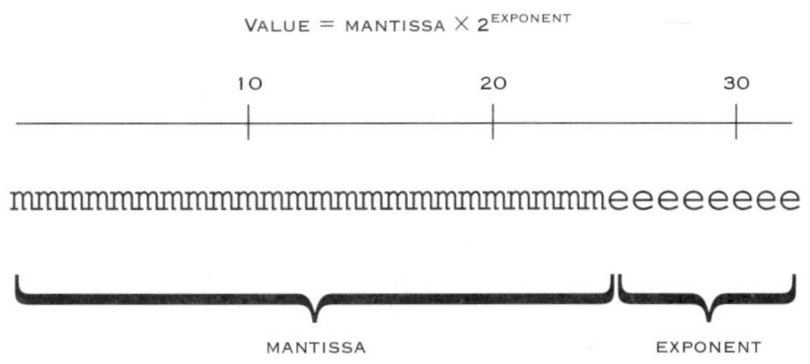

shown in Figure 2.1. The mantissa contains a number between –1.0 and 1.0, and the exponent contains the power of 2 required to scale the number to its actual value.

Real numbers are characterized by two quantities: **precision** and **range**. Precision is the number of significant digits that can be preserved in a number, and range is the difference between the largest and smallest numbers that can be represented. The precision of a real number depends on the number of bits in its mantissa, while the range of the number depends on the number of bits in its exponent. A 24-bit mantissa can represent approximately $\pm 2^{23}$ numbers, or about seven significant decimal digits, so the precision of single-precision real numbers (type `float`) is about seven significant digits. An 8-bit exponent can represent multipliers between 2^{-128} and 2^{127}, so the range of single-precision real numbers is from about 10^{-38} to 10^{38}. Note that the single-precision real data type can represent numbers much larger or much smaller than integers can, but only with seven significant digits of precision.

Similarly, a double-precision real number (type `double`) occupies 64-bits of computer memory, divided into a 53-bit mantissa and an 11-bit exponent. A 53-bit mantissa can represent approximately 15 to 16 decimal digits, so the precision of double-precision real numbers is about 15 significant digits. An 11-bit exponent can represent multipliers between 2^{-1024} and 2^{1023}, so the range of single-precision real numbers is from about 10^{-308} to 10^{308}.

When a value with more than seven digits of precision is stored in a single-precision real variable, *only the most significant seven bits of the number will be preserved*. The remaining information will be lost forever. For example, if the value 12345678.0 is stored in a `float` variable, it will be rounded off to 12345680.0. This difference between the original value and the number stored in the computer is known as **round-off error**. It is important to select a floating-point data type with enough precision to preserve the information needed to solve a particular problem.

A real constant is a literal defining a floating-point constant. It can be distinguished from an integer constant because it contains a decimal point and/or an exponent. If the constant is positive, it may be written either with or without a + sign. No commas may be embedded within a real constant. By default, a real constant is of type `double`, so it is restricted to being between –1.79769313486231570E+308 and +1.79769313486231570E+308.

A real constant *must* have either a decimal point or an exponent, and it may have both. If used, the exponent consists of the letter `E` or `e` followed by a positive or negative integer that specifies the power of ten used when the number is written in scientific notation.

The type of a real constant may be specified by appending either the letter `F` for `float` or the letter `D` for `double`. If there is no appended letter, the constant is of type `double`. The following examples show legal real constants:

```
12.               // Type double
12E2              // Type double
12.0e2            // Type double
3.14159F          // Type float
```

The following constants are not legal real constants:

```
1.2e108F          // Too large for type float
1,234.0           // Embedded comma
1234              // An int constant, not real
```

A real variable is declared with a `float` or `double` declaration statement. For example, the statements

```
float pi = 3.14159F;
double x;
```

declare and initialize a single-precision real variable `pi` and declare a double-precision real variable `x`. The value of variable `x` is undefined.

2.3.3 Boolean Constants and Variables

The `boolean` data type contains one of only two possible values: `true` or `false`. A `boolean` constant can only have one of those two values. Thus, the following are valid `boolean` constants:

```
true
false
```

Note that the words `true` and `false` are reserved. That is, they can only be used as boolean constants. No variable, method, or class may use these names. Boolean constants are rarely used, but boolean expressions and variables are commonly used to control program execution, as we will see in Chapters 3 and 4.

A `boolean` variable is a variable containing a value of the `boolean` data type. It is declared in with a `boolean` declaration statement. For example, the statement

```
boolean test = false;
```

declares a `boolean` variable test, and automatically initializes it to a value of `false`.

2.3.4 Character Constants and Variables

All Java characters and strings use the **Unicode character set**. Unicode is a special coding system in which each character is stored in 16 bits of memory. Since 16 bits are used to represent a character there can be 65,536 possible characters. Unicode assigns a unique number to each character in almost every alphabet used on Earth, including the ideograms used in oriental languages, such as Chinese and Japanese. This support makes it possible to write Java programs that work with any language.

A character constant is a literal representing a *single* Unicode character. The literal is written between single quotes, such as [illegible] or [illegible]. Some important characters are not printable characters, but instead perform control functions. Examples include the carriage return (CR) character, which moves the cursor back to the left-hand end of a line, the line feed (LF) character, which moves the cursor down to the next line, and the tab character, which moves the cursor right by one tab stop. These characters can be represented by special **escape sequences**, as shown in Table 2-3.

A character variable is a variable containing a value of the character data type. It is declared in with a `char` declaration statement. For example, the statements

```
char ch1 = 'A';
char ch2;
```

declare and initialize character variable `ch1`, and declare character variable `ch2`.

TABLE 2-3 Table of Common Escape Sequences

SEQUENCE	COMMENT
`\n`	Newline. Used to position the cursor at the beginning of the next line
`\t`	Horizontal tab. Used to move cursor to next tab stop
`\r`	Carriage return. Used to position the cursor at the beginning of the current line, but not advance to the next line.
`\\`	Backslash. Used to represent the backslash character.
`\'`	Single quote. Used to represent the single quote.
`\"`	Double quote. Used to represent the double quote.
`\u####`	Unicode character specified by sequence number. Used to specify any Unicode character constant. The #### is the hexadecimal representation of the character's sequence number.

2.3.5 Strings

Strings are groups of one or more characters linked together. A string constant is defined in Java by placing the desired characters between double quotes. For example, the following expressions are all valid strings:

```
"This is a string!"
"Line1\nLine2"
"A"
```

Note that escape sequences may be embedded into strings as in the second example above.

A double quote may not appear in the middle of a string, since the double quote character will be interpreted as the end of the string, producing a compile-time error. If a double quote is needed in a string, use the escape sequence `\"` to represent it. For example, the statement

```
System.out.println("She said \"Hello\".");
```

will print out the string

```
She said "Hello".
```

A Java string is fundamentally different from a Java character. A Java character is a *primitive data type*, while a Java string is an *object*. We will learn much more much about strings in Chapter 5, but meanwhile we will use string constants in many input/output (I/O) statements.

2.3.6 Keeping Constants Consistent in a Program

It is important to always keep your physical constants consistent throughout a program. For example, do not use the value 3.14 for π at one point in a program and 3.141593 at another point in the program. Also, you should always write your constants with as much precision as the data type you are using will accept. For example, since the `float` data type has seven significant digits of precision, π should be written as 3.141593, *not* as 3.14!

The best way to achieve consistency and precision throughout a program is to *assign a name to a constant and then to use that name to refer to the constant throughout the program.* If we assign the name `PI` to the constant 3.141593, then we can refer to `PI` by name throughout the program and be certain that we are getting the same value everywhere. Furthermore, assigning meaningful names to constants improves the overall readability of our programs, because a programmer can tell at a glance just what the constant represents.

By convention, the names of Java constants are written in capital letters, with underscore characters between words.

Named constants, or **final variables**, are created using the `final` keyword in a type declaration statement. This keyword means that the value assigned to a name is final and will never change. For example, the following program defines and uses a named constant `PI` containing the value of π to seven significant digits.

```
public class Constant {
   public static void main(String[] args) {

      // Declare constant
      final float PI = 3.14159F;

      // Print out 2*pi
      System.out.println("2*pi = " + 2*PI);
   }
}
```

When this program is executed, the results are:

```
D:\book\java\chap2>java Constant
2*pi = 6.28318
```

Any attempt to modify a final value will produce a compile-time error. For example, the following program attempts to modify the final variable `PI`, producing a compile-time error.

```
public class BadConstant {
   public static void main(String[] args) {

      final float PI = 3.14159F;
      PI = 3.0F;
   }
}
```

```
D:\book\java\rev2\chap2>javac BadConstant.java
BadConstant.java:5: Can't assign a value to a final variable: PI
      PI = 3.0F;
      ^
1 error
```

GOOD PROGRAMMING PRACTICE

Keep your physical constants consistent and precise throughout a program. To improve the consistency and under stand-ability of your code, assign a name to any important constants, and refer to them by that name in the program.

By convention, named constants are written in capital letters, with underscores used to separate the words. This style makes constants stand out from class names, method names, and instance variables. For example, a constant describing the maximum number of values that a program can process might be written as

```
final int MAX_VALUES = 1000;
```

GOOD PROGRAMMING PRACTICE

The names of constants in your program should be in all capital letters, with underscores separating the words.

PRACTICE!

This quiz provides a quick check to see if you have understood the concepts introduced in Sections 2.1 through 2.3. If you have trouble with the quiz, reread the sections, ask your instructor, or discuss the material with a fellow student. The answers to this quiz are found in the back of the book.

Questions 1 through 8 contain a list of valid and invalid constants. State whether or not each constant is valid. If the constant is valid, specify its type. If it is invalid, say why it is invalid.

1. `10.0`
2. `-100,000`
3. `123E-5`
4. `'T'`
5. `'''`
6. `3.14159`
7. `"Who are you?"`
8. `true`

Questions 9 through 11 contain two real constants each. Tell whether or not the two constants represent the same value within the computer:

9. `4650.; 4.65E+3`
10. `-12.71; -1.27E1`
11. `0.0001; 1.0e4`

Questions 12 through 15 contain a list of valid and invalid Java names. State whether or not each name is valid. If it is invalid, say why it is invalid. If it is valid, state what type of item the name represents (assuming that Java conventions are followed).

12. `isVector`
13. `MyNewApp`
14. `2ndChance`
15. `MIN_DISTANCE`

Are the following declarations correct or incorrect? If a statement is incorrect, state why it is invalid.

16. `int firstIndex = 20;`
17. `final short MAX_COUNT = 100000;`
18. `char test = "Y";`
19. Are the following statements legal or illegal? If they are legal, what is their result? If they are illegal, what is wrong with them?

```
int i, j;
final int k = 4;
i = k * k;
j = i / k;
k = i + j;
```

2.4 ASSIGNMENT STATEMENTS AND ARITHMETIC CALCULATIONS

Calculations are specified in Java with an **assignment statement** whose general form is

```
variable_name = expression;
```

The assignment operator calculates the value of the expression to the right of the equals sign and *assigns* that value to the variable named on the left of the equals sign. Note that the equals sign does not mean equality in the usual sense of the word. Instead, it means:

TABLE 2-4 Arithmetic Operators

TYPE	SYMBOL	ALGEBRAIC EXPRESSION	JAVA EXPRESSION
Addition	+	$a + b$	`a + b`
Subtraction	–	$a - b$	`a - b`
Multiplication	*	ab	`a * b`
Division	/	a / b or $\frac{a}{b}$ or $a \div b$	`a / b`
Modulus (Remainder)	%	$a \bmod b$	`a % b`

store the value of `expression` *into location* `variable_name`. For this reason, the equal sign is called the **assignment operator**. A statement such as

```
i = i + 10;
```

is complete nonsense in ordinary algebra, but makes perfect sense in Java. In Java, it means: Take the current value stored in variable `i`, add 10 to it, and store the result back into variable `i`.

The expression to the right of the assignment operator can be any valid combination of constants, variables, parentheses, and arithmetic or boolean operators. The standard arithmetic operators included in Java are given in Table 2-4.

Addition, subtraction, multiplication, and division will be familiar to all readers, but the **modulus** operation may be unfamiliar. The modulus operation calculates the *remainder* left after the division of a whole number has been performed. For example, $23 \div 5$ is 4 with a remainder of 3. Thus so,

```
25 % 5 = 3
```

The five arithmetic operators described in Table 2-4 are **binary operators**. This means that they should occur between and apply to two variables or constants. In addition, the + and – symbols can occur as **unary operators**, which means that they apply to one variable or constant, as shown:

```
+23
-a
```

The binary arithmetic operators are evaluated in order from left to right. Thus the expression $10 + 6 - 4$ will be evaluated in the order $10 + 6 = 16$ and then $16 - 4 = 12$. The unary arithmetic operators are evaluated from right to left. Therefore, the expression `- -z` will be evaluated as `-(-z)`.

2.4.1 Integer Arithmetic

Integer arithmetic is arithmetic involving only integer data. Integer arithmetic always produces an integer result. This is especially important to remember when an expression involves division, since there can be no fractional part in the answer. If the division of two integers is not itself an integer, the computer automatically discards the fractional part of the answer. This behavior can lead to surprising and unexpected answers. For example, integer arithmetic produces the following strange results:

$$\frac{3}{4} = 0 \qquad \frac{4}{4} = 1 \qquad \frac{5}{4} = 1 \qquad \frac{6}{4} = 1$$

$$\frac{7}{4} = 1 \qquad \frac{8}{4} = 2 \qquad \frac{9}{4} = 2$$

Because of this behavior, integers should *never* be used to calculate real-world quantities that vary continuously, such as distance, speed, time, etc. They should only be used for things that are intrinsically integer in nature, such as counters and indices.

PROGRAMMING PITFALLS

Beware of integer arithmetic. Integer division often gives unexpected results.

2.4.2 Floating-Point Arithmetic

Floating-point arithmetic is arithmetic involving floating-point constants and variables. Floating-point arithmetic always produces a floating-point result that is essentially what we would expect. For example, floating-point arithmetic produces the following results:

$$\frac{3.0}{4.0} = 0.75 \qquad \frac{4.0}{4.0} = 1.00 \qquad \frac{5.0}{4.0} = 1.25 \qquad \frac{6.0}{4.0} = 1.50$$

$$\frac{7.0}{4.0} = 1.75 \qquad \frac{8.0}{4.0} = 2.00 \qquad \frac{9.}{4.0} = 2.25 \qquad \frac{1.0}{3.0} = 0.3333333 \quad .$$

However, floating-point numbers do have peculiarities of their own. Because of the finite number of bits used to store a floating-point number, some numbers cannot be represented exactly. For example, the number 1/3 is equal to 0.33333333333. . . . But since the numbers stored in the computer have limited precision, the representation of 1/3 in the computer might be 0.3333333. As a result of this limitation in precision, some quantities that are theoretically equal will not be equal when evaluated by the computer. For example, on some computers

```
3.0 * (1.0 / 3.0) ≠ 1.0,
```

but

```
2.0 * (1.0 / 2.0) = 1.0.
```

Tests for equality must be performed very cautiously when working with real numbers. We will learn how to perform such tests safely in a later Chapter 3.

PROGRAMMING PITFALLS

Beware of floating-point arithmetic. Due to limited precision, two theoretically identical expressions often give slightly different results.

2.4.3 Hierarchy of Operations

Often, many arithmetic operations are combined into a single expression. For example, consider the equation for the distance traveled by an object subjected to a constant acceleration:

```
dist = d0 + v0 * time + 0.5 * acc * time * time;
```

There are four multiplications and two additions in this expression. In such an expression, it is important to know the order in which the operations are evaluated. If addition is evaluated before multiplication, this expression is equivalent to

```
dist = (d0 + v0) * (time + 0.5) * acc * time * time;
```

But if multiplication is evaluated before addition, this expression is equivalent to

```
dist = d0 + (v0 * time) + (0.5 * acc * time * time);
```

These two equations have different results, and we must be able to unambiguously distinguish between them.

To make the evaluation of expressions unambiguous, Java has established a series of rules governing the hierarchy, or order, in which operations are evaluated within an expression. The Java rules generally follow the normal rules of algebra. The order in which the arithmetic operations are evaluated is:

1. The contents of all parentheses are evaluated first, starting from the innermost parentheses and working outward.
2. All multiplications, divisions, and modulus operations are evaluated, working from left to right.
3. All additions and subtractions are evaluated, working from left to right.

Following these rules, we see that the second of our two possible interpretations is correct: The multiplications are performed before the additions.

Note that all of the above operations were applied in order from left to right across an expression. In Java, we say that the **associativity** of the operators is from left to right. Later we will see other operators whose associativity is from right to left.

EXAMPLE 2-1: Assume that the `double` variables `a`, `b`, `c`, `d`, `e`, `f`, and `g` have been initialized to the following values:

```
a = 3.,     b = 2.,     c = 5.,     d = 4
e = 10.,    f = 2.,     g = 3.,
```

Evaluate the following Java assignment statements:

a. `output = a*b+c%d+e/f+g;`
b. `output = a*(b+c)%d+(e/f)+g;`
c. `output = a*(b+c)%(d+e)/f+g;`

SOLUTION

As we can see, the order in which operations are performed has a major effect on the final result of an algebraic expression.

a.	Expression to evaluate:	`output = a*b+c%d+e/f+g;`
	Fill in numbers:	`output = 3.*2.+5.%4.+10./2.+3.`
	First, evaluate multiplication, division, and modulus operations from left to right:	`output = 6.+5. %4.+ 10. / 2.+3.`
		`output = 6.+ 1.+ 10. / 2.+ 3.`
		`output = 6.+ 1.+ 5.+ 3.`
	Now evaluate additions:	`output = 15;`
b.	Expression to evaluate:	`output = a*(b+c)%d+(e/f)+g;`
	Fill in numbers:	`output = 3.*(2.+5.)%4.+(10./2.)+3.;`
	First, evaluate parentheses:	`output = 3. * 7. % 4.+ 5.+ 3.;`
	Evaluate multiplication, division and modulus from left to right:	`output = 21. % 4.+ 5.+ 3.;`
		`output = 1. + 5.+ 3.;`
	Evaluate additions:	`output = 9.;`
c.	Expression to evaluate:	`output = a*(b+c)%(d+e)/f+g;`
	Fill in numbers:	`output = 3.*(2.+5.)%(4.+10.)/2.+3.;`
	First, evaluate parentheses:	`output = 3.*7. %14. /2.+3.;`
	Evaluate multiplication, division, and modulus from left to right:	`output = 21. %14. /2.+3.;`
		`output = 7. /2.+3.;`
		`output = 3.5 +3.;`
	Finally, evaluate addition:	`output = 7.5;`

It is important that every expression in a program be made as clear as possible. Any program of value must not only be written, but also must be maintained and modified when necessary. You should always ask yourself: "Will I easily understand this expression if I come back to it in six months? Can another programmer look at my code and easily understand what I am doing?" If there is any doubt in your mind, use extra parentheses in the expression to make it as clear as possible.

GOOD PROGRAMMING PRACTICE

Use parentheses as necessary to make your equations clear and easy to understand.

If parentheses are used within an expression, then the parentheses must be balanced. That is, there must be an equal number of open parentheses and close parentheses within the expression. It is an error to have more of one type than the other type. Errors of this sort are usually typographical, and they are caught by the Java compiler. For example, the expression

```
(2. + 4.) / 2.)
```

produces an error during compilation because of the mismatched parentheses.

2.4.4 Numeric Promotion of Operands

When an arithmetic operation is performed using two `double` numbers, its immediate result is of type `double`. Similarly, when an arithmetic operation is performed using two `int` numbers, the result is of type `int`. In general, arithmetic operations are only defined between numbers of the same type. For example, the addition of two `double` numbers is a valid operation, and the addition of two `int` numbers is a valid operation, but the addition of a `double` and an `int` is *not* a valid operation. This is true because real numbers and integers are stored in completely different forms in the computer.

What happens if an operation is between a real number and an integer? Expressions containing both real numbers and integers are called **mixed-mode expressions**, and arithmetic involving both real numbers and integers is called **mixed-mode arithmetic**. In the case of an operation between a `double` and an `int`, the `int` is converted by the computer into a `double`, and real arithmetic is used to perform the operation. This automatic conversion is known as **numeric promotion**.

The rules of numeric promotion are designed to preserve as much information as possible during each calculation. The following rules apply to numeric promotion during operations involving binary operators:

1. If either operand is of type `double`, the other operand is converted to `double`.
2. Otherwise, if either operand is of type `float`, the other operand is converted to `float`.
3. Otherwise, if either operand is of type `long`, the other operand is converted to `long`.
4. Otherwise, both operands are converted to type `int`.

The rules governing numeric promotion can be confusing to beginning programmers, and even experienced programmers may trip up on them from time to time. This is especially true when the expression involves division. Consider the following expressions and their results:

	EXPRESSION	RESULT
1.	`1 + 1/4`	`1`
2.	`1.0 + 1/4`	`1.0`
3.	`1 + 1.0/4`	`1.25`

Expression 1 contains only `int`s, so it is evaluated by integer arithmetic. In integer arithmetic, `1 / 4 = 0`, and `1 + 0 = 1`, so the final result is 1 (an `int`). Expression 2 is a mixed-mode expression containing both `double`s and `int`s. However, the first operation to be performed is division, since division comes before addition in the hierarchy of operations. The division is between `int`s, so the result is `1 / 4 = 0`. Next comes an addition between the `double 1.0` and the `int 0`, so Java promotes the `int 0` into a `double 0.0` and then performs the addition. The resulting number is `1.0` (a `double`). Expression 3 is also a mixed-mode expression containing both `double`s and `int`s. The first operation to be performed is a division between a `double` and an `int`, so Java promotes the 4 into a `double 4.0`, and then performs the division. The result is the `double` value `0.25`. The next operation to be performed is an addition between the `int 1` and the `double 0.25`, so Java promotes the integer `1` to a `double 1.0`, and then performs the addition. The resulting number is `1.25` (a `double`).

The following Java program demonstrates these results:

```
// This program illustrates numeric promotion
public class TestPromotion {

   // Define the main method
   public static void main(String[] args) {

      // Demonstrate numeric promotion.
      System.out.println(1 + 1/4);
      System.out.println(1.0 + 1/4);
      System.out.println(1 + 1.0/4);
   }
}
```

When this program is compiled and executed, the results are:

```
D:\book\java\chap2>javac TestPromotion.java
D:\book\java\chap2>java TestPromotion
1
1.0
1.25
```

To summarize,

1. A binary operation between numbers of different types is called a mixed-mode operation.
2. When a mixed-mode operation is encountered, Java promotes one or both of the operands according to the rules specified above, and then performs the operation.
3. The numeric promotion does not occur until two values of different types both appear in the *same* operation. Therefore, it is possible for a portion of an expression to be evaluated in integer arithmetic, followed by another portion evaluated in real arithmetic.

Mixed-mode arithmetic can be avoided by using the *cast operator*, as we will explain in the next section.

PROGRAMMING PITFALLS

Mixed-mode expressions are dangerous, because they are hard to understand and may produce misleading results. Avoid using them whenever possible.

2.4.5 Assignment Conversion and Casting Conversion

Automatic type conversion can also occur when the variable to which the expression is assigned is of a different type than the result of the expression. Such conversion is called **assignment conversion**. There are two possible cases for assignment conversion:

1. The result of an integer expression is assigned to a floating-point variable. This is an example of a **widening conversion**, since any value that can be represented by an integer can also be represented by a floating-point variable (albeit possibly with some loss of precision). *Widening assignment conversions are legal and happen automatically.* For example, the following code is legal and results in a value of 4.0 being stored in `y`.

```
int x = 4;
double y;
y = x;                    // Legal: y = 4.0
```

2. The result of a double expression is assigned to an integer variable. This is an example of a **narrowing conversion**, since the possible range of floating-point values is greater than the possible range of integer values. For example, the floating-point value 1.01E38 cannot be represented as an integer. *Narrowing assignment conversions are illegal and produce a compile-time error.* For example, the following code is illegal and produces a compile time error:

```
int x;
double y = 1.25;
x = y;                    // Illegal!
```

It is possible to explicitly convert or "cast" any numeric type into any other numeric type using a **cast operator**, regardless of whether the conversion involves widening or narrowing. A cast operator is created by placing the desired data type in parentheses before the expression to be converted. For example, the statements

```
int x;
double y = 1.25;
x = (int) y;                  // Legal: x = 1
```

convert the `double` value `1.25` to `int` and store the result in variable `x`. When a floating-point number is converted to an integer, the fractional part of the number is discarded. Thus, the value stored is the integer `1`. Note that we can assign a `double` value to an `int` variable (a narrowing conversion) if we use an explicit cast operator.

The cast operator can be used to make numeric conversions explicit, and thus, it avoids possible confusion associated with mixed-mode arithmetic. The following Java program demonstrates the use of the cast operator:

```
// This program illustrates the cast operator
public class TestCast {

    // Define the main method
    public static void main(String[] args) {

        double x = 3.99, y = 1.1e38;
        System.out.println("(int) x = " + (int) x);
        System.out.println("(int) y = " + (int) y);
    }
}
```

When this program is compiled and executed, the following are the results:

```
(int) x = 3
(int) y = 2147483647
```

Note that the value of `y` was too large to be represented as an `int`, so the cast operator converted it into the largest possible integer. In general, if the value being cast is out of range for the new data type, Java converts it value to the closest possible number in the new data type.

GOOD PROGRAMMING PRACTICE

Use cast operators to avoid mixed-mode expressions and make your intentions clear.

PRACTICE!

This quiz provides a quick check to see if you have understood the concepts introduced in Section 2.4. If you have trouble with the quiz, reread the section, ask your instructor, or discuss the material with a fellow student. The answers to this quiz are found in the back of the book.

1. In what order are the arithmetic operations evaluated if they appear within an arithmetic expression? How do parentheses modify this order?
2. Are the following expressions legal or illegal? If they are legal, what is their result? If they are illegal, what is wrong with them?
   a. `37 / 3`
   b. `37 + 17 / 3`
   c. `28 / 3 / 4`
   d. `28 / 3 / (double) 4`
   e. `(float) 28 / 3 / 4`
   f. `(28 / 3) % 4`
3. Evaluate the following expressions:
   a. `2 + 5 * 2 - 5`
   b. `(2 + 5) * (2 - 5)`
   c. `2 + (5 * 2) - 5`
   d. `(2 + 5) * 2 - 5`

Are the following sets of statements legal or illegal? If they are legal, state the result of the calculations. If they are illegal, state why.

5.
```
int x = 16, y = 3;
double result;
result = x + y/2.0;
```

6.
```
int x = 16, y = 3;
int result;
result = x + y/2.0;
```

2.5 ASSIGNMENT OPERATORS

Java includes several special assignment operators that combine an assignment and a binary operation in a single expression. These assignment operators are convenient shortcuts that reduce the typing required in a program. For example, the assignment statement

```
a = a + 5;
```

can be abbreviated using the addition assignment operator += as

```
a += 5;
```

The += operator adds the value of the variable on the left of the operator to the value to the expression on the right of the operator and stores the result in the variable to the left of the operator.

A similar abbreviation is possible for many other binary operators, including some we have not met yet. Table 2-5 contains a list of the arithmetic assignment operators corresponding to the operators we have seen so far.

TABLE 2-5 Arithmetic Assignment Operators

ASSIGNMENT OPERATOR	SAMPLE EXPRESSION	EXPANDED EXPRESSION	RESULT
Assume: `int a = 3,b = 11;`			
`+=`	`a += 3;`	`a = a + 3;`	6 stored in a
`-=`	`a -= 2;`	`a = a - 2;`	1 stored in a
`*=`	`a *= 4;`	`a = a * 4;`	12 stored in a
`/=`	`a /= 2;`	`a = a / 2;`	1 stored in a
`%=`	`b %= 3;`	`b = b % 3;`	2 stored in b

2.6 INCREMENT AND DECREMENT OPERATORS

Java also includes a unary **increment operator** (++) and a unary **decrement operator** (--). Increment and decrement operators *increase or decrease the value stored in an integer variable by one*. For example, suppose that an integer variable c is to be increased by one. Any of the following statements will perform this operation:

```
c = c + 1;
c += 1;
c++;
```

The increment and decrement operators can be confusing to novice programmers, because they *change the value of a variable without an equals sign* appearing in the expression. However, they are commonly used, because they are so much more compact than the alternative ways of performing the same function.

If the increment or decrement operator is placed *before* a variable, it is called a **preincrement** or **predecrement** operator. The preincrement and predecrement operators cause the variable to be incremented or decremented by one, and then the new value is used in the expression in which it appears. For example, suppose that the variables i and j are defined as shown below. After these statements are executed, the value of i will be 4 and the value of k will be **8**, because the value of i will be incremented *before* the addition is performed.

```
int i = 3, j = 4, k;
k = ++i + j;                    // k = 8
```

If the increment or decrement operator is placed *after* a variable, it is called a **postincrement** or **postdecrement** operator. The postincrement and postdecrement operators cause the old value of the variable to be used in the expression in which it appears, and then the variable to be decremented or decremented by one. For example, suppose that the variables i and j are defined as shown below. After these statements are executed, the value of i will be 4 and the value of k will be **7**, because the value of i will be incremented *after* the addition is performed.

```
int i = 3, j = 4, k;
k = i++ + j;                    // k = 7
```

The operation of these operators is summarized in Table 2-6.

Increment and decrement operators can get very confusing if they are combined in complex expressions, causing unexpected or hard-to-understand results. *Never use more than one of these operators on a single variable in a single expression*—if you do, the resulting expression will be very hard to understand. An example of the misuse of these operators is shown in the following program:

TABLE 2-6 The Increment and Decrement Operators

OPERATOR	SAMPLE EXPRESSION	EXPANDED EXPRESSION RESULT
preincrement	`++a`	Increment a by one and then use the new value of a in the expression in which a is located.
postincrement	`a++`	Use the current value of a in the expression in which a is located, and then increment a by one.
predecrement	`--a`	Decrement a by one and then use the new value of a in the expression in which a is located.
postdecrement	`a--`	Use the current value of a in the expression in which a is located, and then decrement a by one.

```
1  // This program illustrates the mis-use of pre-
2  // incrementing and decrementing
3  public class TestIncrement {
4      // Define the main method
5      public static void main(String[] args) {
6
7        int i = 4, k = 0;
8        k = i-- + 2 * i * ++i;
9        System.out.println( "i = " + i );
10       System.out.println( "k = " + k );
11
12       k = --i + 2 * i * i++;
13       System.out.println( "i = " + i );
14       System.out.println("k = " + k);
15     }
16 }
```

When this program executes, the results are as follows:

```
i = 4
k = 28
i = 4
k = 21
```

In line 8, the value of k was 28, while in line 12, the value of k was 21. The value of i was 4 at the beginning and the end of each statement. Can you determine why the two expressions produced these values of k?

Never write programs containing statements like the ones just shown. The programs will be very prone to errors and difficult to understand!

GOOD PROGRAMMING PRACTICE

Always keep expressions containing increment and decrement operators simple and easy to understand.

2.7 MATHEMATICAL METHODS

In mathematics, a *function* is an expression that accepts one or more input values and calculates a single result from them. Scientific and technical calculations usually require

Figure 2.2. When a mathematical method appears in a Java expression, the compiler generates a call to that method, and then uses the result returned by the method in the original expression. In this case, the compiler generates the call `Math.sin(1.2)`, and uses the result `0.9320390859672` in the original expression.

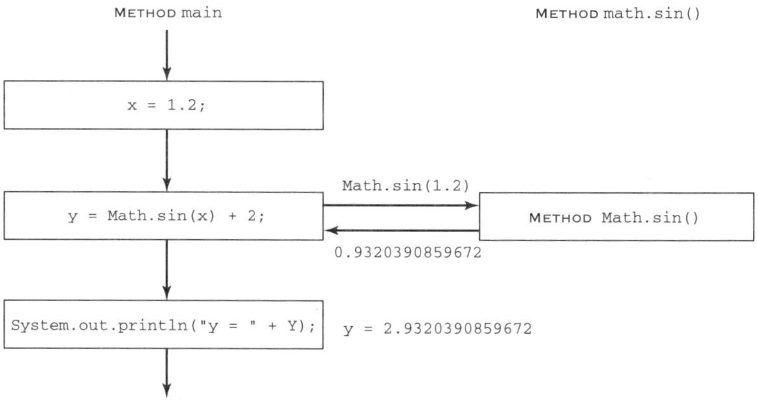

functions that are more complex than the simple addition, subtraction, multiplication, division, and modulus operations that we have discussed so far. Some of these functions are very common and are used in many different technical disciplines. Others are rarer and specific to a single problem or a small number of problems. Examples of very common functions are the trigonometric functions, logarithms, and square roots. Examples of rarer functions include the hyperbolic functions, Bessel functions, and so forth.

The Java language has mechanisms to support both the very common functions and the less common functions. Many of the common ones are implemented as methods of the `Math` class in the `java.lang` package. These methods are automatically available to any Java program. Less common functions are not included in the Java language, but they may be implemented as user-defined methods. User-defined methods are discussed in more detail in Chapter 6.

A Java mathematical method takes one or more input values and calculates a single output value from them. The input values to the method are known as **parameters**; they appear in parentheses immediately after the method name. The output of a mathematical method is a single number, which can be used together with other methods, constants, and variables in Java expressions. When a method name appears in a Java statement, the parameters of the method are passed to the method. The method calculates a result, which is used in place of the method name in the original expression. (See Figure 2.2.)

A list of the mathematical methods in class `Math` is given in Table 2-7. In addition to the methods shown in the table, the class defines two important constants, `Math.PI` (π) and `Math.E` (*e*, the base of the natural logarithms).

Mathematical methods are used by including them in an expression. For example, the method `Math.sin()` can be used to calculate the sine of a number as follows:

```
y = Math.sin(theta);
```

where `theta` is the parameter of the method `sin`. After this statement is executed, the variable `y` contains the sine of the value stored in variable `theta`. Note from Table 2-7 that the trigonometric methods expect their arguments to be in radians. If the variable `theta` is in degrees, then we must convert degrees to radians (180° = π radians) before computing the sine. This conversion can be done in the same statement as the sine calculation:

```
y = Math.sin(theta*(3.141593/180));
```

Alternatively, we could create a named constant containing the conversion factor and refer to that constant when the method is executed:

```
final double DEG_2_RAD = Math.PI / 180;
...
y = Math.sin(theta * DEG_2_RAD);
```

PROGRAMMING PITFALLS

The parameters for all trigonometric functions must be in units of *radians*. It is very common for novice programmers to use degrees by mistake.

2.7.1 Overloaded Methods

The type of parameter required by a method and the type of value returned by it are specified in Table 2-7 for the methods listed there. Some of these methods are **overloaded methods**, which means that more than one method exists with the same name, but with different types of parameters. For example, the absolute value method `Math.abs(x)` is an overloaded method. It really consists of four methods, one each for `double`, `float`, `int`, and `long` parameters. If `x` is a `double`, then the `double` form of this method will be invoked, and the returned value will be of type `double`. If `x` is an `int`, then the integer form of this method will be invoked, and the returned value will be of type `int`, and so forth.

2.7.2 Coercion of Arguments

We can see from Table 2-7 that a method like `Math.sqrt` is only defined for input parameters of type `double`. What happens if this function is invoked with an argument of another type? The answer is that *Java automatically converts arguments of an incorrect type into arguments of the type required by the method*. This process is known as the **coercion of arguments**.

For example, suppose that we execute the following statements:

```
int i = 16;
System.out.println( Math.sqrt(i) );
```

The method `Math.sqrt` is being invoked with an `int` argument, while it needs to have a `double` argument. Therefore, Java automatically converts the integer value 16 into a double-precision floating-point value 16.0 before it passes the value to the method. The resulting output value is a `double` value 4.0.

The behavior is just the same as if we had used an explicit cast to convert the input argument to a `double`:

```
int i = 16;
System.out.println( Math.sqrt( (double) i) );
```

TABLE 2-7 Mathematical Methods

METHOD NAME AND PARAMETERS	METHOD VALUE	PARAMETER	RESULT TYPE	COMMENTS
`Math.abs(x)`	$\lvert x\rvert$	`float`, `double`, `int`, or `long`	same as parameter	Absolute value of x
`Math.acos(x)`	$\cos^{-1} x$	`double`	`double`	Inverse cosine of x for $-1 \le x \le 1$ (results in *radians*)
`Math.asin(x)`	$\sin^{-1} x$	`double`	`double`	Inverse sine of x for $-1 \le x \le 1$ (results in *radians*)
`Math.atan(x)`	$\tan^{-1} x$	`double`	`double`	Inverse tangent of x (results in *radians* in the range $-\pi/2 \le x \le \pi/2$)
`Math.atan2(y,x)`	$\tan^{-1}\frac{y}{x}$	`double`	`double`	Inverse tangent of x (results in *radians* in the range $-\pi \le x \le \pi$)
`Math.ceil(x)`		`double`	`double`	Returns the smallest integer not less than x: `ceil(2.2) = 3` and `ceil(-2.2) = -2`
`Math.cos(x)`	$\cos x$	`double`	`double`	Cosine of x, where x is in radians
`Math.exp(x)`	e^x	`double`	`double`	
`Math.floor(x)`		`double`	`double`	Returns the largest integer not greater than x: `floor(2.2) = 2` and `floor(-2.2) = -3`
`Math.log(x)`	$\log_e x$ or $\ln x$	`double`	`double`	Natural logarithm of x, for $x > 0$
`Math.max(x,y)`		`float`, `double`, `int`, or `long`	same as parameter	Returns the larger of x or y
`Math.min(x,y)`		`float`, `double`, `int`, or `long`	same as parameter	Returns the smaller of x or y
`Math.random()`		none	`double`	Returns a uniformly distributed random value between 0 and 1
`Math.pow(x,y)`	x^y	`double`	`double`	`Math.pow(2,5) = 32` `Math.pow(9, .5) = 3`
`Math.rint(x)`		`double`	`double`	Rounds floating point number to the nearest integer, and returns the result as a floating-point number.
`Math.round(x)`		`double` or `float`	`long` or `int`	Rounds floating point number to the nearest integer, and returns the result as an integer. `round(2.2) = 2` and `round(-2.2) = -2`
`Math.sin(x)`	$\sin x$	`double`	`double`	Sine of x, where x is in radians
`Math.sqrt(x)`	$\sqrt{x}$	`double`	`double`	Square root of x, for $x \ge 0$
`Math.tan(x)`	$\tan x$	`double`	`double`	Tangent of x, where x is in radians

2.8 STANDARD INPUT AND OUTPUT

For a computer program to be useful, there must be some way to read in the data to be processed and to write out the results of the calculations. The process of reading in data and writing out results is known as **input/output (I/O)**. There are many different ways to read in and write out data in a Java program, and we will see some of them in Chapter 5. However, the simplest way to read and write data from a program is through the **standard input** and **standard output** devices of a computer. The standard input device is a special, preopened input channel that is usually connected to the computer's keyboard, so that the program can accept values that a user enters while the program is running. (This is not always true, as the standard input device can be redirected to come from a file or from the output of another program.) The standard output device is a preopened output channel that is usually displayed on the computer's monitor, so that the program can print out results for its user. (The standard output device may also be redirected.) In Java, the data from these devices are *encapsulated* inside objects, which are called **data streams**.

Every Java program has three standard I/O objects: `System.in`, `System.out`, and `System.err`. These objects are ready to read data from or write data to whenever the program begins executing. `System.in` is an object representing the **standard input stream**, which is usually the keyboard. When the program reads input data from this object, the program actually reads values typed by the user at the keyboard. `System.out` is an object representing the **standard output stream**, which is usually the monitor of the computer. When the program writes data to this object, the values are displayed on the monitor. Finally, `System.err` is a special object representing the **standard error stream**. It is a special stream used for displaying severe program errors.

All three standard I/O objects share the characteristic that *they process input or output data one byte at a time*. Sending data to the standard output stream is relatively easy, since there are standard methods that convert the data to be printed into a stream of bytes, and send those bytes to the output stream. We have already seen the `println` method, which performs this function.

However, reading data from the standard input stream is much harder. The standard input stream presents data to the program one byte at a time, and *it is the programmer's responsibility to clump successive bytes together to form meaningful numbers or strings* before attempting to process them. For example, if a user were to type the value 123.4, the program would have to read the characters in one byte at a time and convert the entire string into the appropriate `double` value after all characters had been read. Java includes standard classes to collect and buffer the input bytes until there are enough available to translate into a meaningful number or string. Unfortunately, these classes are relatively complex to use, and they are not discussed in this brief book. A description of how to use them can be downloaded from the book's Web site. In this chapter, we will introduce a single "convenience class" that allows us to read data of any type from the standard input stream.

2.8.1 Using the Standard Output Stream

We have already used the `System.out` object in a number of programs. There are two important methods that we will learn to use with this object: `print` and `println`. The `print` method accepts a single parameter and prints out the value of its parameter on the standard output device. It does *not* send a newline character at the end of the value, so any additional calls to `print` will be displayed on the same line. By contrast, the `println` method accepts a single parameter, and prints out the value of its parameter *followed by a newline character*. Thus `println` terminates the output on a given line.

The behavior of these two methods is illustrated in the following program, which outputs two variables `i` and `j` and a string using both the `print` and `println` methods:

```
1   // This program illustrates the use of
2   // the print and println methods.
3   public class TestOutput {
4      // Define the main method
5      public static void main(String[] args) {
6
7         int i = 1; float j = 1.35F;
8         // Demonstrate output
9         System.out.print( i );
10        System.out.print( j );
11        System.out.print( "String\n" );
12
13        System.out.println( i );
14        System.out.println( j );
15        System.out.println( "String" );
16     }
17  }
```

Line 9 prints out the value "1", line 10 prints out the value "1.35", and line 11 prints out the string "String" followed by a newline character. Since these values were printed with the `print` method, they all appear in consecutive characters on a single line. Line 13 prints out the value "1", line 14 prints out the value "1.35", and line 15 prints out the string "String". Since these values were printed with the `println` method, they all appear consecutive lines. When this program is executed, the results are as follows:

```
11.35String
1
1.35
String
```

Note that there is no space between the values printed out by the `print` method. If you want space between the values, you must explicitly print the spaces. For example, the statements

```
System.out.print( i );
System.out.print( " " );
System.out.print( j );
System.out.print( " " );
System.out.print( "String\n" );
```

produce the output

```
1 1.35 String
```

The + operator has a special meaning when used with strings. If the + operator appears between two strings, it **concatenates** them together into one long string. Furthermore, if data of any other type is combined with a string using the + operator, *that data will automatically be converted into a string and concatenated with the other string*. For example, suppose `v1` is a variable of type `double` containing the value 1.25. Then the statement

```
System.out.println("value = " + v1);
```

converts the contents of variable `v1` into a string and concatenates it with the string `"value = "`. The statement prints out the line

```
value = 1.25
```

The standard error stream works exactly the same way as the standard output stream, except that the object `System.err` is substituted for `System.out`. This data stream is only used for reporting critical errors, so it is rarely used.

2.8.2 Using the Standard Input Stream

The standard input stream is used to read in data from the keyboard or some other specified source. The data in the standard input stream is presented to the program one byte at a time, and the program must combine the bytes after they are read to create the numbers or strings that the program needs to process. Reading data using standard Java methods is very complex, and we will not cover it in this text. Instead, we will use the "convenience class" `StdIn` to read data into our programs.

Class `StdIn` allows Java to accumulate data for you until an entire line is available, and provides methods to translate the data into any primitive data type or into a `String`. For example, if method `readInt()` is called, the line is translated into an integer value, while if method `readDouble()` is called, the line is translated into a double value. (See Figure 2.3.) The `StdIn` class is located in the `chapman.io` package. To use this class, you must import package `chapman.io` into your program with the following statement:

```
import chapman.io.*;
```

`import` statements should be the first noncomment statements in your program, before the class declaration. You must also have the `CLASSPATH` environment variable set as described in the box below.

SETTING THE CLASSPATH ENVIRONMENT VARIABLE

In order to use the special packages supplied with this book, a programmer must first set the `CLASSPATH` environment variable on his or her computer to tell the Java compiler where to look for the packages. This variable must contain the name of the parent directory of the class path structure. For example, if the extra packages appear as subdirectories of directory `c:\packages`, then the `CLASSPATH` variable must include the directory `c:\packages`.

The manner in which the environment variable is set will vary among different types of operating systems. In Windows 95/98, the `CLASSPATH` environment variable is set by including the line

```
set CLASSPATH=c:\packages
```

in the `autoexec.bat` file. If a `CLASSPATH` already exists in the file, add a semicolon (;) followed by `c:\packages` to the end of the existing path.

In Windows NT, the environment variable is set through the System option in the Control Panel. See the Windows NT help system for details.

For Unix systems running the C shell, the class path is set by opening the `.login` file with a text editor and adding the line

```
setenv CLASSPATH $HOME/packages
```

If a `CLASSPATH` already exists in the file, add a colon (:) followed by `$HOME/packages` to the end of the existing path.

For Unix systems running the Bourne or Korn shells, the class path is set by opening the `.profile` file with a text editor and adding the lines

```
CLASSPATH=$HOME/packages
export CLASSPATH
```

If a `CLASSPATH` already exists in the file, add a colon (:) followed by `$HOME/packages` to the end of the existing path.

A `StdIn` object is created by the statement:

```
StdIn in = new StdIn();
```

Figure 2.3. An object of class `StdIn` reads a line one byte at a time from the Standard Input Stream, until an entire line has been read in. It then converts that line into an `int`, `double`, `String`, etc. value, depending on which method was called. The class automatically handles input errors, providing the user a chance to correct them.

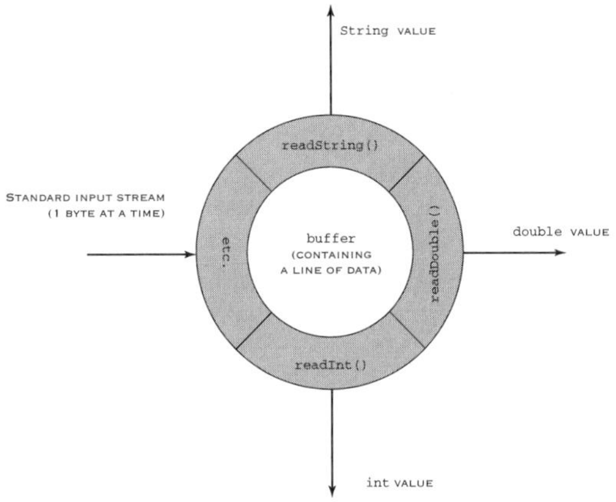

This statement creates a new `StdIn` object and makes the reference `in` refer to that object. The object automatically buffers the data coming out of the standard input stream until an entire line has been read. The `StdIn` methods `readInt()`, `readDouble()`, `readString()`, etc. then convert that line into a value of the appropriate type. The `StdIn` methods are summarized in Table 2-8, shown at the top of the next page.

The following program illustrates the use of these features to read data of various sorts from the standard input stream:

```
// This program tests reading values using class StdIn
import chapman.io.*;
public class ReadStdIn {

   // Define the main method
   public static void main(String[] args) {

      double v1; int i1; boolean test;

      // Create a StdIn object
      StdIn in = new StdIn();

      // Prompt for a double value
      System.out.print("Enter a double value: ");
      v1 = in.readDouble();
      System.out.println("Value = " + v1 );

      // Prompt for an int value
      System.out.print("Enter an int value: ");
      i1 = in.readInt();
      System.out.println("Value = " + i1 );

      // Prompt for a boolean value
      System.out.print("Enter a boolean value: ");
      test = in.readBoolean();
      System.out.println("Value = " + test );
   }
}
```

TABLE 2-8 Methods in Class `StdIn`

METHOD	RESULT
`readBoolean()`	Reads a line and converts it into a `boolean` result. The value is `true` if the input characters are "`true`" disregarding case, and `false` otherwise.
`readByte()`	Reads a line and converts it into a `byte` result.
`readShort()`	Reads a line and converts it into a `short` result.
`readInt()`	Reads a line and converts it into an `int` result.
`readLong()`	Reads a line and converts it into a `long` result.
`readFloat()`	Reads a line and converts it into a `float` result.
`readDouble()`	Reads a line and converts it into a `double` result.
`readString()`	Reads a line and converts it into a `String` result.

Note that this program includes an `import` statement to import the `StdIn` class from the `chapman.io` package. When this program is executed, the results are

```
D:\book\java\chap2>java ReadStdIn
Enter a double value: 45.6
Value = 45.6
Enter an int value: 45.6
Invalid format for integer-try again:
45
Value = 45
Enter a boolean value: true
Value = true
```

Note also that if a user types an incorrectly formatted character string, the `StdIn` class catches the error and allows the user to correct it. In this example, the value "45.6" was not a valid integer, so the `StdIn` object informed the user and gave him or her the chance to correct the error.

PRACTICE!

This quiz provides a quick check to see if you understand the concepts introduced in Sections 2.5 through 2.8. If you have trouble with the quiz, reread the sections, ask your instructor, or discuss the material with a fellow student. The answers to this quiz are found in the back of the book.

Convert the following algebraic equations into Java assignment statements:

1. The equivalent resistance R_{eq} of four resistors R_1, R_2, R_3, and R_4 connected in series:

$$R_{eq} = R_1 + R_2 + R_3 + R_4$$

2. The equivalent resistance R_{eq} of four resistors R_1, R_2, R_3, and R_4 connected in parallel:

$$R_{eq} = \frac{1}{\frac{1}{R_1} + \frac{1}{R_2} + \frac{1}{R_3} + \frac{1}{R_4}}$$

PRACTICE!

3. The period T of an oscillating pendulum:

$$T = 2\pi\sqrt{\frac{L}{g}},$$

where L is the length of the pendulum, and g is the acceleration due to gravity.

4. The equation for damped sinusoidal oscillation:

$$v(t) = V_M e^{-\alpha t} \cos \omega t,$$

where V_M is the maximum value of the oscillation, α is the exponential damping factor, and ω is the angular velocity of the oscillation.

Convert the following Java assignment statements into algebraic equations:

5. The motion of an object in a constant gravitational field:

```
distance = 0.5 * accel * Math.pow(t,2) + vel0 * t + pos0;
```

6. The oscillating frequency of a damped *RLC* circuit:

```
freq = 1 / (2 * Math.PI * Math.sqrt(l * c));
```

7. Energy storage in an inductor:

```
energy = 1. / 2. * inductance * Math.pow(current,2);
```

8. Given the following definitions, decide whether each of the statements shown is legal or illegal? If a statement is legal, what is its result? If it is illegal, tell why it is illegal.

```
double a = 2., b = 3., c;
int i = 3, j = 2, k;
```

a. `c = Math.sin(a * Math.PI));`
b. `k = a / b;`
c. `k = (int) a / b;`
d. `b += a / i;`
e. `c = a / ( j / i ));`

9. After the following statements are executed, what is stored in each of the variables?

```
double a = 2., b = 3., c;
int i = 3, j = 2, k;
k = ++i - j--;
c = i++/j + a/b;
```

2.9 PROGRAM EXAMPLES

In Chapter 2, we have presented the fundamental concepts required to write simple but functional Java programs. We will now present a few examples in which these concepts are used.

EXAMPLE 2-2: TEMPERATURE CONVERSION

Design a Java program that reads an input temperature in degrees Fahrenheit, converts it to an absolute temperature in kelvins, and writes out the result.

SOLUTION

The relationship between temperature in degrees Fahrenheit (°F) and temperature in kelvins (K) can be found in any physics textbook. The relationship is

$$T(\text{in kelvins}) = \left[\frac{5}{9}T(\text{in}^\circ\text{F}) - 32.0\right] + 273.15\,. \tag{2-1}$$

The physics books also give us sample values on both temperature scales, which we can use to check the operation of our program. Two such values are

The boiling point of water:	**212°F**	**373.15**K
and		
The sublimation point of dry ice:	**−110°F**	**194.26**K.

Our program must perform the following steps:

1. Prompt the user to enter an input temperature in °F.
2. Read the input temperature.
3. Calculate the temperature in kelvins from Equation (2-1).
4. Write out the result, and stop.

The resulting program is shown in Figure 2.4.

To test the completed program, we will run it with the known input values given above. Note that user inputs appear in boldface.

```
C:\book\java\chap2>java TempConversion
Enter the temperature in deg Fahrenheit: 212
212.0 deg Fahrenheit = 373.15

C:book\java\chap2>java TempConversion
Enter the temperature in deg Fahrenheit: -110
-110.0 deg Fahrenheit = 194.2611111111111
```

The results of the program match the values from the physics book.

In the previous program, we echoed the input values and printed the output values together with their units. The results of this program only make sense if the units (degrees Fahrenheit and kelvins) are included together with their values. As a general rule, the units associated with any input value should always be printed along with the prompt that requests the value, and the units associated with any output value should always be printed along with that value.

GOOD PROGRAMMING PRACTICE

Always include the appropriate units with any values that you read or write in a program.

```
/*
   Purpose:
     To convert an input temperature from degrees Fahrenheit to
     an output temperature in kelvins.

   Record of revisions:
       Date       Programmer              Description of change
       ====       ==========              =====================
     03/22/98    S. J. Chapman            Original code
*/
import chapman.io.*;
public class TempConversion {

   // Define the main method
   public static void main(String[] args) {

      // Declare variables, and define each variable
      double tempF;          // Temperature in degrees Fahrenheit
      double tempK;          // Temperature in kelvins

      // Create a StdIn object
      StdIn in = new StdIn();

      // Prompt the user for the input temperature.
      System.out.print("Enter the temp in deg Fahrenheit: ");
      tempF = in.readDouble();

      // Convert to kelvins
      tempK = (5. / 9.) * (tempF - 32.) + 273.15;

      // Write out the result.
      System.out.println(tempF + " deg F = " + tempK + " K");
   }
}
```

Figure 2.4. Program to convert degrees Fahrenheit into kelvins

This above program exhibits many of the good programming practices that we have described in this chapter. It includes a data dictionary that defines the meanings of all of the variables in the program, it uses descriptive variable names, and appropriate units are attached to printed values.

EXAMPLE 2-3: ELECTRICAL ENGINEERING: CALCULATING REAL, REACTIVE, AND APPARENT POWER

Figure 2.5 shows a sinusoidal alternating current (AC) voltage source with voltage V supplying a load of impedance $Z\angle\theta\ \Omega$. From simple circuit theory, the rms current I, the real power P, reactive power Q, the apparent power S, and the power factor PF supplied to the load are given by the following equations

$$I = \frac{V}{Z}, \tag{2-2}$$

$$P = VI\cos\theta, \tag{2-3}$$

$$Q = VI\sin\theta, \tag{2-4}$$

$$S = VI, \tag{2-5}$$

$$PF = \cos\theta, \tag{2-6}$$

where V is the rms voltage of the power source in units of volts (V). The units of current are amperes (A), the units of real power are watts (W), the units of reactive power are volt-amperes-reactive (VAR), and of apparent power are volt-amperes (VA). The power factor has no units associated with it.

Figure 2.5. A sinusoidal AC voltage source with voltage V supplying a load of impedance $Z\angle\theta\Omega$

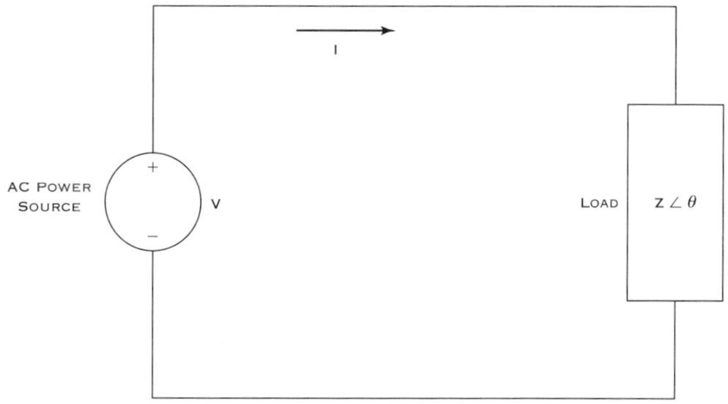

Given the rms voltage of the power source and the magnitude and angle of the impedance Z, write a program that calculates the rms current I, the real power P, the reactive power Q, the apparent power S, and the power factor PF of the load.

SOLUTION

In this program, we need to read in the rms voltage V of the voltage source and the magnitude Z and angle θ of the impedance. The input voltage source will be measured in volts, the magnitude of the impedance Z in ohms, and the angle of the impedance θ in degrees. Once the data is read, we must convert the angle θ into radians for use with the Java trigonometric functions. Next, the desired values must be calculated, and the results must be printed out.

The program must perform the following steps:

1. Prompt the user to enter the source voltage in volts.
2. Read the source voltage.
3. Prompt the user to enter the magnitude and angle of the impedance in ohms and degrees.
4. Read the magnitude and angle of the impedance.
5. Calculate the current I from Equation (2-2).
6. Calculate the real power P from Equation (2-3).
7. Calculate the reactive power Q from Equation (2-4).
8. Calculate the apparent power S from Equation (2-5).
9. Calculate the power factor PF from Equation (2-6).
10. Print out the results, and stop.

The final Java program is shown in Figure 2.6.

This program also exhibits many of the good programming practices that we have described. It includes a variable dictionary defining the uses of all of the variables in the program, it uses descriptive variable names (although the variable names are short, P, Q, S, and PF are the standard accepted abbreviations for the corresponding quantities), and it defines a named constant for the degrees-to-radians conversion factor and then uses that name everywhere throughout the program where required. Also, appropriate units are attached to all printed values.

```
/*
   Purpose:
     To calculate the current, real, reactive, and apparent power,
     and the power factor supplied to a load.

   Record of revisions:
       Date       Programmer          Description of change
       ====       ==========          =====================
     03/22/98    S. J. Chapman        Original code
*/
import chapman.io.*;
public class Power {

   // Define the main method
   public static void main(String[] args) {

      // Declare constants
      final double CONV = Math.PI / 180;    // Degrees to radians

      // Declare variables, and define each variable
      double amps;      // Current in the load (A)
      double p;         // Real power of load (W)
      double pf;        // Power factor of load
      double q;         // Reactive pwr of the load (VA)
      double s;         // Apparent pwr of the load (VAR)
      double theta;     // Impedance angle of the load (deg)
      double volts;     // Rms voltage of the power source (V)
      double z;         // Magnitude of the load impedance (ohms)

      // Create a StdIn object
      StdIn in = new StdIn();

      // Prompt the user for the rms voltage.
      System.out.print("Enter the rms voltage of the source: ");
      volts = in.readDouble();

      // Prompt the user for the magnitude of the impedance
      System.out.print("Enter the magnitude of Z (ohms): ");
      z = in.readDouble();

      // Prompt the user for the angle of the impedance
      System.out.print("Enter the angle of Z (deg): ");
      theta = in.readDouble();

      // Perform calculations
      amps = volts / z;                           // Rms current
      p = volts * amps * Math.cos(theta*CONV); // Real power
      q = volts * amps * Math.sin(theta*CONV); // React. power
      s = volts * amps;                           // App. power
      pf = Math.cos(theta * CONV);               // Power factor

      // Write out the results.
      System.out.println("Voltage      = " + volts + " volts");
      System.out.println("Impedance    = " + z + " ohms at "
                         + theta + " degrees");
      System.out.println("Current      = " + amps + " amps");
      System.out.println("Real Power   = " + p + " W");
      System.out.println("Reactive Pwr = " + q + " VAR");
      System.out.println("Apparent Pwr = " + s + " VA");
      System.out.println("Power Factor = " + pf);
   }
```

Figure 2.6. Program to calculate the real power, reactive power, apparent power, and power factor supplied to a load.

To verify the operation of program `Power`, we will do a sample calculation by hand and compare the results with the output of the program. If the rms voltage V is 120 V, the magnitude of the impedance Z is 5 Ω, and the angle θ is 30°, then the values are as follows:

$$I = \frac{V}{Z} = \frac{120\text{ V}}{5\ \Omega} = 24\text{ A}, \quad (2\text{-}2)$$

$$P = VI\cos\theta = (120\text{ V})(24\text{ A})\cos 30° = 2494\text{ W}, \quad (2\text{-}3)$$

$$Q = VI\sin\theta = (120\text{ V})(24\text{ A})\sin 30° = 1440\text{ VAR}, \quad (2\text{-}4)$$

$$S = VI = (120\text{ V})(24\text{ A}) = 2880\text{ VA}, \quad (2\text{-}5)$$

$$\text{PF} = \cos\theta = \cos 30° = 0.86603\,. \quad (2\text{-}6)$$

When we run program **`Power`** with the specified input data, the results are identical to our hand calculations:

```
C:\book\java\chap2>java Power
Enter the rms voltage of the source: 120
Enter the magnitude of the impedance (ohms): 5
Enter the angle of the impedance (deg): 30
Voltage         = 120.0 volts
Impedance       = 5.0 ohms at 30.0 degrees
Voltage         = 120.0 volts
Current         = 24.0 amps
Real Power      = 2494.1531628991834 W
Reactive Power  = 1439.9999999999998 VAR
Apparent Power  = 2880.0 VA
Power Factor    = 0.8660254037844387
```

EXAMPLE 2-4: CARBON-14 DATING

A radioactive isotope of an element is a form of the element that is not stable. It spontaneously decays into an other element over a period of time. Radioactive decay is an exponential process. If Q_0 is the initial quantity of a radioactive substance at time $t = 0$, then the amount of the substance that will be present at any time t in the future is given by

$$Q(t) = Q_0 e^{-\lambda t} \quad (2\text{-}7)$$

where λ is the radioactive decay constant.

Because radioactive decay occurs at a known rate, it can be used as a clock to measure the time since the decay started. If we know the initial amount of the radioactive material Q_0 present in a sample, and the amount of the material Q left at the current time, we can solve for t in Equation (2-7) to determine how long the decay has been going on. The resulting equation is

$$t_{\text{decay}} = -\frac{1}{\lambda}\log\frac{Q}{Q_0} \quad (2\text{-}8)$$

Equation (2-8) has practical applications in many areas of science. For example, archaeologists use a radioactive clock based on carbon 14 to determine the time that has passed since a once-living thing died. Carbon 14 is continually taken into the body while a plant or animal is living, so the amount of it present in the body at the time of death is assumed to be known. The decay constant λ of carbon 14 is well known to be 0.00012097/year, so if the amount of carbon 14 remaining now can be

Figure 2.7. The radioactive decay of Carbon 14 as a function of time. Notice that 50% of the original carbon 14 is left after about 5730 years have elapsed.

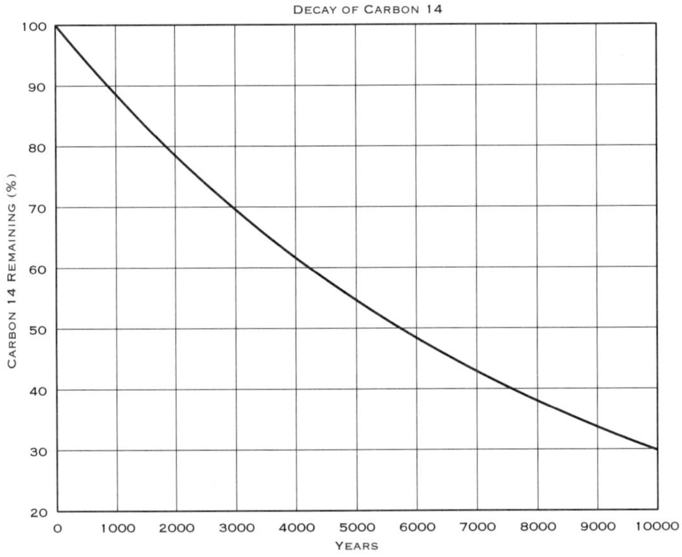

accurately measured, then Equation (2-8) can be used to determine how long ago the thing died.

Write a program that reads the percentage of carbon 14 remaining in a sample, calculates the age of the sample from it, and prints out the result with proper units.

SOLUTION

Our program must perform the following steps:

1. Prompt the user to enter the percentage of carbon 14 remaining in the sample.
2. Read in the percentage.
3. Convert the percentage into the fraction Q/Q_0.
4. Calculate the age of the sample in years using Equation (2-8).
5. Write out the result, and stop.

The resulting code is shown in Figure 2.8.

```
/*
   Purpose:
     To calculate the age of an organic sample from the
     percentage of the original carbon 14 remaining in
     the sample.

     Record of revisions:
       Date        Programmer           Description of change
       ====        ==========           =====================
     03/22/98    S. J. Chapman          Original code
*/
import chapman.io.*;
public class C14Date {
```

Figure 2.8. *(cont.)*

```
        // Define the main method
        public static void main(String[] args) {

        // Declare constants
        final double LAMDA = 0.00012097;    // C14 decay constant (1/year)

        // Declare variables, and define each variable
        double age;        // Age of the sample (years)
        double percent;    // Percentage of carbon 14 remaining
        double ratio;      // Ratio of the Carbon 14 remaining at the time
                           // of the measurement to the original amount
                           // of Carbon 14.

        // Create a StdIn object
        chapman.io.StdIn in = new chapman.io.StdIn();

        // Prompt the user for the percentage of C-14 remaining.
        System.out.print("Enter the percentage of carbon 14 remaining:");
        percent = in.readDouble();

        // Perform calculations
        ratio = percent / 100;                        // Convert to ratio
        age = (-1.0 / LAMDA) * Math.log(ratio);      // Get age in years

        // Tell the user about the age of the sample.
        System.out.println("The age of the sample is " + age + " years.");
    }
}
```

Figure 2.8. Program to calculate the age of a sample from the percentage of carbon 14 remaining in it

To test the completed program, we will calculate the time it takes for half of the carbon 14 to disappear. This time is known as the *half-life* of carbon 14. The program outputs the following (user input appears in boldface):

```
C:\book\java\chap2>java C14Date
Enter the percentage of carbon 14 remaining: 50
The age of the sample is 5729.90973431384 years.
```

The *CRC Handbook of Chemistry and Physics* states that the half-life of carbon 14 is 5730 years, so output of the program agrees with the reference book.

PROFESSIONAL SUCCESS: LEARNING A PROGRAMMING LANGUAGE

You are reading this book because you want to learn how to program in the Java language. What is the best way to learn to program in Java?

The simple answer is: play! The only real way to learn a programming language is to practice, experiment, and play with it. Write programs to try out the features described in this book. When you do it yourself, you will really learn the material.

Don't be afraid to try new things—the worst that could happen is that you make a mistake and your program won't compile. You won't blow the computer up—don't be afraid of it!

2.10 DEBUGGING JAVA PROGRAMS

There is an old saying that the only sure things in life are death and taxes. We can add one more certainty to that list: If you write a program of any significant size, it won't work the first time you try it! Errors in programs are known as *bugs*, and the process of locating and eliminating them is known as *debugging*. Given that we have written a program and it is not working, how do we debug it?

Three types of errors are found in Java programs. The first type of error is a **compile-time error**. Compile-time errors are errors in the Java statement itself, such as spelling errors or punctuation errors. These errors are detected by the compiler during compilation. The second type of error is the **run-time error**. A run-time error occurs when an illegal mathematical operation is attempted during program execution (e.g., attempting to divide by 0). These errors cause exceptions in Java programs. Unless the exception is caught by an exception handler, the program will abort (crash) when the execution occurs. The third type of error is a **logical error**. Logical errors occur when the program compiles and runs successfully, but produces the wrong answer.

The most common mistakes made during programming are *typographical errors*. Some typographical errors create invalid Java statements. These errors produce syntax errors that are caught by the compiler. Other typographical errors occur in variable names. For example, the letters in some variable names might have been transposed. Most of these errors will also be caught by the compiler. However, if one legal variable name is substituted for another legal variable name, the compiler cannot detect the error. This sort of substitution might occur if you have two similar variable names. For example, if variables `vel1` and `vel2` are both used for velocities in the program, then one of them might be inadvertently used instead of the other one at some point. This sort of typographical error will produce a logical error. You must check for that sort of error by manually inspecting the code, since the compiler cannot catch it.

Sometimes is it possible to successfully compile and link the program, but there are run-time errors or logical errors when the program is executed. In this case, there is something wrong with the input data or something wrong with the logical structure of the program (or both). The first step in locating this sort of bug should be to *check the input data to the program*. Verify that the input values are what you expect them to be.

If the variable names seem to be correct and the input data is correct, then you are probably dealing with a logical error. You should check each of your assignment statements

1. If an assignment statement is very long, break it into several smaller assignment statements. Smaller statements are easier to verify
2. Check the placement of parentheses in your assignment statements. It is a very common error to have the operations in an assignment statement evaluated in the wrong order. If you have any doubts as to the order in which the variables are being evaluated, add extra sets of parentheses to make your intentions clear.
3. Make sure that you have initialized all of your variables properly.
4. Be sure that any functions you use are in the correct units. For example, the input to trigonometric functions must be in units of radians, not degrees.
5. Check for possible errors due to integer or mixed-mode arithmetic.

If you are still getting the wrong answer, place `println` statements at various points in your program to print out the results of intermediate calculations. If you can locate the point where the calculations go bad, then you know just where in the code to look for the problem, which is 95% of the battle.

If you still cannot find the problem after completing all of the above steps, explain what you are doing to another student or to your instructor, and let them look at the code. It is very common for a person to see just what he or she expects to see when they look at their own code. Another person can often quickly spot an error that you have overlooked time after time.

All modern compilers have special debugging tools called *symbolic debuggers*. A symbolic debugger is a tool that allows you to walk through the execution of your program one statement at a time, and to examine the values of any variables at each step along the way. Symbolic debuggers allow you to see all of the intermediate results without having to insert a lot of debugging `print` or `println` statements into your code. These debuggers are powerful and flexible, but unfortunately there is a different debugger for each compiler vendor. If you will be using a symbolic debugger in your class, your instructor will introduce you to the debugger appropriate for your compiler and computer.

There is a standard debugger (`jdb`) supplied with the Java Software Development Kit. This debugger is command-line oriented and much less sophisticated than the ones in the commercial packages, but it is free.

PROFESSIONAL SUCCESS: BACK UP YOUR WORK EARLY AND OFTEN

It is very important that you learn to back up your work early and often right from the start of your programming career. Anyone who has programmed for a while can tell you about the frustration of losing parts of a program and having to re-create them.

For example, suppose that you are modifying an existing program to add some new feature, and you discover that the approach you intended to take doesn't work as planned. If you don't have a copy of the original program, you will have to devote extra time removing all of the changes you made and verifying that the program works the way it used to before even thinking about a new way to modify the program.

You should make a habit of saving copies of the programs that you develop at regular "checkpoints" during the development cycle. The checkpoints should be created after any significant changes, when the program is actually functioning. Then, if you lose a file or mess up some additional modification, you can always fall back on the last checkpointed version and start over.

SUMMARY

- Java names may contain any combination of letters, numbers, and underscore characters. They may be of any length, but must begin with a letter.
- Java names may not be the same as any Java keyword. Java keywords are listed in Table 2-1.
- By convention, class names begin with a capital letter, and the first letter of each successive word is also capitalized.
- By convention, local variable and method names begin with a lowercase letter, and the first letter of each successive word is capitalized.

- By convention, named constants (final variables) are written in all capital letters, with underscores used to separate different words.
- Java includes eight primitive data types: `float`, `double`, `char`, `byte`, `short`, `int`, `long`, and `boolean`. Each type is identical on every platform supported by Java.
- Java is a strongly typed language, which means that every variable must be explicitly declared before it is used.
- A Java named constant (also known as a final variable) may be created by prefixing the keyword `final` in front of the type in a declaration statement.
- An assignment statement assigns the value to the right of the equals sign to the variable on the left of the equals sign.
- The preincrement and predecrement operators are placed before a variable, and the postincrement and postdecrement operators are placed after a variable. The preincrement and predecrement operators cause the variable to be incremented or decremented by one, and then the new value is used in the expression in which it appears. The postincrement and postdecrement operators cause the old value of the variable to be used in the expression in which it appears and then the variable to be decremented or decremented by one.
- The cast operator explicitly converts a numeric value of one data type into a numeric value of another data type.
- Java's built-in mathematical methods are contained in the `Math` class of the `java.lang` package. Since this package is automatically imported into any Java program, these methods may be used by simply naming them in a statement. The methods are summarized in Table 2-7.
- Java can include overloaded methods, which are multiple methods with the same name but different parameters. When a programmer uses an overloaded method, the Java compiler decides which actual method to use based on the number and types of the method's arguments. For example, `Math.abs(x)` is an overloaded method.
- Java automatically converts arguments of an incorrect type into arguments of the type required by the method. This process is known as the **coercion of arguments**.
- Every Java program has three standard I/O objects: `System.in`, `System.out`, and `System.err`. These objects either accept or supply data one byte at a time. *It is the programmer's responsibility to clump successive bytes together to form meaningful numbers or strings* before attempting to process them.
- Java input is a very complex subject, and it is described at the book's Web site. For now, the convenience class `StdIn` will be used to read input data and convert it into values of the proper type.
- When the + operator is used between strings, it concatenates them together. If it is used between a string and another operand, it converts the other operand into a string and concatenates the two strings together.

KEY TERMS

assignment conversion
assignment operator
assignment statement
associativity
binary operator
cast operator
coercion of arguments
concatenate
constant
escape sequence
exception
exponent
declaration statement
final
final variable
hierarchy of operations
increment operator error
keyword
mantissa
named constant
narrowing conversion
numeric promotion
overloaded methods
postdecrement
postincrement
precision
predecrement
preincrement
primitive data type
range
reserved
round-off error
standard error stream
standard input stream
standard output stream
String
strongly typed language
variable
unary operator
Unicode character set
widening conversion

APPLICATIONS: SUMMARY OF GOOD PROGRAMMING PRACTICES

Every Java program should be designed so that another person who is familiar with Java can easily understand it. This is very important, since a good program may be used for a long period of time. Over that time, conditions will change, and the program will need to be modified to reflect the changes. The program modifications may be done by someone other than the original programmer. The programmer making the modifications must understand the original program well before attempting to change it.

It is much harder to design clear, understandable, and maintainable programs than it is to simply write programs. To create such clear, understandable, and maintainable programs, a programmer must develop the discipline to properly document his or her work. In addition, the programmer must be careful to avoid known pitfalls along the path to good programs. The following guidelines will help you to develop good programs:

1. Always capitalize the first letter of a class name, and use a lowercase first letter in method and variable names.
2. If a name consists of more than one word, the words should be joined together and each succeeding word should begin with an uppercase letter.
3. Use meaningful variable names whenever possible to make your programs clearer.
4. Create a data dictionary in each program that you write. The data dictionary should explicitly declare and define each variable in the program. Be sure to include the physical units associated with each variable, if applicable.
5. Keep your physical constants consistent and precise throughout a program. To improve the consistency and understandability of your code, assign a name to any important constants, and refer to them by name in the program.
6. The names of constants should be in all capital letters, with underscores separating the words.
7. Use parentheses as necessary to make your equations clear and easy to understand.
8. Use cast operators to avoid mixed-mode expressions and to make your intentions clear.
9. Always keep expressions contianing increment and decrement operators simple and easy to understand.

Problems

1. State whether each of the following Java constants is valid. If valid, state what type of constant it is. If not, state why it is invalid. (If you are not certain, try to compile the constant in a Java program to check your answer.)
   a. `3.141592`
   b. `true`
   c. `-123,456.789`
   d. `+1E-12`
   e. `"Who's coming for dinner?"`
   f. `'Hello'`
   g. `"Enter name:"`
   h. `17.0f`
2. State whether each of the Java names is valid. If a name is not valid, state why the name is invalid. If a name is valid, state what a name of that sort should represent using the Java conventions.
   a. `junk`
   b. `3rd`
   c. `executeAlgorithm`
   d. `timeToIntercept`
   e. `MyMath`
   f. `START_TIME`
3. Which of the following expressions are legal in Java? If an expression is legal, evaluate it.
   a. `5 + 10 % 3 + 2`
   b. `5 + 10 % (3 + 2)`
   c. `23 / (4 / 8)`
4. Which of the following expressions are legal in Java? If an expression is legal, evaluate it.
   a. `((58/4)*(4/58))`
   b. `((58/4)*(4/58.))`
   c. `((58./4)*(4/58.))`
   d. `((58./4*(4/58.))`
5. Assume that the variables `a`, `b`, `c`, `i`, and `j` are initialized as shown in the following code fragment. What is the value of each variable after these statements are executed?

```
int i = 5; j = 2;
double a = 6, b, c;
b = ++i - j--;
j = (int) b / 2;
a += b / j;
```

6. Figure 2.9 shows a right triangle with a hypotenuse of length c and angle θ. From elementary trigonometry, the length of sides *a* and *b* are given by

$$a = c \cos \theta ,$$

and

$$b = c \sin \theta .$$

The following program is intended to calculate the lengths of sides *a* and *b* given the hypotenuse *c* and angle θ. Will this program run? Will it produce the correct result? Why or why not?

```
import chapman.io.*;
public class CalcTriangle {
```

Figure 2.9.

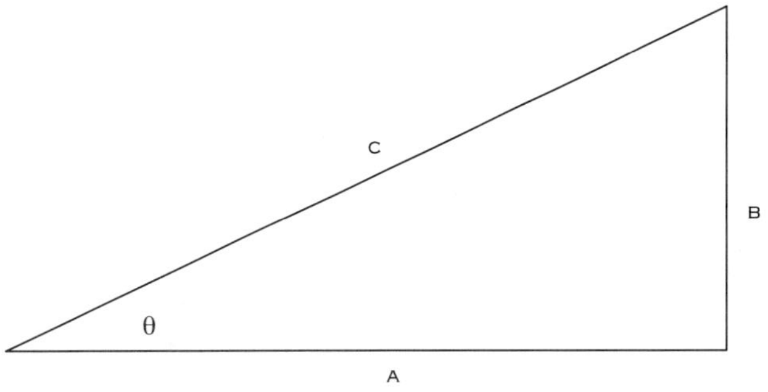


```
    // Define the main method
    public static void main(String[] args) {

       double a, b, c, theta;

       // Create a StdIn object
       StdIn in = new StdIn();

       // Prompt for the hypotenuse
       System.out.print("Enter the length of the hypotenuse c: ");
       c = in.readDouble();

       //Prompt for the angle
       System.out.print("Enter the angle theta in degrees: ");
       theta = in.readDouble();

       // Calculate sides
       a = c * Math.cos( theta );
       b = c * Math.sin( theta );

       // Write results
       System.out.println("Adjacent side = " + a);
       System.out.println("Opposite side = " + b);
    }
}
```

7. Write a Java program that calculates an hourly employee's weekly pay. The program should ask the user for the person's pay rate and the number of hours worked during the week. It should then calculate the total pay from the formula

$$\text{Total Pay} = \text{Hourly Pay Rate} \times \text{Hours Worked}.$$

Finally, it should display the total weekly pay. Check your program by computing the weekly pay for a person earning \$7.50 per hour who worked 39 hours.

8. The potential energy of an object due to its height above the surface of the Earth is given by the equation

$$\text{PE} = mgh, \tag{2-9}$$

where m is the mass of the object, g is the acceleration due to gravity, and h is the height above the surface of the Earth. The kinetic energy of a moving object is given by the equation

$$\text{KE} = \frac{1}{2} mv^2 \tag{2-10}$$

where m is the mass of the object and v is the velocity of the object. Write a Java statement for the total energy (potential plus kinetic) possessed by an object in the earth's gravitational field.

9. Write a Java program that calculates the percentage of carbon 14 that will be left after a given number of years. The program should read the number of years from the standard input stream.

10. If a stationary ball is released at a height h above the surface of the Earth, the velocity of the ball v when it hits the earth is given by the equation

$$v = \sqrt{2gh}\,, \qquad (2\text{-}11)$$

where g is the acceleration due to gravity, and h is the height above the surface of the Earth (assuming no air friction). Write a Java statement for the velocity of the ball when it hits the Earth.

11. **Period of a Pendulum** The period T (in seconds) of an oscillating pendulum is given by the equation

$$T = 2\pi\sqrt{\frac{L}{g}} \qquad (2\text{-}12)$$

where L is the length of the pendulum in meters, and g is the acceleration due to gravity in meters per second squared. Write a Java program to calculate the period of a pendulum of length L. The length of the pendulum will be specified by the user when the program is run. Use good programming practices in your program. (The acceleration due to gravity at the Earth's surface is 9.81 m/s^2. Treat it as a constant in your program.)

12. Write a program to calculate the hypotenuse of a right triangle given the lengths of its two sides. Use good programming practices in your program.

13. The distance between two points (x_1, y_1) and (x_2, y_2) on a Cartesian coordinate plane (Figure 2.10) is given by the equation

$$d = \sqrt{(x_1 - x_2)^2 + (y_1 - y_2)^2}\,. \qquad (2\text{-}13)$$

Write a Java program to calculate the distance between any two points (x_1, y_1) and (x_2, y_2) specified by the user. Use good programming practices in your program. Use the program to calculate the distance between the points (2,3) and (8,–5).

Figure 2.10.

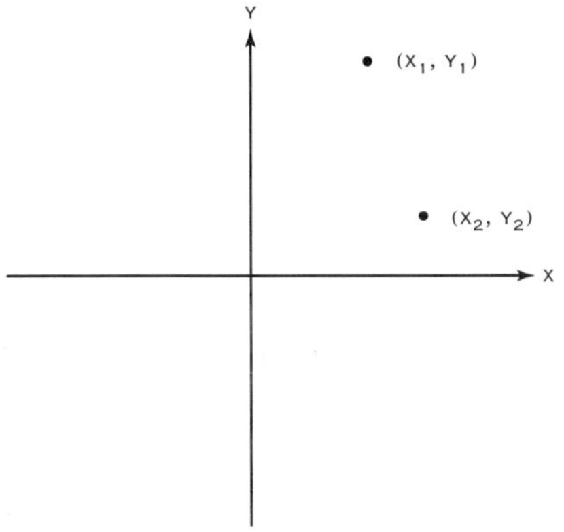

14. **Decibels** Engineers often measure the ratio of two power measurements in *decibels*, or dB. The equation for the ratio of two power measurements in dB is

$$\text{dB} = 10\log_{10}\frac{P_2}{P_1}\,,$$

Figure 2.11. A simplified representation of an AM radio set

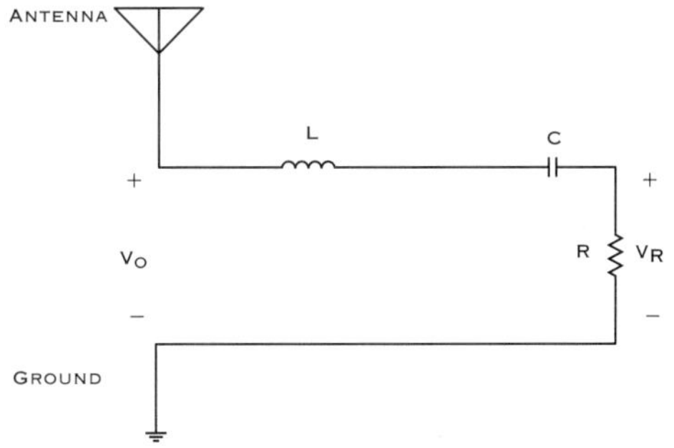

where P_2 is the power level being measured, and P_1 is some reference power level. Assume that the reference power level P_1 is 1 milliwatt, and write a program that accepts an input power P_2 and converts it into dB with respect to the 1 mW reference level.

15. **Hyperbolic cosine** The hyperbolic cosine function is defined by the equation

$$\cosh x = \frac{e^{x} + e^{-x}}{2} .$$

Write a Java program to calculate the hyperbolic cosine of a user-supplied value x. Use the program to calculate the hyperbolic cosine of 3.0.

16. **Radio Receiver** A simplified version of the front end of an AM radio receiver is shown in Figure 2.11. This receiver consists of an *RLC*-tuned circuit containing a resistor, capacitor, and an inductor connected in series. The *RLC* circuit is connected to an external antenna and ground, as shown in the following figure:

The tuned circuit allows the radio to select a specific station out of all the stations transmitting on the AM band. At the resonant frequency of the circuit, essentially all of the signal V_0 appearing at the antenna appears across the resistor, which represents the rest of the radio. In other words, the radio receives its strongest signal at the resonant frequency. The resonant frequency of the LC circuit is given by the equation

$$f_0 = \frac{1}{2\pi\sqrt{LC}} , \qquad (2\text{-}14)$$

where L is inductance in henrys (H) and C is capacitance in farads (F). Write a program that calculates the resonant frequency of this radio set given specific values of L and C. Test your program by calculating the frequency of the radio when $L = 0.1$ mH and $C = 0.25$ nF.

3

Branches and Program Design

In the previous chapter, we developed several complete working Java programs. However, all of the programs were very simple, consisting of a single method containing a series of Java statements, which were executed one after another in a fixed order. Such programs are called *sequential* programs. They read input data, process it to produce a desired answer, print out the answer, and quit. There is no way to repeat sections of the program more than once, and there is no way to selectively execute only certain portions of the program depending on values of the input data.

In the next two chapters, we will introduce a number of Java statements that allow us to control the order in which statements are executed in a program. There are two broad categories of control statements: **selection**, or **branching**, which select specific sections of the code to execute, and **repetition**, which cause specific sections of the code to be repeated. This chapter will deal with selection structures, while Chapter 4 will cover repetition structures.

The operation of selection structures is controlled by `boolean` values (`true` or `false`), so before covering selection structures, we will introduce Java's relational and logical operations, which yield `boolean` results. The results of these operations will be used to control the selection structures.

SECTIONS

OBJECTIVES

After reading this chapter, you should be able to:

- To learn about the program development process and pseudocode
- To understand relational and logical operators
- To learn about the `if` and `switch` structures

Also, with the introduction of selection statements and repetition statements, our programs are going to become more complex, and it will get easier to make mistakes. To help avoid programming errors, we will introduce a formal program design procedure based upon the technique known as **top-down design**. We will also introduce a common algorithm development tool called pseudocode.

3.1 INTRODUCTION TO TOP-DOWN DESIGN TECHNIQUES

Suppose that you are an engineer working in industry and that you need to write a Java program to solve a problem. How do you begin?

When given a new problem, there is a natural tendency to sit down at a keyboard and start programming without "wasting" a lot of time thinking first. It is often possible to get away with this on-the-fly approach to programming for very small problems, such as many of the examples in this book. In the real world, however, problems are larger and more complicated, and a programmer attempting this approach will become hopelessly bogged down. For larger problems, it pays to completely think out the problem and the approach you are going to take to it before writing a single line of code.

We will introduce a formal program design process in this section and then apply that process to major applications developed in the remainder of the book. For some of the simple examples that we will be doing, the design process will seem like overkill. However, as the problems that we solve get larger and larger, the process becomes more and more essential to successful programming.

When I was an undergraduate, one of my professors was fond of saying, "Programming is easy. It's knowing what to program that's hard." His point was forcefully driven home to me after I left the university and began working in industry on larger scale software projects. I found that the most difficult part of my job was to *understand the problem* I was trying to solve. Once I really understood the problem, it became easy to break the problem apart into smaller, more easily manageable pieces with well-defined functions, and then to tackle those pieces one at a time.

In an object-oriented language such as Java, the first step in solving the problem is to break it apart by creating one or more objects that interact with the outside world in well-defined ways. It may be possible to extend a previously defined object for this purpose.

The next step is to determine how each object should interact with the outside world and to define one or more methods to describe that interaction. In general, there should be a separate method for each type of interaction.

There are several analysis techniques in common use to help a programmer decompose a problem into separate objects and methods. These object-oriented programming methods are beyond the scope of this book, since almost all of our examples can be realized as either a single class or a small number of classes. However, these methods may be encountered in advanced programming classes.

Once a problem has been decomposed into appropriate objects and methods, each method must be implemented in a structured way. This is normally done through a process of **top-down design**. Top-down design is the process of starting with a large task and breaking it down into smaller, more easily understandable pieces (subtasks) that perform a portion of the desired task. Each subtask may in turn be subdivided into smaller subtasks if necessary. Once the method is divided into small pieces, each piece can be implemented and tested independently (sometimes as a separate method). We do not attempt to combine the subtasks into a complete method until each of the subtasks has been verified to work properly by itself.

The concept of top-down design is the basis of our formal program design process. We will now introduce the details of the process, which is illustrated in Figure 3.1. The steps involved are:

1. *Clearly state the problem that you are trying to solve.* Programs are usually written to fill some perceived need, but that need may not be articulated clearly by the person requesting the program. For example, a user may ask for a program to solve a system of simultaneous linear equations. This request is not clear enough to allow a programmer to design a program to meet the need; he or she must first know much more about the problem to be solved. Is the system of equations to be solved real or complex? What is the maximum number of equations and unknowns that the program must handle? Are there any symmetries in the equations that might be exploited to make the task easier? The program designer will have to talk with the user requesting the program, and the two of them will have to come up with a clear statement of exactly what they are trying to accomplish. A clear statement of the problem will prevent misunderstandings, and it will also help the program designer to properly organize his or her thoughts. In the example we were just describing, a proper statement of the problem might have been:

Figure 3.1. The program design process used in this book

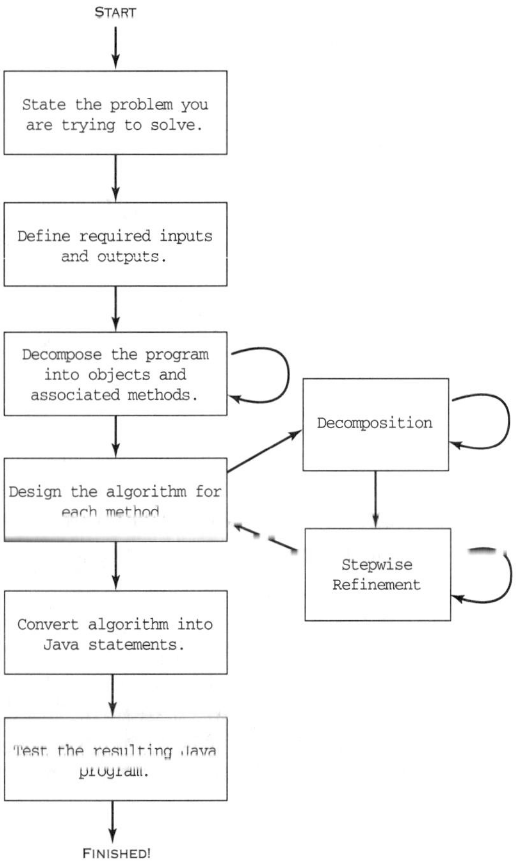

Design and write a program to solve a system of simultaneous linear equations having real coefficients and with up to 20 equations and 20 unknowns.

2. *Define the inputs required by the program and the outputs to be produced by the program.* The inputs to the program and the outputs produced by the program must be specified so that the new program will properly fit into the overall processing scheme. In the previous example, the coefficients of the equations to be solved are probably in some preexisting order, and our new program needs to be able to read them in that order. Similarly, it needs to produce the answers required by the programs that may follow it in the overall processing scheme, and it must write out those answers in the format needed by the programs following it.
3. *Decompose the program into classes and their associated methods.* Define one or more classes, and determine how the classes interact with each other and with the outside world. Define a separate method to implement each interaction.
4. *Design the algorithm that you intend to implement for each method.* An ***algorithm*** is a step-by-step procedure for finding the solution to a problem. It is at this stage in the process that top-down design techniques come into play. The designer looks for logical divisions within a method, and divides the method up into subtasks along those lines. This process is called **decomposition**. If the subtasks are themselves large, the designer can break them up into even smaller subsubtasks. This process continues until the problem has been divided into many small pieces, each of which does a simple, clearly understandable job. (As we shall see in later chapters, these separate pieces can become separate methods.)

   After the problem has been decomposed into small pieces, each piece is further refined through a process called **stepwise refinement**. In stepwise refinement, a designer starts with a general description of what the piece of code should do and then defines the functions of the piece in greater and greater detail until they are specific enough to be turned into Java statements. Stepwise refinement is usually done with **pseudocode**, which will be described in the next section.

   It is often helpful to solve a simple example of the problem by hand during the algorithm development process. If the designer understands the steps that he or she went through in solving the problem by hand, then he or she will be better able to apply decomposition and stepwise refinement to the problem.
5. *Turn the algorithm into Java statements.* If the decomposition and refinement process was carried out properly, this step will be very simple. All the programmer will have to do is to replace pseudocode with the corresponding Java statements on a one-for-one basis.
6. *Test the resulting Java program.* This step is the real killer. The components of the program must first be tested individually, if possible, and then the program as a whole must be tested. When testing a program, we must verify that it works correctly for *all legal input data sets.* It is very common for a program to be written, tested with some standard data set, and released for use, only to find that it produces the wrong answers (or crashes) with a different input data set. If the algorithm implemented in a program includes different branches, we must test all of the possible branches to confirm that the program operates correctly under every possible circumstance.

Large programs typically go through a series of tests before they are released for general use (see Figure 3.2). The first stage of testing is sometimes called **unit testing**. During unit testing, the individual components of the program are tested separately to confirm that they work correctly. After the unit testing is completed, the program goes through a series of *builds* during which the individual components are combined to produce the final program. The first build of the program typically includes only a few of the components. It is used to check the interactions among those components and the functions performed by their associated methods. In successive builds, more and more components are added, until the entire program is complete. Testing is performed on each build, and any errors (bugs) that are detected are corrected before moving on to the next build.

Testing continues even after the program is completed. The first complete version of the program is usually called the **alpha release**. It is used by the programmers and others very close to them in as many different ways as possible, and the bugs discovered during the testing are corrected. When the most serious bugs have been removed from the program, a new version, called the **beta release**, is prepared. The beta release is normally given to friendly outside users who have a need for the program in their normal day-to-day jobs. These users put the program through its paces under many different conditions and with many different input data sets, and they report any

Figure 3.2. A typical testing process for a large program

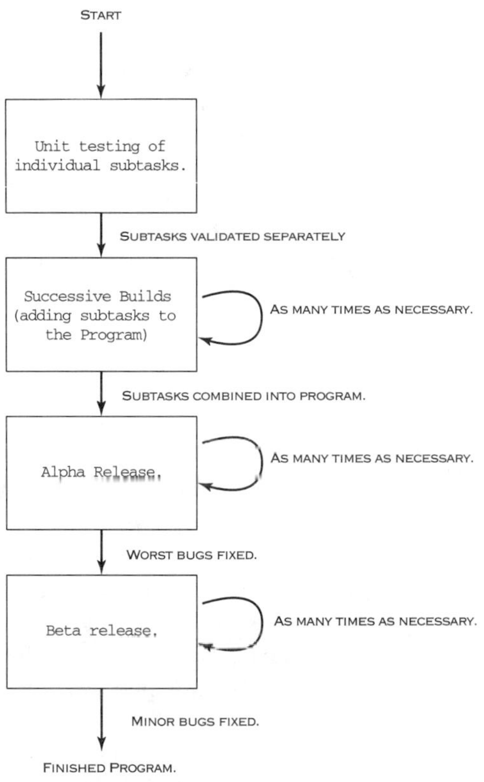

bugs that they find to the programmers. When those bugs have been corrected, the program is ready to be released for general use.

Because the programs in this book are fairly small, we will not go through the sort of extensive testing. However, we will follow the basic principles in testing all of our programs.

The program design process may be summarized as follows:

1. Clearly state the problem that you are trying to solve.
2. Define the inputs required by the program and the outputs to be produced by the program.
3. Decompose the program into classes and their associated methods.
4. Design the algorithm that you intend to implement for each method.
5. Turn the algorithm(s) into Java statements.
6. Test the Java program.

GOOD PROGRAMMING PRACTICE

Follow the steps of the program design process to produce reliable, understandable Java programs.

In a large programming project, the time actually spent programming is surprisingly small. In his book *The Mythical Man-Month,*[1] Frederick P. Brooks, Jr. suggests that in a typical large software project, 1/3 of the time is spent planning what to do (steps 1 through 3), 1/6 of the time is spent actually writing the program (step 4), and fully 1/2 of the time is spent in testing and debugging the program! Clearly, anything that we can do to reduce the testing and debugging time will be very helpful. We can best reduce the testing and debugging time by doing a very careful job in the planning phase and by using good programming practices. Using good programming practices will reduce the number of bugs in the program and will make the ones that do creep in easier to find.

Once a program has been created, it may be used over a lifetime of 20 years or more. Conditions will change during that time, and the program will have to be modified repeatedly over the course of its life. *The cost of the maintenance and modification of a program over its lifetime ususally exceeds the cost required of writing the program in the first place.* The modifications will usually be made by programmers other than the ones who originally wrote the code, and these programmers will be relatively unfamiliar with the program. It is during maintenance that good programming practices really pay off. If a program is well designed, well documented, and uses good programming practices, it will be easier (and cheaper) to modify the program without introducing new bugs.

3.2 USE OF PSEUDOCODE

As a part of the design process, it is necessary to describe the algorithm that you intend to implement. The description of the algorithm should be in a standard form that is easy for both you and other people to understand, and the description should aid you in turn-

[1]*The Mythical Man-Month,* by Frederick P. Brooks Jr., Addison-Wesley, 1974.

ing your concept into Java code. The standard forms that we use to describe algorithms are called **structures**, and an algorithm described using these structures is called a structured algorithm. When the algorithms in a program are implemented in a structured way, the resulting program is called a **structured program**.

The structures used to build algorithms are often described using pseudocode. Pseudocode is a hybrid mixture of Java and English. It is structured like Java, with a separate line for each distinct idea or segment of code, but the descriptions on each line are in English. Each line of the pseudocode should describe its idea in plain, easily understandable English. Pseudocode is very useful for developing algorithms, since it is flexible and easy to modify. It is especially useful, since pseudocode can be written and modified with the same text editor used to write the Java program. That is, no special graphical capabilities are required.

For example, the pseudocode for the algorithm of the `main` method in Example 2-2 is as follows:

```
Prompt user to enter temperature in degrees Fahrenheit
Read temperature in degrees Fahrenheit (tempF)
tempK in kelvins ← (5./9.) * (tempF - 32) + 273.15
Write temperature in kelvins
```

Notice that a left arrow (←) is used instead of an equal sign (=) to indicate that a value is stored in a variable, since this avoids any confusion between assignment and equality. Pseudocode is intended to aid you in organizing your thoughts before converting them into Java code.

3.3 RELATIONAL AND LOGICAL OPERATORS

Relational operators and **logical operators** are two types of operators whose results are `boolean` `true` or `false` values. They are used to control many looping and branching structures in Java, as we shall see later in this chapter and in Chapter 4.

Relational Operators

Relational operators are operators with *two numerical operands* that yield a `boolean` (`true` or `false`) result. The result depends on the *relationship* between the two values being compared, so these operators are called relational. If the relationship expressed by the operator is true, then the result of the operation is `true`; otherwise, the result is `false`. The six relational operators are summarized in Table 3-1.

TABLE 3-1 Relational Operators

OPERATOR	SAMPLE EXPRESSION	MEANING
	Relational operators:	
`>`	`x > y`	`true if` $x > y$
`<`	`x < y`	`true if` $x < y$
`>=`	`x >= y`	`true if` $x \geq y$
`<=`	`x <= y`	`true if` $x \leq y$
	Equality operators:	
`==`	`x == y`	`true if` $x = y$
`!=`	`x != y`	`true if` $x \neq y$

All of these operators associate from left to right. The relational operators >, <, >=, and <= are of equal precedence, below that of the + and – operators. The equality operators == and != are also of equal precedence, just below the relational operators.

The following are some relational operations and their results:

OPERATION	RESULT
`3 < 4`	`true`
`3 <= 4`	`true`
`3 == 4`	`false`
`3 > 4`	`false`
`4 <= 4`	`true`
`'A' < 'B'`	`true`

The last logical expression is `true`, because characters are compared according to their positions in the Unicode character set and `'A'` (sequence number 65) is less than `'B'` (sequence number 66).

The equivalence relational operator is written with two equals signs, while the assignment operator is written with a single equals sign. These are very different operators, and beginning programmers often confuse them. The == symbol is a *comparison* operation that returns a `boolean` result, while the = symbol *assigns* the value of the expression to the right of the equal sign to the variable on the left of the equal sign. It is a very common mistake for beginning programmers to use a single equals sign when trying to do a comparison.

PROGRAMMING PITFALLS

Be careful not to confuse the equivalence relational operator (==) with the assignment operator (=).

In the hierarchy of operations, relational operators are evaluated after all arithmetic operators have been evaluated. Therefore, the following two expressions are equivalent (both are `true`):

```
7 + 3 < 2 + 11
(7 + 3) < (2 + 11)
```

Also, if a comparison is between two operands of differing noncharacter types, the lower ranking type is promoted to the higher ranking type. For example, if a comparison is between a `double` and an `int` value, then the `int` value is promoted to a `double` value before the comparison is performed. Comparisons between numerical data and character data are legal, with the comparison based on the collating sequence of the character:

`4 == 4.`	`true`	(`int` is converted to `double` and comparison is made)
`65 <= 'A'`	`true`	(The sequence value of `'A'` is 65, so 65 <= 65)

Logical Operators

Logical operators are operators with *one or two boolean operands* that yield a boolean (true or false) result. There are six logical operators: && (logical AND), & (boolean logical AND), || (logical OR), | (boolean logical inclusive OR), ^ (boolean logical exclusive OR), and ! (logical NOT). These operators all accept boolean (true or false) operands and return a boolean result.

The general form of a binary logical operation is

$$l_1 \text{ op } l_2 ,$$

where l_1 and l_2 are boolean expressions, variables, or constants, and op is one of the first five of the above binary logical operators.

The results of the operators are summarized in the following *truth tables*, which show the result of each operation for all possible combinations of l_1 and l_2:

TABLE 3-2(a) Truth Table for Binary Logical Operators

l_1	l_2	l_1 && l_2	l_1 & l_2	$l_1 \mid\mid l_2$	$l_1 \mid l_2$	l_1 ^ l_2
false	false	false	false	false	false	false
false	true	false	false	true	true	true
true	false	false	false	true	true	true
true	true	true	true	true	true	false

TABLE 3-2(b) Truth Table for NOT Operator

l_1	$!l_1$
false	true
true	false

To understand these operators, consider the following expressions and their results:

```
(7 > 6) && (2 < 1)        false
(7 > 6) || (2 < 1)        true
!(7 > 6)                  false
```

Remember that relational operators produce a boolean result. In the first case, 7 > 6 is true, and 2 < 1 is false. Thus the result of the logical AND (7 > 6) && (2 < 1) is false, while the result of the logical OR (7 > 6) || (2 < 1) is true. Similarly, since 7 > 6 is true, !(7 > 6) is false.

These operators are most commonly used to combine the results of two or more relational operators to create some test. For example, suppose that we are examining pairs of (x,y) points in a Cartesian plane, and we would like to determine if a point lies in the second quadrant. For a point to lie in the second quadrant, it must have an x value less than zero and a y value greater than zero, as shown inFigure 3.3. A test to determine if the point lies in the second quadrant would be completed as shown:

```
boolean quadrant2;
...
quadrant2 = x < 0 && y > 0;
```

The result of the relational operator `x < 0` is `true` or `false`, and the result of the relational operator `y > 0` is also `true` or `false`. The logical operator then combines these two `boolean` values to calculate the result of the overall expression.

In the hierarchy of operations, logical operators are evaluated *after all arithmetic operations and all relational operators have been evaluated.* Therefore, the `boolean` results of the relational operators are calculated before the AND operator attempts to use them.

Looking at the above truth tables, it appears that the logical AND (`&&`) and the boolean logical AND (`&`) are identical, and that the logical OR (`||`) and the boolean inclusive logical OR (`|`) are identical. Why do these duplicate operators exist? This answer is that *the operators* `&&` *and* `||` *can perform partial evaluations of their operands, while the operators* `&` *and* `|` *always perform full evaluations of their operands.*

If we look at Table 3-2(*a*), we can see that the AND operators will only produce a `true` result of both operands l_1 and l_2 are true. If Java evaluates the first operand l_1 and finds that it is `false`, it already knows that the result of the AND will be `false`. Therefore, there is no need to evaluate l_2. Similarly, we can see that the inclusive OR operators will only produce a `false` result of both operands l_1 and l_2 are `false`. Therefore, if Java evaluates the first operand l_1 and finds that it is `true`, it already knows that the result of the OR will be and `true`, and there is no need to evaluate l_2.

The difference between `&&` and `||` on the one hand, and `&` and `|` on the other hand, is that `&&` and `||` permit decisions to be made if possible after evaluating only `||`, while `&` and `|` always evaluate both operands before making a decision. This difference is illustrated in the following program:

```
// This program illustrates the use of the AND operators
public class TestAnd {
   // Define the main method
   public static void main(String[] args) {

      int i = 10, j = 9;
      boolean test;

      // Demonstrate &&
      test = i > 10 && j++ > 10;
      System.out.println(i);
      System.out.println(j);
      System.out.println(test);

      // Demonstrate &
      test = i > 10 & j++ > 10;
      System.out.println(i);
      System.out.println(j);
      System.out.println(test);
   }
}
```

When this program is executed, the results are as follows.

```
10
9
false
10
10
false
```

Figure 3.3. For a point (x,y) to line in the second quadrant of a Cartesian plane, its x value must be less than zero and its y value must be greater than zero.

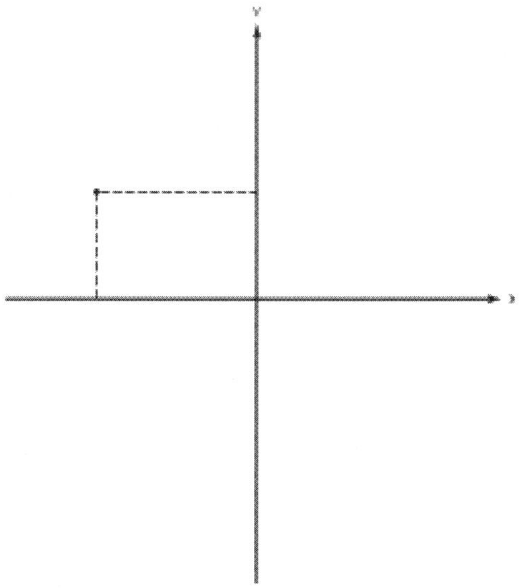

Note that the command `j++` on line 10 is never executed, since the first operand is already false. Therefore, the value of `j` remains 9. The `j++` on line 16 *is* executed, since the `&` operation always evaluates both operands before reaching a decision. Therefore, the value of `j` is increased to 10.

The hierarchy of all operations that we have seen so far is summarized in Table 3-3. In this table, all operators on the same line have equal precedence and are evaluated in the order indicated by the associativity property for that line.

TABLE 3-3 Hierarchy of Operations

OPERATORS	ASSOCIATIVITY	TYPE
`()`	left to right	parentheses
`++ -- + - ! (type)`	right to left	unary
`* / %`	left to right	multiplicative
`+ -`	left to right	additive
`< <= > >=`	left to right	relational
`== !=`	left to right	equality
`&`	left to right	boolean logical AND
`^`	left to right	boolean logical exclusive OR
`|`	left to right	boolean logical inclusive OR
`&&`	left to right	logical AND
`||`	left to right	logical OR
`?:`	right to left	conditional (*described later in this chapter*)
`= += -= *= /= %=`	right to left	assignment

EXAMPLE 3-1 Assume that the following variables are initialized with the values shown, and calculate the result of the specified expressions:

```
var1 = true;
var2 = true;
var3 = false;
```

EXPRESSION	RESULT
a. `! var1`	`false`
b. `var1 | var3`	`true`
c. `var1 && var3`	`false`
d. `var2 ^ var3`	`true`
e. `var1 && var2 || var3`	`true`
f. `var1 | var2 & var3`	`true`
g. `(var1 | var2) & var3`	`false`

The `&` operator is evaluated before the `|` operator in Java. Therefore, the parentheses in part g of this example were required. If they had been absent, the expression in part g would have been evaluated in the order `var1 | (var2 & var3)`.

3.4 SELECTION STRUCTURES

Selection structures are Java statements that permit us to select and execute specific sections of code (called blocks) while skipping other sections of code. They are variations of the `if` structure and the `switch` structure.

The `if` Structure

The `if` structure specifies that a statement (or a block of code) will be executed *if and only if a certain boolean expression is true.* For example, suppose that we were writing a grading program, and we wanted the program to print "Passed" if and only if the student's grade is greater than 70. If the student's grade is less than or equal to 70, then the program will not write out anything. The pseudocode for such a statement would be the following:

```
if the student's grade is greater than 70
   Print "Passed"
```

The Java if structure has the following form:

```
if (boolean_expr)
   statement;
```

or

```
if (boolean_expr) statement;
```

If the boolean expression is `true`, the program executes the statement. If the boolean expression is `false`, then the program skip the statement and executes the next statement after the `if` structure. The Java statement corresponding to the pseudocode above would be

```
if ( grade > 70 )
   System.out.println("Passed");
```

Figure 3.4. An `if` structure consists of a control expression and a controlled statement. The controlled statement will be executed if and only if the control expression evaluates to `true`.

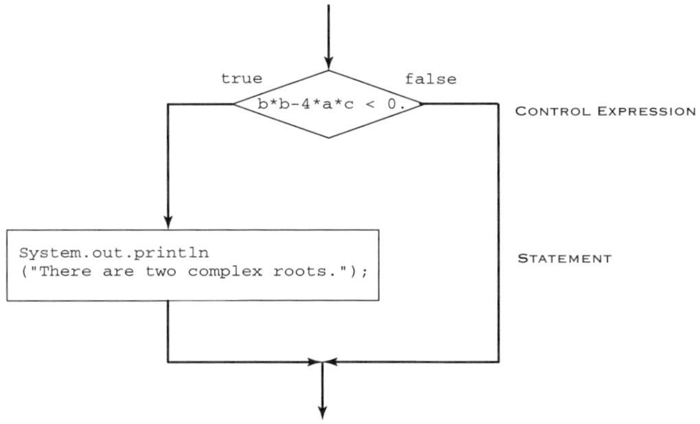

Note that the conditional statement is indented in the `if` structure. This is not a requirement of the Java compiler, since Java ignores *whitespace characters,* such as blanks, tabs, and newlines. However, the indenting makes the program easier for people to read. You should always indent the body of any structure to distinguish it from the surrounding code.

GOOD PROGRAMMING PRACTICE

Always indent the body of any structure by three or more spaces to improve the readability of the code.

As an example of an `if` structure, consider the solution of a quadratic equation of the form

$$ax^2 + bx + c = 0 \,. \tag{3-1}$$

The solution to this equation is

$$x = \frac{-b \pm \sqrt{b^2 - 4ac}}{2a} \,. \tag{3-2}$$

The term $b^2 - 4ac$ is known as the *discriminant* of the equation. If $b^2 - 4ac > 0$, then there are two distinct real roots to the quadratic equation. If $b^2 - 4ac = 0$, then there is a single repeated root to the equation, and if $b^2 - 4ac < 0$, then there are two complex roots to the quadratic equation.

Suppose that we wanted to examine the discriminant of the quadratic equation and tell a user if the equation has complex roots. In pseudocode, the `if` structure to do this would take the following form:

```
If b*b - 4*a*c < 0.
   Write message that equation has two complex roots.
```

In Java, the `if` structure is as follows.

```
if ((b*b - 4*a*c) < 0.)
   System.out.println("There are two complex roots.");
```

This structure is shown pictorially in Figure 3.4.

The `if/else` Structure

In the simple `if` structure, a statement is executed if the controlling `boolean` expression is `true`. If the controlling `boolean` expression is `false`, the statement is skipped.

Sometimes we may want to execute one statement if some condition is `true`, and a different statement if the condition is `false`. This can be accomplished with an `if/else` structure, which takes the following form:

```
if (boolean_expr)
   statement 1;
else
   statement 2;
```

If the `boolean` expression is `true`, then the program executes statement 1 and skips to the statement that follows statement 2. Otherwise, the program executes statement 2.

To illustrate the use of the `if/else` structure, let's reconsider the quadratic equation once more. Suppose that we wanted to examine the discriminant of a quadratic equation and tell a user whether the equation has real or complex roots. If the discriminant is greater than or equal to zero, there are real roots. Otherwise, the equation has complex roots. In pseudocode, this `if/else` structure would take the following form:

```
If (b*b - 4.*a*c) >= 0
   Write message that equation has real roots.
else
   Write message that equation has complex roots.
```

The Java statements to do this are as follows:

```
if ( (b*b - 4.*a*c) >= 0. )
   System.out.println("There are real roots.");
else
   System.out.println("There are complex roots.");
```

This structure is shown pictorially in Figure 3.5. Note that *the statement in the* `else` *clause of an* `if/else` *structure can be another* `if/else` *structure.* This cascading of structures allows us to make more complex selections. An example of cascaded if structures is shown:

```
if (boolean_expr_1)
   statement 1;
else if (boolean_expr_2)
   statement 2;
else
   statement 3;
```

If *boolean_expr_1* is `true`, then the program executes statement 1 and skips to the statement that follows statement 3. Otherwise, the program tests whether *boolean_expr_2* is `true`. If it is, then the program executes statement 2 and skips to the statement that follows statement 3. If both *boolean_expr_1* and *boolean_expr_2* are `false`, then the program executes statement 3. This structure might be more clearly represented as follows:

```
if (boolean_expr_1)
   statement 1;
else
   if (boolean_expr_2)
      statement 2;
   else
      statement 3;
```

Figure 3.5. An `if` structure with an `else` clause. Either statement 1 or statement 2 will be executed, depending on the result of the control expression.

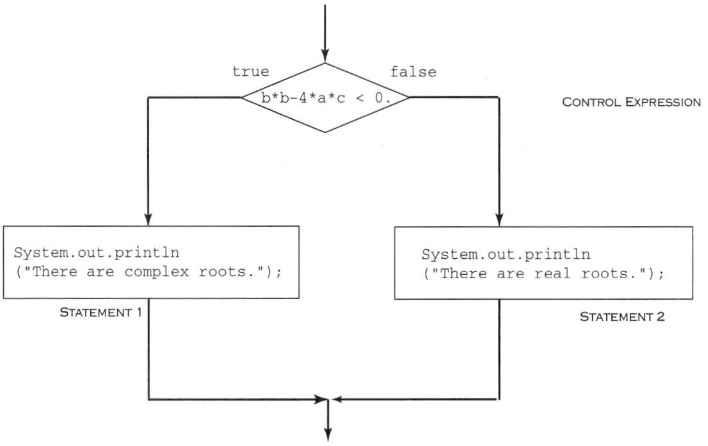

The above structure is perhaps clearer, since it emphasizes the fact that the later `ifs` are inside the `else` clauses of the previous `ifs`. Since Java ignores whitespace, both ways of writing this structure are equivalent.

Any number of `if/else` structures can be cascaded to produce arbitrarily complex selection structures.

To illustrate the use of cascaded `if/else` structures, let's reconsider the quadratic equation once more. Suppose that we wanted to write a program that examines the discriminant of a quadratic equation and tells the user whether the equation has two complex roots, two identical real roots, or two distinct real roots. In pseudocode, this construct would take the following form:

```
If (b*b - 4.*a*c) < 0
   Write message that equation has two complex roots.
else if (b*b - 4.*a*c) == 0.
   Write message that equation has two identical real roots.
else
   Write message that equation has two distinct real roots.
```

The Java statements to do this are as follows:

```
if ((b*b - 4.*a*c) < 0. )
   System.out.println("There are two complex roots.");
else if ( (b*b - 4.*a*c) == 0. )
   System.out.println("There are 2 identical real roots.");
else
   System.out.println("There are 2 distinct real roots.");
```

This cascaded structure is shown pictorially in Figure 3.6.

Executing Multiple Statements in an `if` Structure

The `if` and `if/else` structures are each designed to execute a *single statement* if a particular condition is `true`. What would happen if we needed to execute multiple statements in response to a condition? Java's answer to this problem is the **compound statement**.

A compound statement is a set of statements enclosed between a pair of braces ({ and }). A compound statement can be used anywhere in Java where an ordinary state-

Figure 3.6. A cascaded `if` structure. The second control expression is executed if and only if the first control expression is false.

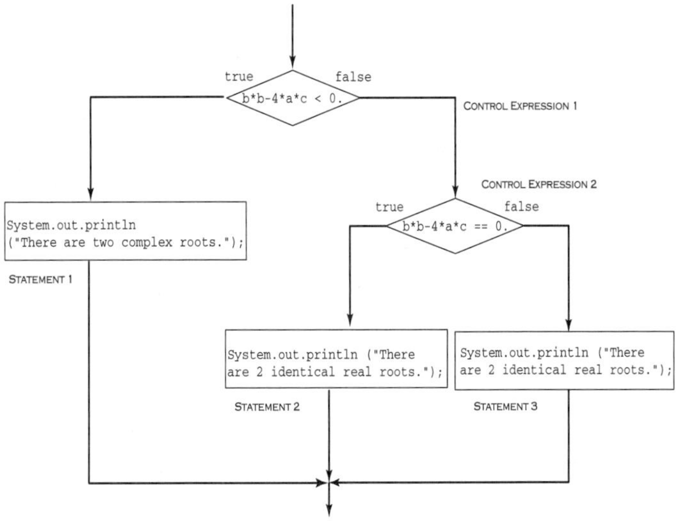

ment is expected. Thus, an `if/else` structure containing multiple lines takes the following form:

```
if (logical_expr){
   statement 1;
   statement 2;
   statement 3;
}
else {
   statement 4;
   statement 5;
   statement 6;
}
```

Examples Using `if` Structures

We will now look at two examples that illustrate the use of `if` structures.

EXAMPLE 3-2: THE QUADRATIC EQUATION

Design and write a program to solve for the roots of a quadratic equation, regardless of type.

SOLUTION

We will follow the design steps outlined earlier in the chapter.

1. **State the problem**. The problem statement for this example is very simple. We want to write a program that will solve for the roots of a quadratic equation, whether they are distinct real roots, repeated real roots, or complex roots.
2. **Define the inputs and outputs**. The inputs required by this program are the coefficients a, b, and c of the quadratic equation

$$ax^2 + bx + c = 0\ . \tag{3-1}$$

The output from the program will be the roots of the quadratic equation, whether they are distinct real roots, repeated real roots, or complex roots.

3. **Decompose the program into classes and their associated methods.** For the next three chapters, this step will be trivial, because every program will have only one class. We will begin learning how to decompose problems into separate classes in Chapter 7, after we learn more about the structure of classes. For now, there will be a single class, and only the `main` method within that class. We will call the class `QuadraticEquation` and make it a subclass of the root class `Object`. (If no other superclass is explicitly specified in the class definition, any new class is automatically a subclass of `Object`.)
4. **Design the algorithm that you intend to implement for each method.** There is only one method in this program. The `main` method can be broken down into three major sections, whose functions are input, processing, and output:

```
Read the input data
Calculate the roots
Write out the roots
```

We will now break each of the above major sections into smaller, more detailed pieces. There are three possible ways to calculate the roots, depending on the value of the discriminant, so it is logical to implement this algorithm with an `if/else` structure. The resulting pseudocode is:

```
Prompt the user for the coefficient a.
Read a
Prompt the user for the coefficient b.
Read b
Prompt the user for the coefficient c.
Read c
discriminant ← b*b - 4. * a * c
if discriminant > 0 {
   x1 ← ( -b + Math.sqrt(discriminant) ) / ( 2 * a )
   x2 ← ( -b - Math.sqrt(discriminant) ) / ( 2 * a )
   Write message that equation has two distinct real roots.
   Write out the two roots.
}
else if discriminant == 0 {
   x1 ← -b / ( 2. * a )
   Write message that equation has two identical real roots.
   Write out the repeated root.
}
else {
   realPart ← -b / ( 2. * a )
   imagPart ← Math.sqrt ( Math.abs( discriminant ) ) / ( 2 * a )
   Write message that equation has two complex roots.
   Write out the two roots.
}
```

5. **Turn the algorithm into Java statements.** The final Java code is shown in Figure 3.7, shown at the top of the next page. Note that the `if` structure is shown in bold face.

```
/*
   Purpose:
     This program solves for the roots of a quadratic equation of the
     form a*x*x + b*x + c = 0. It calculates the answers regardless of
     the type of roots that the equation possesses.

Record of revisions:
       Date     Programmer     Description of change
       ====     ==========     =====================
     3/27/98  S. J. Chapman  Original code

*/
import chapman.io.*;
public class QuadraticEquation {

   // Define the main method
   public static void main(String[] args) {

      // Declare variables, and define each variable
      double a;             // Coefficient of x**2 term of equation
      double b;             // Coefficient of x term of equation
      double c;             // Constant term of equation
      double discriminant;  // Discriminant of the equation
      double imagPart;      // Imag part of equation (for complex roots)
      double realPart;      // Real part of equation (for complex roots)
      double x1;            // 1st soln of equation (for real roots)
      double x2;            // 2nd soln of equation (for real roots)

      // Create a StdIn object
      StdIn in = new StdIn();

      // Prompt the user for the coefficients of the equation
      System.out.println("This program solves for the roots of a");
      System.out.println("quadratic equation of the form ");
      System.out.println("A * X*X + B * X + C = 0.");
      System.out.print("Enter the coefficient A: ");
      a = in.readDouble();
      System.out.print("Enter the coefficient B: ");
      b = in.readDouble();
      System.out.print("Enter the coefficient C: ");
      c = in.readDouble();

      // Calculate discriminant
      discriminant = b*b - 4. * a * c;

      // Solve for the roots, depending on the discriminant
      if ( discriminant > 0. ) {

         // Two real roots...
         x1 = ( -b + Math.sqrt(discriminant) ) / ( 2 * a );
         x2 = ( -b - Math.sqrt(discriminant) ) / ( 2 * a );
         System.out.println("This equation has 2 real roots:");
         System.out.println("X1 = " + x1 + ", X2 = " + x2);
      }

      else if ( discriminant == 0. ) {

         // One repeated root
         x1 = ( -b ) / ( 2 * a );
         System.out.println("This equation has 2 identical real roots:");
         System.out.println("X1 = X2 = " + x1);
      }
```

Figure 3.7. *(cont.)*

```
         else {
            // Complex roots...
            realPart = ( -b ) / ( 2. * a );
            imagPart = Math.sqrt( Math.abs ( discriminant ) ) / ( 2 * a );
            System.out.println("This equation has complex roots:");
            System.out.println("X1 = " + realPart + " +i " + imagPart);
            System.out.println("X2 = " + realPart + " -i " + imagPart);
         }
      }
   }
```

Figure 3.7. Program to solve for the roots of a quadratic equation

6. **Test the program**. Next, we must test the program using real input data. Since there are three possible paths through the program, we must test all three paths before we can be certain that the program is working properly. From Equation (3-2), it is possible to verify the solutions to the following equations:

$$x^2 + 5x + 6 = 0, \qquad x = -2, \text{ and } x = -3,$$

$$x^2 + 4x + 4 = 0, \qquad x = -2,$$

$$x^2 + 2x + 5 = 0, \qquad x = -1 \pm i2.$$

If this program is compiled and run three times with the above coefficients, the results are as follows (user inputs are shown in bold face):

```
C:\book\java\chap3>jave QuadraticEquation
This program solves for the roots of a quadratic
equation of the form A * X*X + B * X + C = 0.
Enter the coefficient A: 1
Enter the coefficient B: 5
Enter the coefficient C: 6
This equation has two real roots:
X1 = -2.0, X2 = -3.0

This program solves for the roots of a quadratic
equation of the form A * X*X + B * X + C = 0.
Enter the coefficient A: 1
Enter the coefficient B: 4
Enter the coefficient C: 4
This equation has two identical real roots:
X1 = X2 = -2.0

This program solves for the roots of a quadratic
equation of the form A * X*X + B * X + C = 0.
Enter the coefficient A: 1
Enter the coefficient B: 2
Enter the coefficient C: 5
This equation has complex roots:
X1 = -1.0 +i 2.0

X2 = -1.0 -i 2.0
```

The program gives the correct answers for our test data in all three possible cases.

EXAMPLE 3-3: EVALUATING A FUNCTION OF TWO VARIABLES

Write a Java program to evaluate a function $f(x,y)$ for any two user-specified values x and y. The function $f(x,y)$ is defined as follows:

$$f(x, y) = \begin{cases} x + y & x \geq 0 \text{ and } y \geq 0 \, , \\ x + y^2 & x \geq 0 \text{ and } y < 0 \, , \\ x^2 + y & x < 0 \text{ and } y \geq 0 \, , \\ x^2 + y^2 & x < 0 \text{ and } y < 0 \, . \end{cases}$$

SOLUTION

The function $f(x,y)$ is evaluated differently depending on the signs of the two independent variables x and y. To determine the proper equation to apply, it will be necessary to check for the signs of the x and y values supplied by the user. We will use the following steps to solve the problems:

1. **State the problem**. This problem statement is very simple: Evaluate the function $f(x,y)$ for any user-supplied values of x and y.
2. **Define the inputs and outputs**. The inputs required by this program are the values of the independent variables x and y. The output from the program will be the value of the function $f(x,y)$.
3. **Decompose the program into classes and their associated methods**. Again, there will be a single class and only the `main` method within that class. We will call the class `Eval` and make it a subclass of the root class `Object`.
4. **Design the algorithm that you intend to implement for each method**. The `main` method can be broken down into three major sections, whose functions are input, processing, and output:

   ```
   Read the input values x and y
   Calculate f(x,y)
   Write out f(x,y)
   ```

   We will now break each of the above major sections into smaller, more detailed pieces. There are four possible ways to calculate the function $f(x,y)$, depending on the values of x and y, so it is logical to implement this algorithm with a four-branched `if/else` structure.

   ```
   Prompt the user for thc value x.
   Read x.
   Prompt the user for the value x.
   Read y.
   Echo the input coefficients.
   if x >= 0 and y >= 0
      fun ← x + y
   else if x ≥ 0 and y < 0
      fun ← x + y*y
   else if x < 0 and y ≥ 0
      fun ← x*x + y
   else
      fun ← x*x + y*y
   Write out f(x,y).
   ```

5. **Turn the algorithm into Java statements**. The final Java code is shown in Figure 3.8.

```
/*
   Purpose:
     This program solves the function f(x,y) for user-specified x and
          y, where f(x,y) is defined as:

                       _
                      |
                      | x + y             x >= 0 and y >= 0,
                      | x + y*y           x >= 0 and y < 0,
          f(x,y)  =   | x*x + y           x < 0 and y >= 0,
                      | x*x + y*y         x < 0 and y < 0.
                      |_

   Record of revisions:
       Date      Programmer          Description of change
       ====      ==========          =====================
     03/27/98  S. J. Chapman         Original code
*/
import chapman.io.*;
public class Eval {

// Define the main method
   public static void main(String[] args) {

   // Declare variables, and define each variable
   double x;              // First independent variable
   double y;              // Second independent variable
   double fun;            // Resulting function

   // Create a StdIn object
   StdIn in = new StdIn();

   // Prompt the user for the coefficients of the equation
   System.out.print("Enter x: ");
   x = in.readDouble();
   System.out.print("Enter y: ");
   y = in.readDouble();

   // Calculate the function f(x,y) based on the sign of x and y
   if ( (x >= 0) && (y >= 0) )
      fun = x + y;
   else if ( (x >= 0) && (y < 0) )
      fun = x + y*y;
   else if ( (x < 0) && (y >= 0) )
      fun = x*x + y;
   else
      fun = x*x + y*y;

   // Write the value of the function.
   System.out.println("The value of the function is: " + fun);
   }
}
```

Figure 3.8. Program `Eval` from Example 3-3

6. **Test the program.** Next, we must test the program using real input data. Since there are four possible paths through the program, we must test all four paths before we can be certain that the program is working properly. To test all

four possible paths, we will execute the program with the four sets of input values $(x,y) = (2,3), (-2,3), (2,-3)$, and $(-2,-3)$. Calculating by hand, we see that

$$f(2, 3) = 2 + 3 = 5 ,$$
$$f(2, -3) = 2 + (-3)^2 = 11 ,$$
$$f(-2, 3) = (-2)^2 + 3 = 7 ,$$
$$f(-2, -3) = (-2)^2 + (-3)^2 = 13 .$$

If this program is compiled, and then run four times with the previously stated values, the results are as follows:

```
C:\book\java\chap3>java Eval
Enter x: 2
Enter y: 3
The value of the function is: 5.0

C:\book\java\chap3>java Eval
Enter x: 2
Enter y: -3
The value of the function is: 11.0

C:\book\java\chap3>java Eval
Enter x: -2
Enter y: 3

The value of the function is: 7.0

C:book\java\chap3>java Eval
Enter x: -2
Enter y: -3
The value of the function is: 13.0
```

The program gives the correct answers for our test values in all four possible cases.

Testing for Equality in `if` Structures

A common problem with `if` statements occurs when *floating-point* (`float` *and* `double`) *variables are tested for equality.* Because of small round-off errors during floating-point arithmetic operations, two numbers that theoretically should be equal may differ by a tiny amount, and the test for equality will fail. This failure can cause a program to execute the wrong statement in an `if/else` structure, producing a subtle and hard-to-find bug. For example, consider the quadratic equation program of Example 3-2. In that program, we concluded that a quadratic equation had two identical real roots if the discriminant, $b^2 - 4ac = 0$. Depending on the coefficients of the equation, round-off errors might cause the discriminant to be a very small nonzero number, say 10^{-14}, when it should theoretically be zero. If this happens, then the test for identical real roots would fail.

When working with floating-point variables, it is often a good idea to replace a test for equality with a test for *near equality.* For example, instead of testing to see if the `double` value `x` is equal to `10.`, you might test to see if `|x - 10.| < 1e-10`. Any value of `x` between 9.9999999999 and 10.0000000001 will satisfy the latter test, so round-off error will not cause problems. The Java statement,

```
if ( x == 10. )
```

would be replaced by

```
if ( Math.abs(x - 10.) <= 1e-10 )
```

GOOD PROGRAMMING PRACTICE

When working with floating-point values, it is a good idea to replace test for equality with tests for near equality to avoid improper results due to cumulative round-off errors.

Nested `if` Structures

The `if` structure is very flexible. Since any number of `if` structures can be cascaded in an `if/else...if/else...if/else` structure, where each succeeding `if` lies in the `else` clause of the preceding `if`, it is possible to implement any desired selection construct. When `if` structures lie inside other if structures, they are said to be **nested**.

Nested `if` structures are very useful and flexible, but they are also a very common source of bugs in Java programs. This happens because novice programmers often make mistakes in the use of `else` clauses. *The Java compiler always associates an* `else` *statement with the immediately preceding* `if` *statement,* unless told to otherwise by the proper use of braces ({}). Bugs can occur when using `else` clauses with nested `if` structures, because the Java compiler associates the `else` clause with a different `if` than the programmer expected it to.

Let's take a simple example to illustrate this problem. Suppose that we want to test two variables `x` and `y` to determine if they are both greater than zero. If they are, we will print out the string `"x and y are > 0"`. If `x` is `<= 0`, we would like to print out the string `"x <= 0"` instead. A programmer might attempt to implement this function as follows:

```
if ( x > 0)
   if (y > 0)
      System.out.println("x and y are > 0");
else
   System.out.println("x <= 0");
```

At first glance, this code seems to be saying that if `x > 0`, then we will test to see if `y > 0` and print the string if true, while if `x <= 0`, we will print out `"x <= 0"`. In other words, the `else` clause seems to be associated with the `if (x > 0)` statement. However, *the Java compiler associates the* `else` *with the immediately preceding* `if` *statement,* so the compiler actually interprets the structure as follows:

```
if ( x > 0)
   if (y > 0)
      System.out.println("x and y are > 0");
   else
      System.out.println("x <= 0");
```

The code will actually print out `"x <= 0"` when `x > 0` and `y <= 0`! This sort of error is known as the **dangling-else problem**.

The dangling-else problem can be avoided by using braces to force the `else` to be associated with the proper `if`.

```
if ( x > 0) {
   if (y > 0)
      System.out.println("x and y are > 0");
}
else
   System.out.println("x <= 0");
```

The braces indicate to the compiler that the second `if` is inside the body of the first `if`, and that the `else` is matched with the first `if`. (If the `else` had been *inside* the braces, it would have been matched with the second `if`).

GOOD PROGRAMMING PRACTICE

Use braces in nested if structures to make your intentions clear and avoid dangling `else` clauses.

Another common error with cascaded `if/else` structures occurs when a programmer fails to carefully consider the order of tests that he or she is performing. For example, suppose that we want to write out messages depending on the current temperature in degrees Fahrenheit as follows:

```
temp < 60       "cold"
temp > 70       "warm"
temp > 90       "hot"
```

One possible structure might be the following:

```
if (temp < 60)
   System.out.println("cold");
else if (temp > 70)
   System.out.println("warm");
else if (temp > 90)
   System.out.println("hot");
```

This structure will compile and execute, but *it will not work properly.* If the temperature is greater than 90, we would like to print out "hot". However, the statement `if (temp > 70)` will be executed first, and since it is true, the message "warm" will be printed out. *The test* `if (temp > 90)` *will never be executed.* This is an example of a logical error in a program.

If the statements are restructured as follows, they work correctly:

```
if (temp > 90)
   System.out.println("hot");
else if (temp > 70)
   System.out.println("warm");
else if (temp < 60)
   System.out.println("cold");
```

PROGRAMMING PITFALLS

Carefully consider the order in which your perform tests in your `if` structures to avoid creating incorrect branches.

The Conditional Operator

The **conditional operator** (`?:`) is essentially a compact `if/else` structure. It is a **ternary operator**, which means that it takes three arguments that together form a conditional expression. The first argument is a `boolean` expression, whose result must be either `true` or `false`. The second argument, which follows the question mark (`?`), is the value of the operation if the condition is `true`. The third argument, which follows the colon (`:`), is the value of the operation if the condition is `false`. For example, the `if/else` structure

```
if ( grade > 70 )
   System.out.println( "Passed");
else
   System.out.println( "Failed");
```

could also be written with the conditional operator as follows:

```
System.out.println( grade > 70 ? "Passed" : "Failed");
```

The precedence of the conditional operator is very low, ranking after all relational and logical operators and before assignment operators, so the expression `grade > 70` is evaluated as a `boolean` result before the conditional operator is evaluated.

The `switch` Structure

The `switch` structure is another form of selection. It permits a programmer to select a particular code block to execute based on the value of a single integer or character expression. The general form of a `switch` structure is as follows:

```
switch (switch_expr) {

   case case_selector_1:
      Statement 1;          //
      Statement 2;          // Block 1
      ...                   //
      break;
   case case_selector_2:
      Statement 1;          //
      Statement 2;          // Block 2
      ...                   //
      break;
   ...
   default:
      Statement 1;          //
      Statement 2;          // Block n
      ...                   //
      break;
}
```

When a `switch` structure is encountered, Java evaluates the value of *switch_expr*, and execution jumps to the case whose selector matches the value of the expression. The program then executes statements in order from that point on until a `break` statement is encountered. When a `break` statement is encountered, execution skips to the first statement after the end of the `switch`.

If the value of *switch_expr* does not match any case expression, the execution will jump to the `default` case if such a case is present. If one is not present, execution continues with the first statement after the end of the `switch`.

For example, consider the simple `switch` structure in the following program:

```
// This program tests the switch structure
import chapman.io.*;
public class TestSwitch {

   // Define the main method
   public static void main(String[] args) {
```

```
            int a;

            // Create a StdIn object
            StdIn in = new StdIn();

            // Prompt for an integer
            System.out.print("Enter an integer: ");
            a = in.readInt();

            // Switch
            switch (a) {
               case 1:
                  System.out.println("Value is 1");
                  break;
               case 2:
                  System.out.println("Value is 2");
                  break;
               default:
                  System.out.println("Other than 1 or 2");
                  break;
            }

            System.out.println("After switch");
         }
      }
```

If a value of 1 is read into this program, the switch expression a will be 1, so the statements from case 1 until the break will be executed, followed by the first statement after the switch. Similarly, if a value of 2 is read into this program, the switch expression a will be 2, so the statements from case 2 until the break will be executed, followed by the first statement after the switch. If any value other than 1 or 2 is read in, then the statements after default until the break will be executed, followed by the first statement after the switch.

```
C:\book\java\chap3>java TestSwitch
Enter an integer: 1
Value is 1
After switch

C:\book\java\chap3>java TestSwitch
Enter an integer: 2
Value is 2
After switch

C:\book\java\chap3>java TestSwitch
Enter an integer: 3
Other than 1 or 2
After switch
```

Note that the break statements are necessary for the proper operation of the switch structure. If they are missing, then execution will continue from the selected case through to the end of the switch. If the switch structure in the previous program did not have the break statements:

```
switch (a) {
   case 1:
      System.out.println("Value is 1");
   case 2:
```

```
            System.out.println("Value is 2");
         default:
            System.out.println("Other than 1 or 2");
      }
```

then the results would have been

```
C:\book\java\chap3>java TestSwitch
Enter an integer: 1
Value is 1
Value is 2
Other than 1 or 2
After switch

C:\book\java\chap3>java TestSwitch
Enter an integer: 2
Value is 2
Other than 1 or 2
After switch

C:\book\java\chap3>java TestSwitch
Enter an integer: 3
Other than 1 or 2
After switch
```

Also, note that the case selectors must be *mutually exclusive.* It is an error for the same selector to appear more than once in a single structure.

GOOD PROGRAMMING PRACTICE

The `switch` structure may be used to select among *mutually exclusive* options based on the results of a single integer or character expression.

The `switch` structure is never really necessary, since any selection that can be represented by a `switch` can also be represented by a cascaded `if/else` structure. For example, the `switch` in the previous program could be rewritten as follows:

```
if (a == 1)
   System.out.println("Value is 1");
else if (a == 2)
   System.out.println("Value is 2");
else
   System.out.println("Other than 1 or 2");
```

The `switch` structure is just a limited form of the cascaded `if/else` structure. However, some programmers prefer the `switch` structure for stylistic reasons.

PROGRAMMING PITFALLS

Be sure to include `break` statements in each case of a `switch` structure, so that only the statements in that case are executed when the case is selected.

PRACTICE!

This quiz provides a quick check to see if you have understood the concepts introduced in Sections 3.1 through 3.4. If you have trouble with the quiz, reread the sections, ask your instructor, or discuss the material with a fellow student. The answers to this quiz are found in the back of the book.

1. Suppose that the `double` variables `a`, `b`, and `c` contain the values `-10.`, `0.1`, and `2.1`, respectively, and that the `boolean` variable `b1`, `b2`, and `b3` contain the values `true`, `false`, and `false`, respectively. Is each of the following expressions legal or illegal? If an expression is legal, what will its result be?
   a. `a > b || b > c;`
   b. `(!a) || b1`
   c. `b1 & !b2`
   d. `a < b == b < c`
   e. `b1 || b2 && b3`
   f. `b1 | b2 && b3`

Write Java statements that perform the following functions:

2. If `x` is greater than or equal to zero, then assign the square root of `x` to variable `sqrtX` and print out the result. Otherwise, print out an error message about the argument of the square root function, and set `sqrtX` to zero.
3. A variable `fun` is calculated by dividing variable `numerator` by variable `denominator`. If the absolute value of `denominator` is less than 1.0e-30, write "Divide by zero error." Otherwise, calculate and print out `fun`.
4. The cost per mile for a rented vehicle is $0.50 for the first 100 miles, $0.30 for the next 200 miles, and $0.20 for all miles in excess of 300 miles. Write Java statements that determine the total cost and the average cost per mile for a given number of miles (stored in variable `distance`).

Examine the following Java statements. Are they correct or incorrect? If they are correct, what is output by them? If they are incorrect, what is wrong with them?

5.
```
if (volts > 125)
     System.out.println("WARNING: High voltage on line.");
if (volts < 105)
   System.out.println("WARNING: Low voltage on line.");
else
   System.out.println("Line voltage is within tolerances.");
```

6.
```
double i = 3., j = 5., k;
k = i > j ? i / j : j / i;
System.out.println("k = " + k);
```

7.
```
double a = 2 * Math.PI;
double b;
switch( a ) {
   case 1:
      b = Math.sqrt(a);
      break;
   case 2:
      b = Math.pow(a,3);
      break;
   default
      b = a;
      break;
}
```

PRACTICE!

8.
```
if (temperature > 37.)
   System.out.println("Human body temperature exceeded.");
else if (temperature > 100.)
   System.out.println("Boiling point of water exceeded.");
```

PROFESSIONAL SUCCESS: RAPID PROTOTYPING

The program design process described in this chapter is very useful for creating well thought-out working programs. However, the process of decomposition and stepwise refinement can sometimes be very difficult. Suppose that you break a program down into classes and methods, but you don't know how to implement the function to be performed by one of the methods. What do you do then?

If you don't know how to perform some function, write a quick prototype program that performs *only* that function, and play with it until you get satisfactory results. Once you clearly understand what you want to do, go back to the stepwise refinement process of the original program and continue from there.

In the real world, such *rapid prototyping* is often faster and more efficient than trying to get everything exactly right on paper before writing the first line of code. This is especially true when the program is performing a function that you have never done before.

However, be careful not to incorporate the rapid prototype code directly into your final program. The prototype is a learning experience. When you rewrite the code for the final program, you will do a much better job because you now know exactly what you want to do.

SUMMARY

- Pseudocode is a hybrid mixture of Java and English used to express programming thoughts without worrying about the details of Java syntax.
- The relational operators (>, >=, <, <=, ==, !=) compare two numerical operands, and produce a `boolean` result based on that comparison.
- The logical operators (&&, &, ||, |, ^, !) accept one or two `boolean` operands and produce a `boolean` result based on the values of the operands. The results of the operations are given in Table 3-2.
- The hierarchy of Java operations is summarized in Table 3-3.
- Selection structures are Java statements that permit a program to select and execute specific sections of code (statements or compound statements) while skipping other sections of code.
- The `if` selection structure executes a code block if its condition is `true`, and skips the code block if its condition is `false`.
- The `if/else` selection structure executes one code block if its condition is `true`, and another code block if its condition is `false`.
- The Java compiler always associates an `else` clause with the most recent `if`, unless braces are used to force the association of the `else` clause with a specific `if` statement.
- The `switch` selection structure allows a program to select a set of statements to execute based on the value of an integer control expression.

- The conditional operator (`?:`) is essentially a compact `if/else` structure that selects the value of one of two possible expressions, depending on the result of a `boolean` control expression.

APPLICATIONS: SUMMARY OF GOOD PROGRAMMING PRACTICES

The following guidelines introduced in this Chapter will help you to develop good programs:

1. Follow the steps of the program design process to produce reliable, understandable Java programs.
2. Always indent the body of any structure by three or more spaces to improve the readability of the code.
3. When working with floating-point values, it is a good idea to replace tests for equality with tests for near equality to avoid improper results due to cumulative round-off errors.
4. Use braces in nested `if` structures to make your intentions clear, and avoid dangling `else` clauses.
5. The `switch` structure may be used to select among *mutually exclusive* options, based on the results of a single integer or character expression.
6. Be sure to include `break` statements in each `case` of a `switch` structure so that only the statements within that `case` are executed when the case is selected.

KEY TERMS

algorithm
alpha release
beta release
`break` statement
compound statement
`if` structure
`if/else` structure
logical operator
null statement
pseudocode
relational operator
repetition
selection
structure
structured program
`switch` structure
top-down design
unit testing

Problems

1. The tangent function is defined as $\tan\theta = \sin\theta / \cos\theta$. This expression can be evaluated to solve for the tangent as long as the magnitude of $\cos\theta$ is not too near to zero. (If $\cos\theta$ is 0, evaluating the equation for $\tan\theta$ will produce the non-numerical value `Inf.`) Assume that θ is given in degrees, and write Java statements to evaluate $\tan\theta$ as long as the magnitude of $\cos\theta$ is greater than or equal to 10^{-20}. If the magnitude of $\cos\theta$ is less than 10^{-20}, write out an error message instead.
2. Which of the following expressions are legal in Java? If an expression is legal, evaluate it.
   a. `5.5 >= 5`
   b. `20 > 20`
   c. `!(6 > 5)`
   d. `15 <= 'A'`
   e. `true > false`
   f. `35 / 17. > 35 / 17`
   g. `17.5 && (3.3 > 2)`
3. The following Java statements are intended to alert a user to dangerously high oral thermometer readings (values are in degrees Fahrenheit). Are they in the correct order? If they are not, explain why and correct them.

```
if ( temp < 97.5 )
   System.out.println("Temperature below normal");
else if ( temp > 97.5 )
   System.out.println("Temperature normal");
else if ( temp > 99.5 )
   System.out.println("Temperature slightly high");
else if ( temp > 103.0 )
   System.out.println("Temperature dangerously high");
```

4. The cost of sending a package by an express delivery service is \$10.00 for the first two pounds and \$3.75 for each pound or fraction thereof over two pounds. If the package weighs more than 70 pounds, a \$10.00 excess weight surcharge is added to the cost. No package over 100 pounds will be accepted. Write a program that accepts the weight of a package in pounds and computes the cost of mailing the package. Be sure to handle the case of overweight packages.
5. Modify program `QuadraticEquation` to treat test for near equality instead of equality. The discriminant should be considered to be equal to zero if $|b^2 - 4ac| < 10^{-14}$.
6. The inverse sine method `Math.asin(x)` is only defined for the range $-1.0 \le$ `x` ≤ 1.0. If `x` is outside this range, an error will occur when the function is evaluated. The following Java statements calculate the inverse sine of a number if it is in the proper range and print an error message if it is not. Assume that `x` and `inverseSine` are both type `double`. Is the code correct or incorrect? If it is incorrect, explain why and correct it.

```
if ( Math.abs(x) <= 1. )
   inverseSine = Math.asin(x);
else
   System.out.println(x + " is out of range!");
```

7. What is wrong with the following code segment?

```
switch (n) {
   case 1:
      System.out.println("Number is 1");
   case 2:
      System.out.println("Number is 2");
      break;
   default:
      System.out.println("Number is not 1 or 2");
      break;
}
```

8. Write a Java program to evaluate the function

$$y(x) = \ln\frac{1}{1-x}$$

for any user-specified value of x, where ln is the natural logarithm (logarithm to the base e). Note that the natural logarithm is only defined for positive values; when an illegal value of x is entered, tell the user and terminate the program.

9. **Refraction** When a ray of light passes from a region with an index of refraction n_1 into a region with a different index of refraction n_2, the light ray is bent. (See Figure 3.9.) The angle at which the light is bent is given by *Snell's Law*

$$n_1 \sin\theta_1 = n_2 \sin\theta_2 , \qquad (3\text{-}20)$$

where θ_1 is the angle of incidence of the light in the first region, θ_2 is the angle of incidence of the light in the second region, and n_1 and n_2 are the indices of refraction for the first and second regions respectively. Using Snell's Law, it is possible to predict the angle of incidence of a light ray in Region 2 if the angle of incidence θ_1 in Region 1 and the indices of refraction n_1 and n_2 are known. The equation to perform this calculation is

Figure 3.9. A ray of light bends as it passes from one medium into another one. (a) If the ray of light passes from a region with a low index of refraction into a region with a higher index of refraction, the ray of light bends more towards the vertical. (b) If the ray of light passes from a region with a high index of refraction into a region with a lower index of refraction, the ray of light bends away from the vertical.

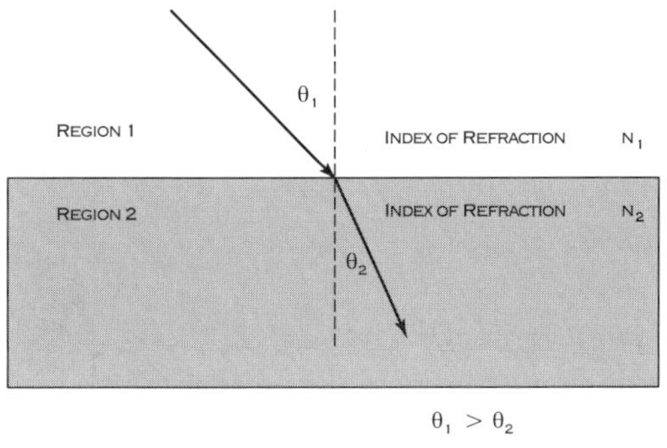

(a)

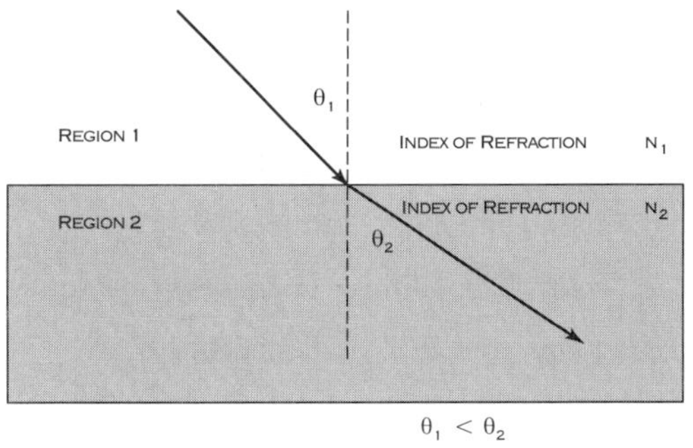

(b)

$$\theta_2 = \sin^{-1}\left(\frac{n_1}{n_2}\sin\theta_1\right). \tag{3-21}$$

Write a Java program to calculate the angle of incidence (in degrees) of a light ray in Region 2 given the angle of incidence θ_1 in Region 1 and the indices of refraction n_1 and n_2. (*Note*. If $n_1 > n_2$, then for some angles θ_1, Equation (3-20) will have no real solution, because the absolute value of the quantity $\left(\frac{n_1}{n_2}\sin\theta_1\right)$ will be greater than 1.0. When this occurs, all light is reflected back into Region 1, and no light passes into Region 2 at all. Your program must be able to recognize and properly handle this condition.) Test your program by running it for the following two cases:

a. $n_1 = 1.0$, $n_2 = 1.7$, and $\theta_1 = 45°$.

b. $n = 1.7$, $n_2 = 1.0$, and $\theta_1 = 45°$.

4

Repetition Structures

This chapter focuses on a different type of structure that allows us to control the order in which Java statements are executed: **repetition structures**, or **loops**. Repetition structures are Java structures that permit us to execute a sequence of statements more than once. There are three basic forms of repetition structures: **`while`** loops, **`do/while`** loops, and **`for`** loops. The first two types of loops repeat a sequence of statements an indefinite number of times until some specified control condition becomes `false`. In contrast, the `for` loops repeat a sequence of statements a specified number of times, and the number of repetitions is known before the loop starts.

SECTIONS

OBJECTIVES

After reading this chapter, you should be able to:

- To learn how to use `while` and `do/while` loops
- To learn how to use `for` loops
- To learn how to use the `continue` and `break` statements
- To learn how to create formatted output using the `chapman.io.Fmt` class

4.1 THE `while` LOOP

A `while` **loop** is a statement or block of statements that are repeated indefinitely as long as some condition is satisfied. The general form of a `while` loop in Java is

```
while ( boolean_expr )
   statement;
```

or, with a compound statement,

```
while ( boolean_expr ) {
   statement 1;
   statement 2;
   ...
}
```

When a `while` loop is encountered, Java evaluates the `boolean` expression. If *boolean_expr* is `true`, Java executes the statement(s) in the loop body. It then evaluates *boolean_expr* again. If it is still `true`, the statement is executed again. This process is repeated until the expression becomes `false`. When the expression becomes `false`, execution skips to the first statement after the loop. Note that if *boolean_expr* is `false` the first time the expression is evaluated, the statement will never be executed at all! The operation of a `while` loop is illustrated in Figure 4.1.

As an example of `while` loops, consider the following code:

```
k = 3;
while (k < 5) {
   System.out.print(" " + k);
   k++;
}
```

When this loop is first executed, `k` is equal to 3. The boolean expression `k < 5` is `true`, so the statements in the block are executed, writing out `k` and incrementing its value to 4. Then the boolean expression evaluated again. Since `k < 5` is still `true`, the statements in the block are executed again, writing out `k` and incrementing its value to 5. Then the `boolean` expression evaluated again. This time, `k < 5` is `false`, so execution skips to the first statement following the loop. The resulting output is as follows:

```
3 4
```

On the other hand, the following loop is never executed at all, because the `boolean` expression is `false` when loop is first reached:

```
k = 5;
while ( k < 5 ) {
   System.out.print(" " + k);
   k++;
}
```

Figure 4.1. The structure of a `while` loop. Note that the boolean expression is evaluated *first*, and the statement(s) are only executed if the expression is `true`.

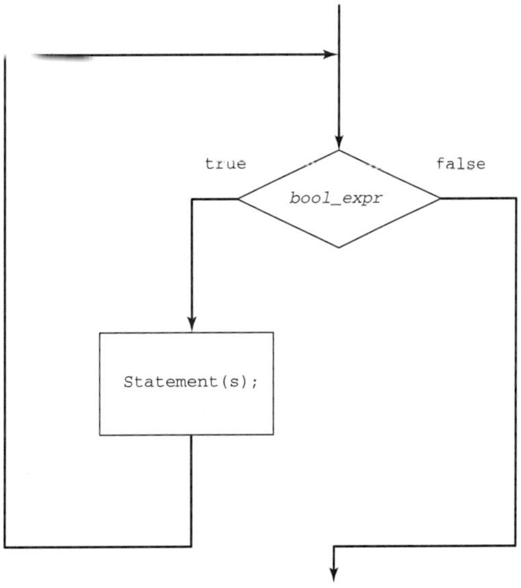

The way that `while` loops are defined in Java produces a nasty trap for programmers, one that you will probably stumble into several times as you learn the language. A typical Java `while` loop might be written as follows:

```
x = 4;
while ( x < 10 )
   x += 4;
```

Note that *there is no semicolon after the* `while`. That is, the `while` structure is not terminated until after the loop body. If we make a mistake and place a semicolon after the `while`, the statements would be:

```
x = 4;
while ( x < 10 );
   x += 4;
```

Unfortunately, this error creates a disaster. A semicolon by itself represents a **null statement**, which is a statement that does nothing. Recall that Java will happily accept multiple statements on a single line. In this case, the semicolon will create a null statement as the body of the loop, and the original statement `x += 4;` will be *outside* of the loop. In other words, it is as though we wrote the following code:

```
x = 4;
while ( x < 10 )
   ;
x += 4;
```

The Java compiler will compile this program with no warnings, and when the program executes, it will go into an infinite loop! Since the null statement never changes `x`, `x` will always be less than 10, and the loop will execute forever.

What makes this bug particularly dangerous is that the compiler gives no warning that anything is wrong. The program executes and runs, but produces an infinite loop. And of course, the error is very hard to spot unless you are specifically looking for it.

PROGRAMMING PITFALLS

Adding a semicolon after a `while` statement can produce a logical error. Java will compile and execute the program, but the program may go into an infinite loop.

We will now show an example statistical analysis program that is implemented using a `while` loop.

EXAMPLE 4-1: STATISTICAL ANALYSIS

It is very common in science and engineering to work with large sets of numbers, each of which is a measurement of some particular property that we are interested in. A simple example would be the grades on the first test in a college course. Each grade would be a measurement of how much a particular student has learned in the course to date.

Much of the time, we are not interested in looking closely at every single measurement that we make. Instead, we want to summarize the results of a set of measurements with a few numbers that tell us a lot about the overall data set. Two such numbers are the *average*, or *arithmetic mean*, and the *standard deviation* of the set of measurements. The average or arithmetic mean of a set of numbers is defined as

$$\bar{x} = \frac{1}{N} \sum_{i=1}^{N} x_i , \qquad (4\text{-}1)$$

where x_i is sample i out of N samples. The standard deviation of a set of numbers is defined as

$$s = \sqrt{\frac{N\sum_{i=1}^{N} x_i^2 - \left(\sum_{i=1}^{N} x_i\right)^2}{N(N-1)}}. \tag{4-2}$$

Standard deviation is a measure of the amount of scatter on the measurements; the greater the standard deviation is, the more scattered the points in the data set are.

Implement an algorithm that reads in a set of measurements and calculates the mean and the standard deviation of the input data set.

SOLUTION

This program must be able to read in an arbitrary number of measurements and then calculate the mean and standard deviation of those measurements. We will use a `while` loop to accumulate the input measurements before performing the calculations.

When all of the measurements have been read, we must have some way of telling the program that there are no more data to enter. For now, we will assume that all the input measurements are either positive or zero, and we will use a negative input value as a *flag* to indicate that there are no more data to read. If a negative value is entered, then the program will stop reading input values and will calculate the mean and standard deviation of the data set.

1. **State the problem**. Since we assume that the input numbers must be positive or zero, a proper statement of this problem would be to *calculate the average and the standard deviation of a set of measurements, assuming that all of the measurements are either positive or zero and that we do not know in advance how many measurements are included in the data set. A negative input value will mark the end of the set of measurements.*
2. **Define the inputs and outputs**. The inputs required by this program are an unknown number of positive or zero `double` numbers. The outputs from this program are a printout of the mean and the standard deviation of the input data set. In addition, we will print out the number of data points input to the program, since this is a useful check that the input data was read correctly.
3. **Decompose the program into classes and their associated methods.** Again, there will be a single class, and only the `main` method within that class. We will call the class `Stats` and make it a subclass of the root class `Object`.
4. **Design the algorithm that you intend to implement for each method**. The `main` method can be broken down into three major sections, whose functions are accumulating input data, processing, and output:

```
Accumulate the input data
Calculate the mean and standard deviation
Write out the mean, standard deviation, and number of points
```

The first major step of the program is to accumulate the input data. To do this, we will have to prompt the user to enter the desired numbers. When the numbers are entered, we will have to keep track of the number of values entered, plus the sum and the sum of the squares of those values. The pseudocode for these steps is as follows:

```
Initialize n, sumX, and sumX2 to 0
Prompt user for first number
Read in x
while (x >= 0)
   ++n;
   sumX ← sumX + x;
   sumX2 ← sumX2 + x*x;
   Prompt user for next number
   Read in next x
End of while
```

Note that we have to read in the first value before the `while` loop starts so that the `while` loop can have a value to test the first time it executes.

Next, we must calculate the mean and standard deviation. The pseudocode for this step is just the Java versions of Equations (4-1) and (4-2) and is the following:

```
xBar ← sumX / n
stdDev ← Math.sqrt((n*sumX2 - sumX*sumX) / (n*(n-1)))
```

Finally, we must write out the results:

```
Write out the mean value xBar
Write out the standard deviation stdDev
Write out the number of input data points n
```

5. **Turn the algorithm into Java statements**. The final Java program is shown in Figure 4.2, with the `while` loop shown in boldface.

6. **Test the program**. To test this program, we will calculate the answers by hand for a simple data set and then compare the answers to the results of the program. If we used three input values: 3, 4, and 5, then the mean and standard deviation would be the following:

$$\bar{x} = \frac{1}{N}\sum_{i=1}^{N} x_i = \frac{1}{3}\,12 = 4$$

$$s = \sqrt{\frac{N\sum_{i=1}^{N} x_i^2 - \left(\sum_{i=1}^{N} x_i\right)^2}{N(N-1)}} = 1$$

When the foregoing values are entered into the program, the results are as follows, with the input in boldface:

```
C:\book\java\chap4>java Stats
Enter first value: 3
Enter next value: 4
Enter next value: 5
Enter next value: -1
The mean of this data set is: 4.0
The standard deviation is: 1.0
The number of data points is: 3
```

The program gives the correct answers for our test data set.

```
/*
   Purpose:
      To calculate mean and the standard deviation of an input
      data set containing an arbitrary number of input values.

  Record of revisions:
      Date        Programmer      Description of change
      ====        ==========      =====================
     3/29/98      S. J. Chapman   Original code
*/
import chapman.io.*;
public class Stats {

   // Define the main method
   public static void main(String[] args) {

      // Declare variables, and define each variable
      int n = 0;              // The number of input samples.
      double stdDev = 0; // The standard deviation of the input samples.
      double sumX = 0;    // The sum of the input values.
      double sumX2 = 0;   // The sum of the squares of the input values.
      double x = 0;       // An input data value.
      double xBar = 0;    // The average of the input samples.

      // Create a StdIn object
      StdIn in = new StdIn();

      // Get first input value
      System.out.print("Enter first value: ");
      x = in.readDouble();

      // while loop to accumulate input values.
      while (x >= 0) {

         // Accumulate sums.
         ++n;
         sumX += x;
         sumX2 += x*x;

         // Read next value
         System.out.print("Enter next value: ");
         x = in.readDouble();
      }

      // Calculate the mean and standard deviation
      xBar = sumX / n;
      stdDev = Math.sqrt((n * sumX2 - sumX*sumX) / (n * (n-1)));

      // Tell user
      System.out.println("The mean of this data set is: " + xBar);
      System.out.println("The standard deviation is: " + stdDev);
      System.out.println("The number of data points is: " + n);
   }
}
```

Figure 4.2. Program to calculate the mean and standard deviation of a set of nonnegative real numbers.

In the previous example, we failed to follow the design process completely. This failure has left the program with a fatal flaw! Did you spot it?

We have failed because *we did not completely test the program for all possible types of inputs.* Look at the example once again. If we enter either no numbers or only one number, then we will be dividing by zero in the equations! The division-by-zero error will cause the program to produce an `Inf` (standing for infinity) result. We need to modify the program to detect this problem, inform the user of it, and stop gracefully.

A modified version of the program called `Stats1` is shown in Figure 4.3. Here, we check to see if there are enough input values before performing the calculations. If not, the program will print out an intelligent error message and quit. Test the modified program for yourself.

```
/*
   Purpose:
      To calculate mean and the standard deviation of an input
      data set containing an arbitrary number of input values.

   Record of revisions:
      Date        Programmer              Description of change
      ====        ==========              =====================
    3/29/98     S. J. Chapman             Original code
  1. 3/29/98    S. J. Chapman             Correct divide-by-0 error if 0 or
                                          1 input values given.
*/
import chapman.io.*;
public class Stats1 {

   // Define the main method
   public static void main(String[] args) {

      // Declare variables, and define each variable
      int n = 0;                // The number of input samples.
      double stdDev = 0;        // The standard deviation of the input samples.
      double sumX = 0;          // The sum of the input values.
      double sumX2 = 0;         // The sum of the squares of the input values.
      double x = 0;             // An input data value.
      double xBar = 0;          // The average of the input samples.

      // Create a StdIn object
      StdIn in = new StdIn();

      // Get first input value
      System.out.print("Enter first value: ");
      x = in.readDouble();

      // while loop to accumulate input values.
      while (x >= 0) {

         // Accumulate sums.
         ++n;
         sumX += x;
         sumX2 += x*x;

         // Read next value
         System.out.print("Enter next value: ");
         x = in.readDouble();
      }

      // Check to see if we have enough input data.
      if (n < 2)
         System.out.println("At least 2 values must be entered!");
      else {
         // There is enough information, so
         // calculate the mean and standard deviation
         xBar = sumX / n;
         stdDev = Math.sqrt((n * sumX2 - sumX*sumX) / (n * (n-1)));
```

Figure 4.3. *(cont.)*

```
            // Tell user.
            System.out.println("The mean of this data set is: " + xBar);
            System.out.println("The standard deviation is: " + stdDev);
            System.out.println("The number of data points is: " + n);
        }
    }
}
```

Figure 4.3. A modified statistical analysis program that avoids the divide-by-zero problems inherent in program `Stats`.

4.2 THE `do/while` LOOP

There is another form of the `while` loop in Java, called the **`do/while` loop**. The `do/while` structure has the following form:

```
do {
   Statement 1;
   ...
   Statement n;
} while( boolean_expr )
```

In this loop, statements 1 through *n* will be executed, and then the `boolean` expression will be tested. If the `boolean` expression is `true`, then statements 1 through *n* will be executed again. This process will be repeated until the `boolean` expression becomes `false`. When the bottom of the loop is reached and the `boolean` expression is `false`, the program will execute the first statement after the end of the loop. The operation of a `do/while` loop is illustrated in Figure 4.4.

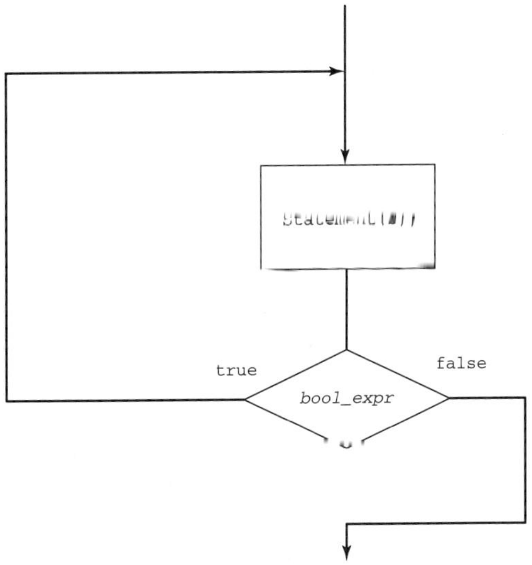

Figure 4.4. The structure of a `do/while` loop. Note that the statement(s) are executed *first*, and then boolean expression is evaluated. The statement(s) in this loop will be executed *at least once* even if the boolean expression is false before the loop starts.

The major difference between the `while` loop and the `do/while` loop is that *the test for the* `while` *loop is executed at the* ***top*** *of the loop, while the test for the* `do/while` *loop is executed at the* ***bottom*** *of the loop.* Because of this difference, the statements in a `do/while` loop will always be executed at least once.

GOOD PROGRAMMING PRACTICE

Use a `while` or `do/while` loop to repeat a set of statements indefinitely until a condition becomes `false`. Use a `while` loop in cases where you wish to perform the loop repetition test at the top of the loop, and use a `do/while` loop in cases where you wish to perform the loop repetition test at the bottom of the loop.

4.3 THE `for` LOOP

In the Java language, a loop that executes a block of statements a specified number of times is called a **`for` loop**. The `for` loop structure has the form

```
for ( index = initExpr; continueExpr; incrementExpr )
   Statement;
```

or

```
for ( index = initExpr; continueExpr; incrementExpr ) {
   Statement 1;
   ...
   Statement n;
}
```

where the expressions within the parentheses control the operation of the loop. The *`index`* is an integer whose value varies each time the loop is executed. This value is initialized to the value of *`initExpr`* when the loop first starts to execute. The *`continueExpr`* is then evaluated. If this expression is `true`, then the statement(s) in the body of the loop are executed. After the statement(s) are executed, the *`incrementExpr`* is executed to increment the value of the loop index, and *`continueExpr`* is reevaluated. If the expression is still `true`, then the statement(s) are executed again. This process is repeated until the *`continueExpr`* is `false`, at which time execution continues with the first statement following the loop. The operation of a `for` loop is illustrated in Figure 4.5.

A simple example of the `for` loop is shown below.

```
for ( int i = 1; i <= 2; i++ )
   System.out.println("i = " + i);
```

The index variable in this loop is `i`. When the loop first starts, `i` is initialized to 1, and the expression `i <= 2` is evaluated. Since this expression is `true`, the print statement is executed. After the statement is executed, `i++` is evaluated, increasing `i` to 2, and the expression `i <= 2` is evaluated again. Since this expression is still `true`, the print statement is executed again. After the statement is executed, `i++` is evaluated, increasing `i` to 3, and the expression `i <= 2` is evaluated a third time. The expression is

Figure 4.5. The structure of a `for` loop

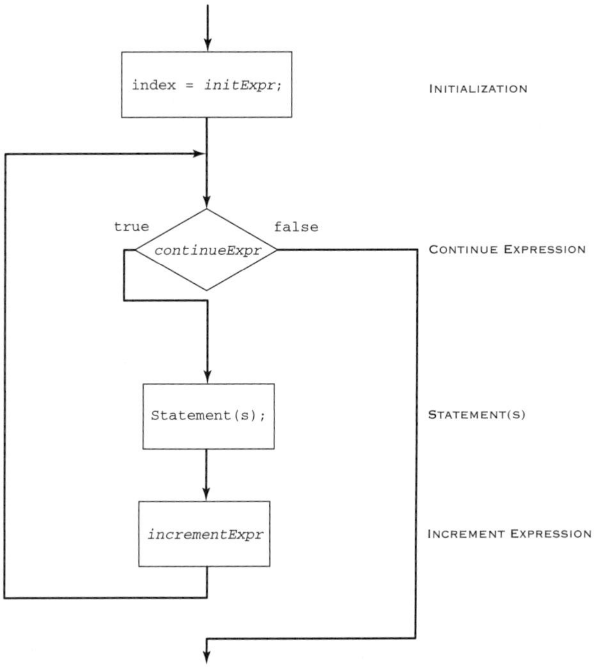

`false` this time, so execution transfers to the first statement after the loop. The output of this loop is the following:

```
i = 1
i = 2
```

Let's look at a number of specific examples to make the operation of the `for` loop clearer. First, consider the following example:

```
for ( int count = 1; count <= 10; count++ ) {
   Statement 1;
   ...
   Statement n;
}
```

In this case, statements 1 through *n* will be executed 10 times. The index variable `count` will be 1 the first time, 2 the second time, and so on. The index variable will be 10 on the last pass through the statements. At the end of the tenth pass, the index variable `count` will be increased to 11. Since the expression `count <= 10` is now `false`, control will transfer to the first statement after the loop.

Second, consider the following example:

```
for ( int count = 1; count <= 10; count += 2 ) {
   Statement 1;
   ...
   Statement n;
}
```

In this case, statements 1 through *n* will be executed five times. The index variable count will be 1 the first time, 3 the second time, and so on. The index variable will be 9 on the fifth and last pass through the statements. At the end of the fifth pass, the index variable `count` will be increased to 11. Since the expression `count <= 10` is `false`, control will transfer to the first statement after the loop.

Third, consider the following example:

```
for ( int count = 10; count <= 1; count++ ) {
   Statement 1;
   ...
   Statement n;
}
```

Here, *statements 1 through* n *will never be executed*, since the expression `count <= 1` is `false` when the continue expression is first evaluated. Instead, control will transfer to the first statement after the loop.

Finally, consider the example:

```
for ( int count = 3; count >= -3; count -= 2 ) {
   Statement 1;
   ...
   Statement n;
}
```

In this case, statements 1 through *n* will be executed 4 times. The index variable `count` will be 3 the first time, 1 the second time, –1 the third time, and –3 the fourth time. At the end of the fourth pass, the index variable `count` will be decreased to –5. Since the expression `count >= -3` is now `false`, control will transfer to the first statement after the loop.

EXAMPLE 4-2: THE FACTORIAL FUNCTION

We will illustrate the operation of a `for` loop by creating one to calculate the value of factorial function. The factorial function is defined as

$$\mathrm{n!} = \begin{Bmatrix} 1 & \mathrm{n} = 0 \\ \mathrm{n} \times (\mathrm{n}-1) \times (\mathrm{n}-2) \times \ldots \times 3 \times 2 \times 1 & \mathrm{n} > 0 \end{Bmatrix}. \tag{4-3}$$

The Java code to calculate `n` factorial for positive value of `n` would be the following:

```
nFactorial = 1;
for (int count = 1; count <= n; count++)
   nFactorial *= count;
```

Suppose that we wish to calculate the value of `5!`. If `n` is 5, this loop will be executed 5 times, with the variable `count` taking on values of 1, 2, 3, 4, and 5 in the successive loops. The resulting value of `nFactorial` will be $1 \times 2 \times 3 \times 4 \times 5 = 120$.

EXAMPLE 4-3: CALCULATING THE DAY OF YEAR

The *day of year* is the number of days (including the current day) that have elapsed since the beginning of a given year. It is a number in the range 1 to 365 for ordinary years, and 1 to 366 for leap years. Write a Java program that accepts a day, month, and year and calculates the day of year corresponding to that date.

SOLUTION

To determine the day of year, this program will need to sum up the number of days in each month preceding the current month, and add that to the number of elapsed days in the current month. A `for` loop will be used to perform this sum. Since the number of days in each month varies, it is necessary to determine the correct number of days to add for each month. A `switch` structure will be used to determine the proper number of days to add for each month.

During a leap year, an extra day must be added to the day of year for any month after February. This extra day accounts for the presence of February 29 in the leap year. Therefore, to perform the day of year calculation correctly, we must determine which years are leap years. In the Gregorian calendar, leap years are determined by the following rules:

1. Years evenly divisible by 400 are leap years.
2. Years evenly divisible by 100 but *not* by 400 are not leap years.
3. All years divisible by 4 but *not* by 100 are leap years.
4. All other years are not leap years.

We will use the modulus operator (%) to determine whether a year is evenly divisible by a given number. If the result of the modulus operation is zero, then the year is evenly divisible.

1. **State the problem**. Write a Java program that accepts a day, month, and year and calculates the day of year corresponding to that date, taking into account leap years.
2. **Define the inputs and outputs**. The inputs required by this program are the month, day, and year to be translated. The output from this program is the day of year corresponding to that date.
3. **Decompose the program into classes and their associated methods**. Again, there will be a single class, and only the `main` method within that class. We will call the class `DayOfYear` and make it a subclass of the root class `Object`.
4. **Design the algorithm that you intend to implement for each method**. The `main` method can be broken down into four major sections as follows:

```
Read the month, day, and year
Check to see if this is a leap year
Add up the days until now in current year
Write out the day of year
```

The first major step of the program is to read the month, day, and year. To do this, we will have to prompt the user to enter the desired numbers. The pseudocode for this step is the following:

```
Prompt user for the month
Read in month
Prompt user for the day
Read in day
Prompt user for the year
Read in year
```

Next, we must determine if this is a leap year by using the algorithm previously described. The pseudocode for this step is:

```
if year is divisible by 400
   leapDay ← 1;
else if year is divisible by 100
   leapDay ← 0;
else if year is divisible by 4
   leapDay ← 1;
else
   leapDay ← 0;
```

Then, we must add up the days in the month so far plus all the days in all months before the current one. The pseudocode for this step is:

```
dayOfYear ← day;
for ( int i = 1; i <= month-1; i++ ) {

   switch (i) {
      case 1: case 3: case 5: case 7: case 8:
      case 10: case 12:
         dayOfYear ← dayOfYear + 31;
         break;
      case 4: case 6: case 9: case 11:
         dayOfYear ← dayOfYear + 30;
         break;
      case 2:
         dayOfYear ← dayOfYear + 28 + leapDay;
      }
   }
}
```

Finally, we must write out the results:

```
Write out the current day of year
```

5. **Turn the algorithm into Java statements**. The final Java program is shown in Figure 4.6.

```
/*
   Purpose:
      This program calculates the day of year corresponding to a speci-
        fied date. It illustrates the use for loops and the switch struc-
        ture.

   Record of revisions:
      Date       Programmer       Description of change
      ====       ==========       =====================
     3/30/98     S. J. Chapman    Original code
*/
import chapman.io.*;
public class DayOfYear {

   // Define the main method
   public static void main(String[] args) {

      // Declare variables, and define each variable
      int day;               // Day (dd)
      int dayOfYear;         // Day of year
      int leapDay;           // Extra day for leap year
      int month;             // Month (mm)
      int year;              // Year (yyyy)
```

Figure 4.6. *(cont.)*

```
      // Create a StdIn object
      StdIn in = new StdIn();

      // Get day, month, and year to convert
      System.out.println("This program calculates the day of year");
      System.out.print("given the current date. ");
      System.out.print("Enter current month (1-12): ");
      month = in.readInt();
      System.out.print("Enter current day(1-31): ");
      day = in.readInt();
      System.out.print("Enter current year (yyyy): ");
      year = in.readInt();

      // Check for leap year, and add extra day if necessary
      if ( year % 400 == 0 )
         leapDay = 1;          // Years divisible by 400 are leap years
      else if ( year % 100 == 0 )
         leapDay = 0;          // Other centuries are not leap years
      else if ( year % 4 == 0 )
         leapDay = 1;          // Otherwise every 4th year is a leap year
      else
         leapDay = 0;          // Other years are not leap years

      // Calculate day of year
      dayOfYear = day;
      for ( int i = 1; i <= month-1; i++ ) {

         // Add days in months from January to last month
         switch (i) {
            case 1: case 3: case 5: case 7: case 8:
            case 10: case 12:
               dayOfYear = dayOfYear + 31;
               break;
            case 4: case 6: case 9: case 11:
               dayOfYear = dayOfYear + 30;
               break;
            case 2:
               dayOfYear = dayOfYear + 28 + leapDay;
         }
      }

      // Tell user
      System.out.println("Day         = " + day);
      System.out.println("Month       = " + month);
      System.out.println("Year        = " + year);
      System.out.println("day of year = " + dayOfYear);
   }
}
```

Figure 4.6. A program to calculate the equivalent day of year from a given day, month, and year.

6. **Test the program**. To test this program, we must test the program with both dates that are in leap years and dates that are not in leap years. We will use the following known results to test the program:

   1. Year 1999 is not a leap year. January 1 must be day of year 1, and December 31 must be day of year 365.

2. Year 2000 is a leap year. January 1 must be day of year 1, and December 31 must be day of year 366.
3. Year 2001 is not a leap year. March 1 must be day of year 60, since January has 31 days, February has 28 days, and this is the first day of March.

Using these test data, the program produces the following results, with inputs appearing in boldface:

```
C:\book\java\chap4>java DayOfYear
This program calculates the day of year
given the current date.  Enter current month (1-12): 1
Enter current day(1-31):     1
Enter current year (yyyy):  1999
Day           = 1
Month         = 1
Year          = 1999
day of year = 1

C:\book\java\chap4>java DayOfYear
This program calculates the day of year
given the current date.  Enter current month (1-12): 12
Enter current day(1-31):     31
Enter current year (yyyy):  1999
Day           = 31
Month         = 12
Year          = 1999
day of year = 365

C:\book\java\chap4>java DayOfYear
This program calculates the day of year
given the current date.  Enter current month (1-12): 1
Enter current day(1-31):     1
Enter current year (yyyy):  2000
Day           = 1
Month         = 1
Year          = 2000
day of year = 1

C:\book\java\chap4>java DayOfYear
This program calculates the day of year
given the current date.  Enter current month (1-12): 12
Enter current day(1-31):     31
Enter current year (yyyy):  2000
Day           = 31
Month         = 12
Year          = 2000
day of year = 366

C:\book\java\chap4>java DayOfYear
This program calculates the day of year
given the current date.  Enter current month (1-12): 3
Enter current day(1-31):     1
Enter current year (yyyy):  2001
Day           = 1
Month         = 3
Year          = 2001
day of year = 60
```

The program gives the correct answers for our test dates in all five test cases.

EXAMPLE 4-4: STATISTICAL ANALYSIS

Implement an algorithm that reads in a set of measurements and calculates the mean and the standard deviation of the input data set. Values in the data set can be positive, negative, or zero.

SOLUTION

This program must be able to read in an arbitrary number of measurements, and then calculate the mean and standard deviation of those measurements. Each measurement can be positive, negative, or zero.

Since we cannot use a data value as a flag this time, we will ask the user for the number of input values, and then use a `for` loop to read in those values. This program is shown in Figure 4.7. Note that the `while` loop has been replaced by a `for` loop. Verify its operation for yourself by finding the mean and standard deviation of the following 5 input values: 3., –1., 0., 1., and –2.

```
/*
   Purpose:
      To calculate mean and the standard deviation of an input data set,
      where each input value can be positive negative,or zero.

   Record of revisions:
      Date        Programmer          Description of change
      ====        ==========          =====================
     3/31/98     S. J. Chapman        Original code
*/
import chapman.io.*;
public class Stats2 {

   // Define the main method
   public static void main(String[] args) {

      // Declare variables, and define each variable
      int i;                  // Loop index
      int n = 0;              // The number of input samples.
      double stdDev = 0;     // The standard deviation of the input samples.
      double sumX = 0;        // The sum of the input values.
      double sumX2 = 0;       // The sum of the squares of the input values.
      double x = 0;           // An input data value.
      double xBar = 0,        // The average of the input samples.

      // Create a StdIn object
      StdIn in = new StdIn();

      // Get the number of points to input
      System.out.print("Enter number of points: ");
      n = in.readInt();

      // Check to see if we have enough input data.
      if (n < 2)
         System.out.println("At least 2 values must be entered!");

      else {

         // Loop to read input values.
         for (i = 1; i <= n; i++) {
```

Figure 4.7. *(cont.)*

```
            // Read values
            System.out.print("Enter number: ");
            x = in.readDouble();

            // Accumulate sums.
            sumX += x;
            sumX2 += x*x;
         }

         // Calculate the mean and standard deviation
         xBar = sumX / n;
         stdDev = Math.sqrt((n * sumX2 - sumX*sumX) / (n * (n-1)));

         // Tell user.
         System.out.println("The mean of this data set is:" + xBar);
         System.out.println("The standard deviation is: " + stdDev);
         System.out.println("The number of data points is: " + n);
      }
   }
}
```

Figure 4.7. Modified statistical analysis program that works with both positive and input values.

Details of Operation

Now that we have seen examples of a `for` loop in operation, we will examine some of the important details required to use `for` loops properly.

1. It is not necessary to indent the body of the `for` loop as we have just shown. The Java compiler will recognize the loop even if every statement in it starts in column 1. However, the code is much more readable if the body of the `for` loop is indented, so you should always indent the bodies of your loops.

GOOD PROGRAMMING PRACTICE

Always indent the body of a `for` loop by three or more spaces to improve the readability of the code.

2. The index variable of a `for` loop *must not be modified anywhere within the* `for` *loop*. Since the index variable is used to control the repetitions in the loop, changing it could produce unexpected results. In the worst case, modifying the index variable could produce an *infinite loop* that never completes. Consider the following example:

```
for ( i = 1; i <= 10; i++) {
   System.out.println("i = " + i);
   i = 5;
}
```

If `i` is reset to 5 every time through the loop, the loop will never end, because the index variable can never be greater than 10! This loop will run forever unless the program containing it is killed. You should *never* modify the loop index within the body of a `for` loop.

PROGRAMMING PITFALL

Never modify the value of a `for` loop index variable while inside the loop.

3. Never use `float` or `double` variables as the index variable in a `for` loop. If you do, roundoff errors can sometimes cause unexpected results. For example, consider the following program code:

```
// This program tests a double index in a for loop
public class TestDoubleIndex

   // Define the main method
   public static void main(String[] args) {

      // Set up for loop
      for ( double i = 0.1; i < 1.0; i += 0.1) {
         System.out.println("i = " + i);
      }
   }
}
```

This loop *should* execute nine times, with the results being `i` = 0.1, 0.2, ..., 0.9. In the next pass through the loop, `i` would be 1.0, so the condition `i < 1.0` should be `false`, and the loop would terminate. However, when we execute this program the results are as follows:

```
C:\book\java\chap4>java TestDoubleIndex
i = 0.1
i = 0.2
i = 0.30000000000000004
i = 0.4
i = 0.5
i = 0.6
i = 0.7
i = 0.7999999999999999
i = 0.8999999999999999
i = 0.9999999999999999
```

The loop really executed *ten times,* because round-off errors prevented `i` from being exactly equal to 1.0 on the tenth pass. This sort of problem can never happen with integer loop indexes.

GOOD PROGRAMMING PRACTICE

Always use integer variables as `for` loop indexes.

4. It is a very common error to use a comma instead of a semicolon to separate the control statements in a `for` structure. This is a syntax error that will be caught by the compiler. For example, the loop

```
for (j = 1; j <= 3, j++) {
   . . .
}
```

will produce the following compiler error:

```
TestFor.java:10: ';' expected.
          for (j = 1; j \<= 3, j++) {
                                ^

1 error
```

PROGRAMMING PITFALL

Be sure to separate the control statements in a `for` structure with a semicolon.

5. Placing a semicolon after the `for` in a `for` loop produces a logical error. A typical Java `for` loop may be written as follows:

   ```
   for (j = 1; j <= 10; j++)
      statement;
   ```

   Note that *there is no semicolon after the* `for`—the `for` structure is not terminated until after the loop body. If we make a mistake and place a semicolon after the `for`, the statements would be the following:

   ```
   for (j = 1; j <= 10; j++);
      statement;
   ```

   Unfortunately, this creates a serious error *that is not detected by the compiler.* A semicolon by itself represents a **null statement**, which is a statement that does nothing. As with the `while` loop, the semicolon will make a null statement the body of the loop, and the original `statement` will be outside of the loop. In other words, it is as though we wrote the following:

   ```
   for (j = 1; j <= 10; j++)
      ;
   statement;
   ```

   The Java compiler will compile this program with no warnings, and when the program executes, it will increment `j` from 1 to 10 while doing nothing, and afterwards execute `statement` one time only! This is the type of bug known as a logical error.

   What makes this bug particularly dangerous is that the compiler gives no warning that anything is wrong. The program executes and runs, but produces the wrong answer. And of course, the error is very hard to spot unless you are specifically looking for it.

PROGRAMMING PITFALL

Adding a semicolon after a `for` statement can produce a logical error. Java will compile and execute the program, but the program will produce incorrect results.

6. It is possible to design `for` loops that count down as well as up. The following `for` loop executes 3 times with `j` being 3, 2, and 1 in the successive loops:

   ```
   for (j = 3; j >= 1; j--) {
      . . .
   }
   ```

The continue and break Statements

There are two additional statements that can be used to control the operation of loops: **continue** and **break**.

If the continue statement is executed in the body of a loop, the execution of the body will stop and control will be returned to the top of the loop. The loop index will be incremented, and execution will resume again if the continuation condition is still true. An example of the continue statement in a for loop is shown in the following code:

```
// This program tests the continue statement
public class TestContinue {

   // Define the main method
   public static void main(String[] args) {

      int i;

      // Set up for loop
      for ( i = 1; i <= 5; i++) {
         if ( i == 3 ) continue;
         System.out.println("i = " + i);
      }
      System.out.println("End of loop!");
   }
}
```

When this program is executed, the output is:

```
C:\book\java\chap4>java TestContinue
i = 1
i = 2
i = 4
i = 5
End of loop!
```

Note that the continue statement was executed on the iteration when i was 3 and that control returned to the top of the loop without executing the output statement.

If the break statement is executed in the body of a loop, the execution of the body will stop and control will be transferred to the first executable statement after the loop. An example of the break statement in a for loop is shown in the following code:

```
// This program tests the break statement
public class TestBreak {

   // Define the main method
   public static void main(String[] args) {

      int i;

      // Set up for loop
      for ( i = 1; i <= 5; i++) {
         if ( i == 3 ) break;
         System.out.println("i = " + i);
      }
      System.out.println("End of loop!");
   }
}
```

When this program is executed, the output is:

```
C:\book\java\chap4>java TestBreak
i = 1
i = 2
End of loop!
```

Note that the `break` statement was executed on the iteration when `i` was 3, and that control transferred to the first executable statement after the loop without executing the output statement.

Both the `continue` and `break` statements work with `while`, `do/while`, and `for` loops.

Nesting Loops

It is possible for one loop to be completely inside another loop. If one loop is completely inside another one, the two loops are called **nested loops**. The following example shows two nested `for` loops used to calculate and write out the product of two integers:

```
// This program tests nested for loops
public class NestedFor {

   // Define the main method
   public static void main(String[] args) {

      int i, j, product;

      for ( i = 1; i <= 3; i++) {
         for ( j = 1; j <= 3; j++) {
            product = i * j;
            System.out.println(i + " * " + j + " = " + product);
         }
      }
   }
}
```

In this example, the outer `for` loop will assign a value of 1 to index variable `i`, and then the inner `for` loop will be executed. The inner `for` loop will be executed 3 times with index variable `j` having values 1, 2, and 3. When the entire inner `for` loop has been completed, the outer `for` loop will assign a value of 2 to index variable `i`, and the inner `for` loop will be executed again. This process repeats until the outer `for` loop has executed three times. The resulting output is

```
1 * 1 = 1
1 * 2 = 2
1 * 3 = 3
2 * 1 = 2
2 * 2 = 4
2 * 3 = 6
3 * 1 = 3
3 * 2 = 6
3 * 3 = 9
```

Note that the inner `for` loop executes completely before the index variable of the outer `for` loop is incremented.

If `for` loops are nested, they must have independent index variables. Otherwise, the inner loop would be modifying the loop variable of the outer loop, causing the outer loop to behave improperly.

Labeled break and continue Statements

If a `break` statement appears in a nested loop structure, it breaks out of the *innermost loop* only. The outer loops will continue to execute. For example, consider the following program:

```
public class TestBreak2 {
   // Define the main method
   public static void main(String[] args) {

      int i, j, product;

      for ( i = 1; i <= 3; i++) {
         for ( j = 1; j <= 3; j++) {
            product = i * j;
            if ( j == 3 ) break;
            System.out.println(i + " * " + j + " = " + product);
         }
      }
   }
}
```

This `break` statement is executed when `j` is 3. When the `break` statement is executed, only the *innermost* loop will terminated. The outer loop continues to execute, producing the following results:

```
1 * 1 = 1
1 * 2 = 2
2 * 1 = 2
2 * 2 = 4
3 * 1 = 3
3 * 2 = 6
```

To break out of more than one level of a nested structure, we must use a **labeled break statement**. We must place a **label** on the loop that we wish to break out of, and then specify that label in the `break` statement.

For example, suppose that we wish to modify the previous program to break out of both loops when the `break` statement is executed. We would place a label on the outer loop, and then refer to that label in the break statement:

```
public class TestBreak3 {

   // Define the main method
   public static void main(String[] args) {

      int i, j, product;

      outer: for (i = 1; i <= 3; i++) {
         for (j = 1; j <= 3; j++) {
            product = i * j;
            if (j == 3) break outer;
            System.out.println(i + " * " + j + " = " + product);
         }
      }
      System.out.println("Outside nested loops.");
   }
}
```

When this program is executed, the results are

```
C:\book\java\chap4>java TestBreak3
1 * 1 = 1
1 * 2 = 2
Outside nested loops.
```

The `break` statement executed when `j` was equal to 3, and the program broke out of both loops at that time.

Similarly, if a `continue` statement appears in a nested loop structure, it breaks out of the *innermost* loop only. The outer loops will continue to execute. For example, consider the following program:

```
public class TestContinue2 {

   // Define the main method
   public static void main(String[] args) {

      int i, j, product;

      for ( i = 1; i <= 3; i++) {
         for ( j = 1; j <= 3; j++) {
            product = i * j;
            if ( j == 2 ) continue;
            System.out.println(i + " * " + j + " = " + product);
         }
      }
   }
}
```

This `continue` statement is executed when `j` is 2. When the `continue` statement is executed, the remaining statements in the *innermost* loop will be skipped, and execution continues at the top of the inner loop. The outer loop remains totally unaffected. The results of this program are

```
1 * 1 = 1
1 * 3 = 3
2 * 1 = 2
2 * 3 = 6
3 * 1 = 3
3 * 3 = 9
```

To cause a higher level loop to continue, we must use a **labeled `continue` statement**. We will place a label on the loop that we wish to continue and then specify that label in the `continue` statement. For example, the following program causes the *outer* loop to continue each time that `j` reaches 2 in the inner loop:

```
public class TestContinue3 {

   // Define the main method
   public static void main(String[] args) {

      int i, j, product;

      outer: for ( i = 1; i <= 3; i++) {
         inner: for ( j = 1; j <= 3; j++) {
            product = i * j;
            if ( j == 2 ) continue outer;
            System.out.println(i + " * " + j + " = " +
            product);
         }
      }
   }
}
```

When this program is executed, the results are

```
1 * 1 = 1
2 * 1 = 2
3 * 1 = 3
```

Compare this result to the result of the previous example, where the innermost loop was affected by the `continue` statement.

GOOD PROGRAMMING PRACTICE

Use labeled `break` or `continue` statements to break out of or continue outer loops in a nested loop structure.

PRACTICE!

This quiz provides a quick check to see if you have understood the concepts introduced in Sections 4.1 through 4.3. If you have trouble with the quiz, reread the sections, ask your instructor, or discuss the material with a fellow student. The answers to this quiz are found in the back of the book.

Examine the control parameters of the following `for` loops, and determine how many times each loop will be executed:

1. `for (index = 7; index <= 10; index++)`
2. `for (j = 7; j <= 10; j--)`
3. `for (index = 1; index <= 10; index += 10)`
4. `for (k = 1; k < 10; k++)`
5. `for (counter = -2; counter <= 10; counter += 2)`
6. `for (time = -2; time >= -10; time--)`
7. `for (i = -10; i <= -7; i -= 3)`

Examine the following loops, and determine the value in `ires` at the end of each of the loops. Assume that `ires`, `index`, and all loop variables are integers. How many times does each loop execute?

8.
```
ires = 0;
for (index = 1; index <= 10; index++)
   ires++;
```

9.
```
ires = 0;
for (index = 1; index <= 10; index++)
   ires += index;
```

10.
```
ires = 0; index = 0;
while ( ires < 12 )
   ires += ++index;
```

11.
```
ires = 0; index = 0;
while ( index < 5 )
   ires += ++index;
```

12.
```
ires = 0; index = 0;
do {
   ires += ++index;
} while ( index < 5 );
```

13.
```
ires = 0;
for (index = 1; index <= 6; index++) {
   if ( index == 3 )
      continue;
   ires += index;
}
```

PRACTICE!

14.
```
ires = 0;
for (index = 1; index <= 6; index++) {
   if ( index == 3 )
      break;
   ires += index;
}
```

15.
```
ires = 0;
for (index1 = 1; index1 <= 5; index1++) {
   for (index2 = 1; index2 <= 5; index2++) {
      ires++;
   }
}
```

16.
```
ires = 0;
for (index1 = 1; index1 <= 5; index1++) {
   for (index2 = index1; index2 <= 5; index2++) {
      ires++;
   }
}
```

17.
```
ires = 0;
loop1: for (index1 = 1; index1 <= 5; index1++) {
   loop2: for (index2 = 1; index2 <= 5; index2++) {
      if ( index2 == 3 )
         break loop1;
      ires++;
   }
}
```

18.
```
ires = 0;
loop1: for (index1 = 1; index1 <= 5; index1++) {
   loop2: for (index2 = 1; index2 <= 5; index2++) {
      if ( index2 == 3 )
         break loop2;
      ires++;
   }
}
```

Examine the following Java statements and state whether they are valid. If they are invalid, indicate the reason why they are invalid.

19.
```
loop1: for (i = 1; i <= 10; i++ ) {
   loop2: for (i = 1; i <= 10; i++ ) {
      ...
   }
}
```

20.
```
x = 10;
while ( x > 0);
   x -= 3;
```

21.
```
ires = 0;
for (i = 1; i <= 10; i++ ) {
   ires += i--;
}
```

4.4 FORMATTING OUTPUT DATA

One serious limitation of the standard Java API is that there is no convenient way to format numbers for display. When Java prints out a number, it displays *all nonzero significant digits* of the number. while this is suitable under some circumstances, it is not particularly useful at other times.

For example, suppose that we were interested in creating a program that reads in the prices of a series of purchases and calculates the average price. If the average price is written out with a statement like

```
System.out.println("Average price = " + ave);
```

the result might be

```
Average price = $3.766666666666667
```

This output is not very useful, since monetary amounts are only significant to the nearest cent.

A similar problem happens if we are trying to create tables of information. If the values in the table contain different numbers of significant digits, then the columns of data in the table will not line up. For example, the following program calculates a table of the square roots and cube roots of all integers from 0 to 10.

```
public class SquareCubeRoot {
   public static void main(String[] args) {

      // Declare variables, and define each variable
      int i;                 // Loop index
      double cubeRoot;       // Cube root of index
      double squareRoot;     // Square root of index

      // Print title
      System.out.println("Table of Square and Cube Roots:");

      // Calculate and print values
      for ( i = 0; i <= 10; i++ ) {
         squareRoot = Math.sqrt(i);
         cubeRoot = Math.pow(i,1./3.);
         System.out.println(i + "  " + squareRoot + "  " + cubeRoot);
      }
   }
}
```

When this program is executed, the results are:

```
Table of Square and Cube Roots:
 0  0.0  0.0
 1  1.0  1.0
 2  1.4142135623730951  1.2599210498948732
 3  1.7320508075688772  1.4422495703074083
 4  2.0  1.5874010519681994
 5  2.23606797749979  1.7099759466766968
 6  2.449489742783178  1.8171205928321397
 7  2.6457513110645907  1.912931182772389
 8  2.8284271247461903  2.0
 9  3.0  2.080083823051904
10  3.1622776601683795  2.154434690031884
```

This output looks terrible! It would be completely unacceptable for any real task.

Although Java does not include a standard formatting method, *it is possible to create a class* that will format numbers nicely. The `chapman.io` package that accompanies this

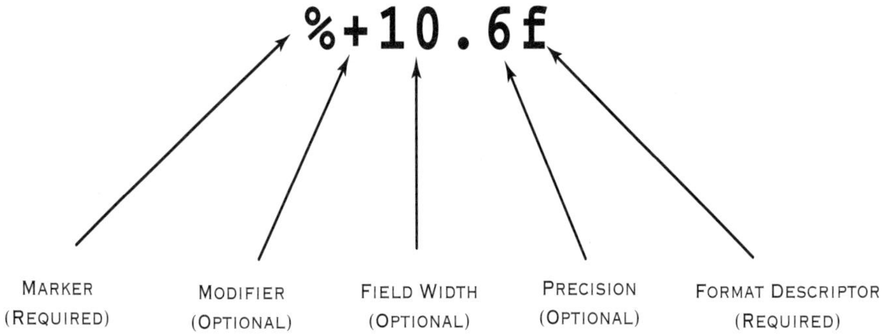

Figure 4.8. The components of a format descriptor

TABLE 4-1 The Components of a Format Descriptor

COMPONENT	STATUS	MEANING
Marker (%)	Required	Marks the beginning of a format descriptor.
Modifier	Optional	A one-character value that modifies the behavior of the descriptor:
		+ Display a leading + for positive numbers
		0 + Show leading zeros
		- Left-align the number in the field space
		space Add a space in front of positive numbers
		# Use "alternative" format. For floating-point numbers, this modifier displays trailing zeros. For octal or hexadecimal numbers, it adds a "0" or "0x" to the front of the number.
Field Width	Optional	The number of characters to use to display the value.
Precision	Optional	A period followed by the number of places to display after the decimal point.
Format Descriptor	Required	The basic format for the value to be displayed, as shown in Table 4-2.

book includes such a class, called `Fmt`. This class contains methods that are modelled after the `printf` and `sprintf` functions of the C language and are used in a similar fashion.

The `Fmt.printf` method formats an output value according to a format string and then prints that value on the standard output stream. The `Fmt.sprintf` method is similar, except that it returns the formatted value to the calling method as a string.

The `Fmt.printf` method has the form

```
Fmt.printf (format, value);
```

where `value` is the value to be printed, and `format` is a string describing how to print the value. For example, the average price just described might be printed with the following statement:

```
Fmt.printf("Average price = $ %5.2f \n", ave);
```

The output of this statement would be:

```
Average price = $ 3.77
```

The **format string** must contain a format descriptor (which is marked by an initial % character), and may contain any number of characters before and after the format descriptor. All of the characters before and after the format descriptor will be printed unchanged, while the format descriptor is replaced by the value being printed out. The descriptor serves as a "model," describing how the value should be printed. For example, the descriptor `%5.2f` states that a *floating-point value* should be printed out in a field that is *five characters wide*, with *two characters after the decimal point*.

A `format` string may also contain **escape sequences**, which are convenient ways to include unprintable characters in the output stream. For example, the sequence `\n` in the foregoing example represents the **newline character**, which positions the cursor to the beginning of a new line.

The components of a format descriptor are shown in Figure 4.8, and the meanings of each component are given in Tables 4-1 and 4-2. A list of common escape sequences is given in Table 4-3.

TABLE 4-2 Format Descriptors

DESCRIPTOR	MEANING
`%f`	Display a floating-point number in fixed-point format (`12.345`).
`%e, %E`	Display a floating-point number in exponential format. The `%e` descriptor produces a lower-case `e` for the exponent marker (`1.234500e+001`), while the `%E` descriptor produces a upper-case `E` for the exponent marker (`1.234500E+001`).
`%g, %G`	Display a floating-point number in general format, which is fixed-point format for relatively small numbers and exponential format for extremely large or extremely small numbers. The `%g` descriptor produces a lower-case `e` for the exponent marker (if used), `while` the `%G` descriptor produces a upper-case `E` for the exponent marker (if used).
`%d, %i`	Display an integer in decimal format.
`%x`	Display an integer in hexadecimal format.
`%o`	Display an integer in octal format.
`%c`	Display a character.
`%s`	Display a string.

TABLE 4-3 Common Escape Sequences

DESCRIPTOR	MEANING
`\n`	Newline. Positions the cursor at the beginning of the next line.
`\t`	Horizontal tab. Moves the cursor to the next tab stop.
`\r`	Carriage return. Positions the cursor at the beginning of the *current* line.
`\\`	Backslash. Used to print the backslash character.
`\'`	Single quote. Used to print the single quote character.
`\"`	Double quote. Used to print the double quote character.
`\u####`	Unicode character. Used to display the Unicode character whose location in the character set is ####, where #### is a hexadecimal value between `0000` and `FFFF`.

EXAMPLE 4-5: If i, pi, and e are initialized as shown, then the following statements will produce the indicated results. Can you explain why each result is produced?

```
int i = 12345;
double pi = 3.14159265358979;
double e = -1.602e-19;
```

EXPRESSION	RESULT
a. `Fmt.printf("i = %d\n", i);`	`i = 12345`
b. `Fmt.printf("i = %10d\n", i);`	`i =      12345`
c. `Fmt.printf("i = %010d\n", i);`	`i = 0000012345`
d. `Fmt.printf("i = %-10d\n", i);`	`i = 12345`
e. `Fmt.printf("pi = %f\n", pi);`	`pi = 3.141593`
f. `Fmt.printf("pi = %+8.4f\n", pi);`	`pi = +3.1416`
g. `Fmt.printf("pi = %12.4e\n", pi);`	`pi = 3.1416e+000`
h. `Fmt.printf("e = %12.4e\n", e);`	`e = -1.6020e-019`

Each `Fmt.printf` method call can format one and only one value. If you need to print multiple values on a line, then you will have to make multiple calls to the method. This fact is illustrated by Example 4-6.

EXAMPLE 4-6: GENERATING A TABLE OF INFORMATION Write a program to generate a table containing the square roots and cube roots of all integers between 0 and 10. Use formatted output to generate a neat table with five places after the decimal point.

SOLUTION

This program will be similar to class SquareCubeRoot, except that the output will use `Fmt.printf` statements to format the data. The resulting program is as follows:

```
import chapman.io.*;
public class SquareCubeRoot1 {
   public static void main(String[] args) {

      // Declare variables, and define each variable
      int i;                 // Loop index
      double cubeRoot;       // Cube root of index
      double squareRoot;     // Square root of index

      // Print title
      System.out.println("Table of Square and Cube Roots:");

      // Calculate and print values
      for ( i = 0; i <= 10; i++ ) {
         squareRoot = Math.sqrt(i);
         cubeRoot = Math.pow(i,1./3.);
         Fmt.printf("%5d",i);
         Fmt.printf("     %7.5f",squareRoot);
         Fmt.printf("     %7.5f\n",cubeRoot);
      }
   }
}
```

When this program is executed, the results are as follows and are much nicer than before:

```
D:\book\java\chap4>java SquareCubeRoot1
Table of Square and Cube Roots:
      0       0.00000       0.00000
      1       1.00000       1.00000
      2       1.41421       1.25992
      3       1.73205       1.44225
      4       2.00000       1.58740
      5       2.23607       1.70998
      6       2.44949       1.81712
      7       2.64575       1.91293
      8       2.82843       2.00000
      9       3.00000       2.08008
     10       3.16228       2.15443
```

GOOD PROGRAMMING PRACTICE

Use the formatted output method `Fmt.printf` to create neat tabular output in your programs.

4.5 EXAMPLE PROBLEM

EXAMPLE 4-7: PHYSICS—THE FLIGHT OF A BALL

If we assume negligible air friction and ignore the curvature of the earth, a ball that is thrown into the air from any point on the earth's surface will follow a parabolic flight path (see Figure 4.9*a*). The height of the ball at any time *t* after it is thrown is given by

$$y(t) = y_0 + v_{y0} + t + \frac{1}{2} g\, t^2 , \tag{4-4}$$

where y_0 is the initial height of the object above the ground, v_{y0} is the initial vertical velocity of the object, and g is the acceleration due to the earth's gravity. The horizontal distance (range) traveled by the ball as a function of time after it is thrown is given by

$$x(t) = x_0 + v_{x0}\, t , \tag{4-5}$$

where x_0 is the initial horizontal position of the ball on the ground and v_{x0} is the initial horizontal velocity of the ball.

If the ball is thrown with some initial velocity v_0 at an angle of θ degrees with respect to the earth's surface, then the initial horizontal and vertical components of velocity will be the following

$$v_{x0} = v_0 \cos \theta , \tag{4-6}$$

$$v_{y0} = v_0 \sin \theta . \tag{4-7}$$

Assume that the ball is initially thrown from position $(x_0, y_0) = (0,0)$ with an initial velocity v of 20 meters per second at an initial angle of θ degrees. Design, write, and test a program that will determine the horizontal distance traveled by the ball from the time it was thrown until it touches the ground again. The program should do this for all angles θ from 0 to 90° in 1° steps. Determine the angle θ that maximizes the range of the ball.

Figure 4.9. (a) When a ball is thrown upwards, it follows a parabolic trajectory. (b) The horizontal and vertical components of a velocity vector v at an angle θ with respect to the horizontal.

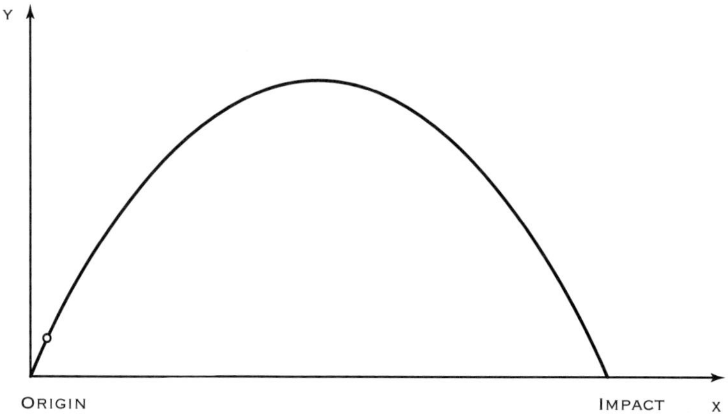

(a)

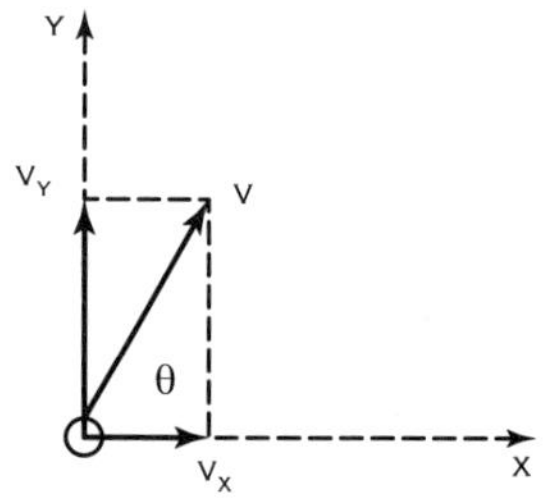

(b)

SOLUTION

In order to solve this problem, we must determine an equation for the range of the thrown ball. We can do this by first finding the time that the ball remains in the air, and then finding the horizontal distance that the ball can travel during that time.

The time that the ball will remain in the air after it is thrown may be calculated from Equation 4-4. The ball will touch the ground at the time t for which $y(t) = 0$. Remembering that the ball will start from ground level ($y(0) = 0$), and solving for t, we get

$$y(t) = y_0 + v_{y0}t + \frac{1}{2}g\,t^2, \qquad (4\text{-}4)$$

$$0 = 0 + v_{y0}t + \frac{1}{2}g\,t^2,$$

$$0 = \left(v_{y0} + \frac{1}{2}g\,t\right)t.$$

So, the ball will be at ground level at time $t_1 = 0$ (when we threw it), and at time

$$t_2 = -\frac{2v_{y0}}{g}.$$

The horizontal distance that the ball will travel in time t_2 is found as follows:

$$\text{range} = x(t_2) = x_0 + v_{x0}t_2, \tag{4-5}$$

$$\text{range} = 0 + v_{x0}\left(-\frac{2\,v_{y0}}{g}\right)$$

$$\text{range} = -\frac{2\,v_{x0}\,v_{y0}}{g}.$$

We can substitute Equations (4-6) and (4-7) for v_{x0} and v_{y0} to get an equation expressed in terms of the initial velocity v and initial angle θ:

$$\text{range} = -\frac{2(v_0 \cos\theta)(v_0 \sin\theta)}{g}$$

$$\text{range} = -\frac{2v_0^2}{g}\cos\theta\,\sin\theta \tag{4-8}$$

From the problem statement, we know that the initial velocity v_0 is 20 meters per second and that the ball will be thrown at all angles from 0° to 90° in 1° steps. Finally, any elementary physics textbook will tell us that the acceleration due to the earth's gravity is –9.81 meters per second squared.

Now let's apply our design technique to this problem.

1. **State the problem.** A proper statement of this problem would be: *Calculate the range that a ball would travel when it is thrown with an initial velocity of* v_0 *at an initial angle* θ. *Calculate this range for a* v_0 *of 20 meters per second and all angles between 0° and 90°, in 1° increments. Determine the angle* θ *that will result in the maximum range for the ball. Assume that there is no air friction.*
2. **Define the inputs and outputs.** As the problem is defined, no inputs are required. We know from the problem statement what v_0 and θ will be, so there is no need to read them in. The outputs from this program will be a table showing the range of the ball for each angle θ and the angle θ for which the range is maximum.
3. **Decompose the program into classes and their associated methods.** This step will be trivial until we learn how to write objects in Chapter 7. For now, there will be a single class, and only the `main` method within that class. We will call the class `Ball`.
4. **Design the algorithm that you intend to implement for each method.** There is only one method in this program. The `main` method can be broken down into the following major steps:

```
for theta = 0 to 90 degrees in 1 degree steps {
   Calculate the range of the ball for each angle theta.
   Determine if this theta yields the maximum range so far.
   Write out the range as a function of theta.
}
WRITE out the theta yielding maximum range
```

A `for` loop is appropriate for this algorithm, since we are calculating the range of the ball for a specified number of angles. We will calculate the range for each value of θ and compare each range with the maximum range found so far to determine which angle yields the maximum range. Note that the trigonometric functions work in radians, so the angles in degrees must be converted to radians before the range is calculated. The detailed pseudocode for this algorithm is as follows:

```
Declare variables.
Initialize v0 to 20 meters/second.
for theta = 0 to 90 degrees in 1 degree steps {
   radian ← theta * DEG_TO_RAD;   (Convert degrees to radians)
   angle ← (-2. * v0**2 / gravity) * sin(radian) * cos(radian);
   Write out theta and range.
   if range > max_range then
      max_range ← range;
      max_degrees ← theta;
   }
}
Write out max_degrees, max_range.
```

5. **Turn the algorithm into Java statements**. The final Java program is shown in Figure 4.10.

```
/*
   Purpose:
     To calculate distance traveled by a ball thrown at a specified angle
       theta and at a specified velocity V0 from a point on the surface
       of the earth, ignoring the effects of air friction and the earth's
       curvature.

   Record of revisions:
      Date        Programmer      Description of change
      ====        ==========      =====================
     3/31/98    S. J. Chapman   Original code
*/
import chapman.io.*;
public class Ball {

   // Define the main method
   public static void main(String[] args) {

      // Declare variables, and define each variable
      double gravity = -9.81; // Accel. due to gravity (m/s/s)
      final double DEG_TO_RAD = Math.PI / 180; // Deg ==> rad conv. factor
      int maxDegrees = 0;     // angle at which the max rng occurs (deg)
      double maxRange = 0;    // Max range for the ball at vel v0 (m)
      double range;            // Range of the ball at a given angle (m)
      double radian;           // Angle of throw (radians)
      int theta;               // Angle of throw (deg)
      double v0 = 20;          // Velocity of the ball (m/s)

      for (theta = 0; theta <= 90; theta++) {

         // Get angle in radians
         radian = theta * DEG_TO_RAD;

         // Calculate range in meters.
         range = (-2*v0*v0/gravity) * Math.sin(radian)* Math.cos(radian);
```

Figure 4.10. *(cont.)*

```
            // Write out the range for this angle.
            Fmt.printf("Theta = %2d deg; ", theta);
            Fmt.printf("Range = %8.4f meters\n", range);

            // Compare the range to the previous maximum range. If this
            // range is larger, save it and the angle at which it occurred.
            if (range > maxRange) {
               maxRange = range;
               maxDegrees = theta;
            }
         }

         // Skip a line, and then write out the maximum range and
         // the angle at which it occurred.
         Fmt.printf("Max range = %8.4f ", maxRange);
         Fmt.printf("at %2d degrees.\n", maxDegrees);
      }
   }
```

Figure 4.10. Program `Ball` to determine the angle which maximizes the range of a thrown ball

The degrees-to-radians conversion factor is always a constant, so in the program it is declared as a named constant, and all references to the constant within the program use that name. The acceleration due to gravity at sea level can be found in any physics text. It is it is about 9.81 m/s^2, directed downward.

6. **Test the program**. To test this program, we will calculate the answers by hand for a few of the angles and compare the results with the output of the program:

$$\theta = 0°: \quad \text{range} = -\frac{2(20^2)}{-9.81} = \cos 0 \sin 0 = 0 \text{ meters}$$

$$\theta = 5°: \quad \text{range} = -\frac{2(20^2)}{-9.81} \cos\left(\frac{5\pi}{180}\right) \sin\left(\frac{5\pi}{180}\right) = 7.080 \text{ meters}$$

$$\theta = 40°: \quad \text{range} = -\frac{2(20^2)}{-9.81} \cos\left(\frac{40\pi}{180}\right) \sin\left(\frac{40\pi}{180}\right) = 40.16 \text{ meters}$$

$$\theta = 45°: \quad \text{range} = -\frac{2(20^2)}{-9.81} \cos\left(\frac{45\pi}{180}\right) \sin\left(\frac{45\pi}{180}\right) = 40.77 \text{ meters}$$

When program `Ball` is executed, a 90-line table of angles and ranges is produced. To save space, only a portion of the table is reproduced:

```
Theta =  0 deg; Range =   0.0000 meters
Theta =  1 deg; Range =   1.4230 meters
Theta =  2 deg; Range =   2.8443 meters
Theta =  3 deg; Range =   4.2621 meters
Theta =  4 deg; Range =   5.6747 meters
Theta =  5 deg; Range =   7.0805 meters
Theta =  6 deg; Range =   8.4775 meters
Theta =  7 deg; Range =   9.8643 meters
Theta =  8 deg; Range =  11.2390 meters
Theta =  9 deg; Range =  12.6001 meters
Theta = 10 deg; Range =  13.9458 meters
...
```

```
Theta = 40 deg; Range =  40.1553 meters
Theta = 41 deg; Range =  40.3779 meters
Theta = 42 deg; Range =  40.5514 meters
Theta = 43 deg; Range =  40.6754 meters
Theta = 44 deg; Range =  40.7499 meters
Theta = 45 deg; Range =  40.7747 meters
Theta = 46 deg; Range =  40.7499 meters
Theta = 47 deg; Range =  40.6754 meters
Theta = 48 deg; Range =  40.5514 meters
Theta = 49 deg; Range =  40.3779 meters
Theta = 50 deg; Range =  40.1553 meters
...
Theta = 80 deg; Range =  13.9458 meters
Theta = 81 deg; Range =  12.6001 meters
Theta = 82 deg; Range =  11.2390 meters
Theta = 83 deg; Range =   9.8643 meters
Theta = 84 deg; Range =   8.4775 meters
Theta = 85 deg; Range =   7.0805 meters
Theta = 86 deg; Range =   5.6747 meters
Theta = 87 deg; Range =   4.2621 meters
Theta = 88 deg; Range =   2.8443 meters
Theta = 89 deg; Range =   1.4230 meters
Theta = 90 deg; Range =   0.0000 meters
Max range =  40.7747 at 45 degrees.
```

The program output matches our hand calculation for the angles calculated above to the four-digit accuracy of the hand calculation. Note that the maximum range occurred at an angle of 45°.

4.6 MORE ON DEBUGGING JAVA PROGRAMS

It is much easier to make a mistake when writing a program containing selection structures and loops than it is when writing simple sequential programs. Even after going through the full design process, a program of any size is almost guaranteed not to be completely correct the first time it is used.

Programs with many levels of `if` structures, `for` loops, etc. will contain many nested layers of braces (`{}`). One of the most common problems is to have *mismatched braces* within a program. A Java compiler will always catch this error, but the error message may not be very informative. For example, consider a portion of the program `Ball` from Figure 4.9. Suppose that by accident we leave out the opening brace of the `for` loop:

```
for (theta = 0; theta <= 90; theta++) // { missing

   // Get angle in radians
   radian = theta * deg2Rad;

   // Calculate range in meters.
   range = (-2*v0*v0/gravity) * Math.sin(radian) * Math.cos(radian);

   // Write out the range for this angle.
   System.out.print ("Theta = " + theta + " deg; ");
   System.out.println ("Range = " + range + " meters");

   // Compare the range to the previous maximum range. If this
   // range is larger, save it and the angle at which it occurred.
   if (range > max_range) {
```

```
            max_range = range;
            max_degrees = theta;
        }
    }

    // Skip a line, and then write out the maximum range and
    // the angle at which it occurred.
    System.out.println (" ");
```

If we compile this modified program with the Java compiler, the result is as follows:

```
C:\book\java\chap4>javac Ball.java
Ball.java:51: Type expected.
      System.out.println (" ");
                           ^

1 error
```

The compiler knows that something is wrong, but the resulting error message seems to be complete nonsense. In fact, the compiler has interpreted the closing brace of the `for` loop as the closing brace of the `main` method. It thinks that the next statement is a new method definition, and it says that the definition has the wrong syntax. The true error was many lines away from the point that the compiler reported an error, and it had a very different cause. If you get a problem like this, you must inspect your program *very* carefully for syntax problems such as missing braces.

Suppose that we have built and compiled a program, and after that program executes, we find that the output values are in error when it is tested. How do we go about finding the bugs and fixing them?

The best approach to locating the error is to use a symbolic debugger, if one is supplied with your compiler. You must ask your instructor or else check with your system's manuals to determine how to use the symbolic debugger supplied with your particular compiler and computer.

An alternative approach to locating the error is to insert output statements into the code to print out important variables at key points in the program. When the program is run, the output statements will print out the values of the key variables. These values can be compared with the ones you expect, and the places where the actual and expected values differ will serve as a clue to help you locate the problem. For example, to verify the operation of a `for` loop, the following output statements could be added to the program:

```
System.out.println("At loop1: ist, ien = " + ist + ", " + ien);
loop1: for (i = ist; i <= ien; i++) {
   System.out.println("In loop1: i = ", + i);
   ...
}
System.out.println("loop1 completed");
```

When the program is executed, its output listing will contain detailed information about the variables controlling the `for` loop and just how many times the loop was executed. Similar output statements could be used to debug the operation of an `if` structure:

Once you have located the portion of the code in which the error occurs, you can take a look at the specific statements in that area to locate the problem. A list of some common errors follows. Be sure to check for them in your code.

1. *If the problem is in an* `if` *structure, check to see if you used the proper relational operator in your logical expressions.* Did you use > when you really

intended >=, etc.? Logical errors of this sort can be very hard to spot, since the compiler will not give an error message for them.

2. *Another common problem with* `if` *statements occurs when floating-point* (`float` *and* `double`) *variables are tested for equality.* Because of small round-off errors during floating-point arithmetic operations, two numbers that theoretically should be equal will differ by a tiny amount, and the test for equality will fail. Instead of testing for equality, you shold test for *near equality,* as described in Chapter 3.
3. *Most errors in* `for` *loops involve mistakes with the loop parameters.* If you add output statements to the `for` loop as previously shown, the problem should be fairly clear. To reduce the risk of errors, ask yourself the following questions: Did the `for` loop start with the correct value? Did it end with the correct value? Did it increment at the proper step? If not, check the parameters of the `for` loop closely. You will probably spot an error in the control parameters.
4. Errors in `while` and `do/while` loops are usually related to errors in the logical expression used to control their function. These errors may be detected by examining the test expression of the loop with output statements. Errors can also be caused by using a `while` where a `do/while` is required and vice versa. In other words, confirm whether you want to test for the loop condition at the beginning or at the end of each loop.

SUMMARY

- A `while` loop executes a block of statements repeatedly until its `boolean` control expression becomes `false`. The control expression test occurs before the loop executes.
- A `do/while` loop executes a block of statements repeatedly until its `boolean` control expression becomes `false`. The control expression test occurs after the loop executes.
- A `for` loop executes a block of statements a specified number of times.
- The `continue` statement causes execution of the remaining statements within the body of a loop to be skipped, and execution resumes at the top of the loop. If the loop is a `for` loop, the loop increment expression is executed.
- The `break` statement causes execution of the remaining statements within the body of a loop to be skipped, and execution resumes at the next statement following the end of the loop.
- If `while`, `do/while`, or `switch` structures are nested, a `break` or `continue` statement applies to the *innermost* structure containing the statement.
- A labeled `break` or `continue` statement applies to the particular structure with that label, even if it is not the innermost one.

PROFESSIONAL SUCCESS: THE JOYS OF OBSFUCATION

When I taught at various universities in the 1970s and 1980s, computer science students used to have competitions to see who could write the most difficult to understand programs. They would create working programs in C, and challenge other students to figure out what the programs did. These students got a great deal of ego gratification from writing programs that no one else could understand, but in the meantime they learned *horrible* programming practices. I have learned through bitter experience that such students often do not make good employees in industry.

A properly designed program should be understandable to any professional working in the same field, and it should be possible for such a person to modify the program without excessive pain. Java goes a long way to ensuring that by eliminating pointer arithmetic, by adding boolean data types, and by rigidly enforcing the rules of the language. However, you must do your part, too. Don't get excessively "cute" in the design of your programs. Keep things simple and well commented. This is especially important when working with loops. Make the control parameters for your loops as simple as possible, and perform your calculations within the body of the loop.

For example, the following two program fragments both read `int`s from the standard input stream and display them, until a value greater than or equal to 10 is entered.

```
StdIn in = new StdIn();
System.out.println ("Enter values:");
for (int i = in.readInt(); i < 10; i = in.readInt() ) {
   System.out.println(" i = " + i);
}
```

and

```
int i;
StdIn in = new StdIn();
System.out.println("Enter values: ");
do {
   i = in.readInt();
   System.out.println(" i = " + i);
} while(i < 10);
```

The first program fragment uses a `for` loop and performs `readInt()` calls to get each new value of `i`. Each time that the loop is executed, a value is read from the standard input stream and displayed if it is less than 10. This code compiles and works, but it is a misuse of the `for` structure. The `for` loop is designed to execute a block of code a finite number of times. Here, the loop is executing an indefinite number of times, and using a complicated loop control expression that involves reading data from the standard input stream. This is *very bad* programming practice.

The second program fragment implements the same function as a `do/while` loop. It is much clearer, and it is using the language as it is intended to be used.

APPLICATION: SUMMARY OF GOOD PROGRAMMING PRACTICES

The following guidelines introduced in this Chapter will help you to develop good programs:

1. Always indent the body of any structure by three or more spaces to improve the readability of the code.
2. Use a `while` or `do/while` loop to repeat a set of statements indefinitely until a condition becomes `false`. Use a `while` loop in cases where you wish to perform the loop repetition test at the top of the loop, and use a `do/while` loop in cases where you wish to perform the loop repetition test at the bottom of the loop.
3. Use labeled `break` or `continue` statements to break out of or continue outer loops in a nested loop structure.
4. Always use integer variables as `for` loop indexes.
5. Use the formatted output method `Fmt.printf` to create neat tabular output in your programs.

KEY TERMS

`break` statement
`continue` statement
`do/while` loop
`Fmt` class
`for` loop
format descriptor
label
labeled `break` statement
labeled `continue` statement
nested loops
null statement
`while` loop

Problems

1. Write the Java statements to calculate and print out the squares of all the even integers between 0 and 50, inclusive. Create a neat table of youre results.
2. Write a Java program to evaluate the equation $y(x) = x^2 - 3x + 2$ for all values of x between -1 and 3, inclusive, in steps of 0.1.
3. Write a Java program to calculate the factorial function, as defined in Example 4-2. Be sure to handle the special cases of 0! and of illegal input values.
4. What is the difference in behavior between a `continue` statement and a `break` statement?
5. Modify the program `Stats1` to use the `do/while` structure instead of the `while` structure currently in the program.
6. Modify the program `Stats1` to use method `Fmt.printf` for output. Display the average and standard deviation with four digits after the decimal place.
7. What is wrong with each of the following code segments?

   a.
```
x = 5;
while (x >= 0)
   x++;
```
   b.
```
x = 1;
while (x <= 5);
   x++;
```
   c.
```
for (x = 0.1; x < 1.0; x += 0.1)
   System.out.println("x = " + x);
```
   d.
```
switch (n) {
case 1:
   System.out.println("Number is 1");
case 2:
   System.out.println("Number is 2");
   break;
default:
   System.out.println("Number is not 1 or 2");
   break;
}
```

8. What does the following program do?

```
public class Print {

   // Define the main method
   public static void main(String[] args) {

      for ( int i = 1; i <= 10; i++ ) {
         for ( int j = i; j <= 10; j++ ) {
            System.out.print("*");
         }
         System.out.println();
      }
   }
}
```

9. Examine the following `for` statements and determine how many times each loop will be executed. Assume that all loop index variables are integers.

   a. `for (range = -32768; range <= 32767; range++)`

   b. `for (j = 100; j >= 1; j -= 10)`

   c. `for (k = 2; k <= 3; k += 4)`

   d. `for (i = -4; i <= -7; i++)`

   e. `for (x = -10; x <= 10; x -= 10)`

10. Examine the following `for` loops. Determine the value of `ires` at the end of each of the loops and also the number of times each loop executes. Assume that all variables are integers.

   a.
```
ires = 0;
for (index = -10; index <= 10; index++)
   ires++;
}
```

   b.
```
ires = 0;
loop1: for (idx1 = 1; idx1 <= 20; idx1 += 5) {
   if (idx1 <= 10) continue;
   loop2: for (idx2=idx1; idx2 <= 20; idx2 += 5) {
      ires = ires + idx2;
   }
}
```

   c.
```
ires = 0;
loop1: for (idx1 = 10; idx1 >= 4; idx1 -= 2) {
   loop2: for (idx2 = 2; idx2 <= idx1; idx2 += 2) {
      if (idx2 > 6) break loop2;
      ires = ires + idx2;
   }
}
```

11. Examine the following `while` loops. Determine the value of `ires` at the end of each of the loops and the number of times each loop executes. Assume that all variables are integers.

   a.
```
ires = 1;
loop1: do {
   ires = 2 * ires;
} while (ires / 10 == 0);
```

   b.
```
ires = 2
loop2: while (ires <= 512) {
   ires = ires * ires;
   if (ires == 128) break loop2;
}
```

12. Modify the program `Ball` from Example 4-7 to read in the acceleration due to gravity at a particular location. Have the program calculate the maximum range that the ball will travel for that acceleration. After modifying the program, run it with accelerations of -9.8 m/s^2, -9.7 m/s^2, and -9.6 m/s^2. What effect does the reduction in gravitational attraction

have on the range of the ball? What effect does the reduction in gravitational attraction have on the best angle θ at which to throw the ball?

13. Write a program to calculate π from the following infinite series:

$$\pi = 4 - \frac{4}{3} + \frac{4}{5} - \frac{4}{7} + \frac{4}{9} - \ldots$$

Print a table showing the value of π approximated by 1 term, 2 terms, etc. from the series. The table should have three columns, showing the number of terms used, the approximate value of π, and the difference between the approximate value and the actual value. How many terms of this series are needed to get three significant digits of accuracy (3.14)?

14. Program `DayOfYear` in Example 4-3 calculates the day of year associated with any given month, day, and year. As written, this program does not check to see if the data entered by the user is valid. It will accept nonsense values for months and days, and it will do calculations with them to produce meaningless results. Modify the program so that it checks the input values for validity before using them. If the inputs are invalid, the program should tell the user what is wrong and quit. The year should be a number greater than zero, the month should be a number between 1 and 12, inclusive, and the day should be a number between 1 and a maximum that depends on the month, inclusive. Use a `switch` structure to implement the bounds checking performed on the day.

15. **Current Through a Diode** The current flowing through the semiconductor diode shown in Figure 4.11 is given by the equation

$$i_D = I_0\left(e^{\frac{qv_D}{kT}} - 1\right), \tag{4-9}$$

where i_D = the voltage across the diode, in volts
v_D = the current flow through the diode, in amps
I_0 = the leakage current of the diode, in amps
q = the charge on an electron, 1.602×10^{-19} coulombs
k = Boltzmann's constant, 1.38×10^{-23} joule/K
T = temperature, in kelvins (K)

The leakage current I_0 of the diode is 2.0 μA. Write a computer program to calculate the current flowing through this diode for all voltages from −1.0 V to + 0.7 V, inclusive, in 0.1 V steps. Repeat this process for the following temperatures of 75 °F, and 100 °F, and 125 °F. Use the program of Example 2-2 to convert the temperatures from °F to kelvins.

16. **Tension on a Cable** A 200 pound object is to be hung from the end of a rigid 8-foot horizontal pole of negligible weight, as shown in Figure 4.12. The pole is attached to a wall by a pivot and is supported by an 8-foot cable that is attached to the wall at a higher point. The tension on this cable is given by the equation

$$T = \frac{W \cdot l_c \cdot l_p}{d\sqrt{l_p^2 - d^2}} \tag{4-10}$$

where T is the tension on the cable, W is the weight of the object, l_c is the length of the cable, l_p is the length of the pole, and d is the distance along the pole at which the cable is attached. Write a program to determine the distance d at which to attach the cable to the pole in order to minimize the tension on the cable. To do this, the program should calculate the tension on the cable at 0.1 foot intervals from d = 1 foot to d = 7 feet, inclusive, and should locate the position d that produces the minimum tension.

17. **Bacterial Growth** Suppose that a biologist performs an experiment in which he or she measures the rate at which a specific type of bacterium reproduces asexually in different culture media. The experiment shows that in Medium A the bacteria reproduce once every

Figure 4.11. A semiconductor diode

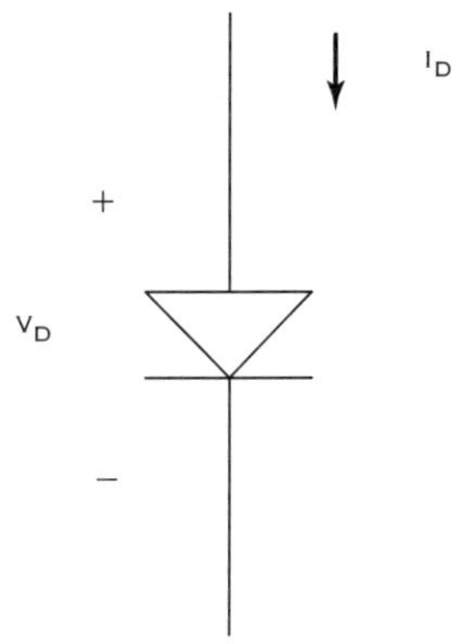

60 minutes, and in Medium B the bacteria reproduce once every 90 minutes. Assume that a single bacterium is placed on each culture medium at the beginning of the experiment. Write a Java program that calculates and writes out the number of bacteria present in each culture at intervals of three hours from the beginning of the experiment until 24 hours have elapsed. How do the numbers of bacteria compare on the two media after 24 hours?

18. **Decibels** Engineers often measure the ratio of two power measurements in *decibels*, or dB. The equation for the ratio of two power measurements in decibels is

$$\mathrm{dB} = 10 \log_{10}\frac{P_2}{P_1}, \tag{4-11}$$

where P_2 is the power level being measured and P_1 is some reference power level. Assume that the reference power level P_1 is 1 watt (W), and write a program that calculates the decibel level corresponding to power levels between 1 W and 20 W, inclusive, in 0.5 W steps.

Figure 4.12. A 200 pound weight suspended from a rigid bar supported by a cable

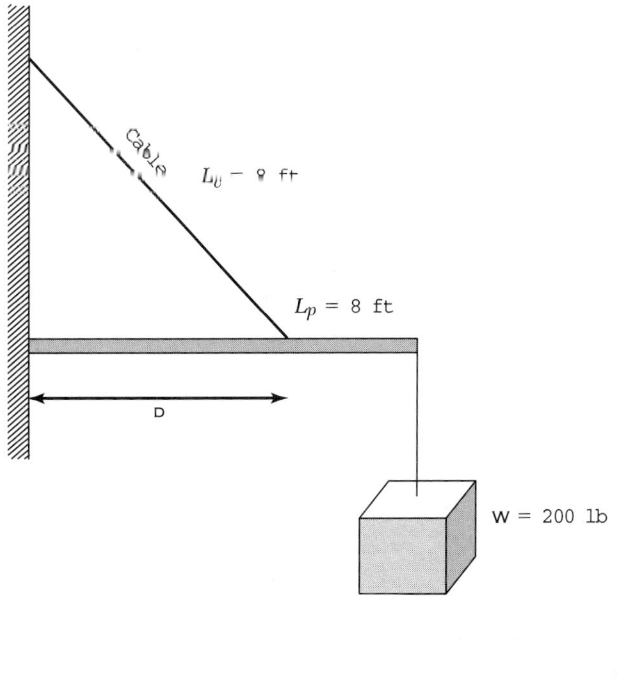

19. **Infinite Series** Trigonometric functions are usually calculated on computers by using a *truncated infinite series.* An *infinite series* is an infinite set of terms that together add up to the value of a particular function or expression. For example, one infinite series used to evaluate the sine of a number is

$$\sin x = x - \frac{x^3}{3!} + \frac{x^5}{5!} - \frac{x^7}{7!} + \frac{x^9}{9!} + \ldots, \tag{4-12a}$$

or

$$\sin x = \sum_{n=1}^{\infty} (-1)^{n-1} \frac{x^{2n-1}}{(2n-1)!}, \tag{4-12b}$$

where x is in units of radians.

Since a computer does not have enough time to add an infinite number of terms for every sine that is calculated, the infinite series is *truncated* after a finite number of terms. The number of terms that should be kept in the series is just enough to calculate the function to the precision of the floating point numbers on the computer on which the function is being evaluated. The truncated infinite series for $\sin x$ is

$$\sin x = \sum_{n=1}^{N} (-1)^{n-1} \frac{x^{2n-1}}{(2n-1)!}, \tag{4-13}$$

where N is the number of terms to retain in the series.

Write a Java program that reads in a value for x in degrees and then calculates the sine of x using the sine intrinsic function. Next, calculate the sine of x using Equation 3-13, with $N = 1, 2, 3, \ldots, 10$. Compare the true value of $\sin x$ with the values calculated using the truncated infinite series. How many terms are required to calculate $\sin x$ to the full accuracy of your computer?

20. **Geometric Mean** The *geometric mean* of a set of numbers x_1 through x_n is defined as the nth root of the product of the numbers as follows:

$$\text{geometric mean} = \sqrt[n]{x_1 x_2 x_3 \ldots x_n}. \tag{4-14}$$

Write a Java program that will accept an arbitrary number of positive input values and calculate both the arithmetic mean (i.e., the average) and the geometric mean of the numbers. Use a `while` loop to get the input values, and stop receiving inputs if the user enters a negative number. Test your program by calculating the average and geometric means of the four numbers 10, 5, 2, and 5.

21. **RMS Average** The *root-mean-square,* or *rms, average* is another way of calculating a mean for a set of numbers. The rms average of a series of numbers is the square root of the arithmetic mean of the squares of the numbers:

$$\text{rms average} = \sqrt{\frac{1}{N}\sum_{i=1}^{N} x_i^2}. \tag{4-15}$$

Write a Java program that will accept an arbitrary number of positive input values and calculate the rms average of the numbers. Prompt the user for the number of values to be entered, and use a `for` loop to read in the numbers. Test your program by calculating the rms average of the four numbers 10, 5, 2, and 5.

22. **Harmonic Mean** The *harmonic mean* is yet another way of calculating a mean for a set of numbers. The harmonic mean of a set of numbers is given by the following equation:

$$\text{harmonic mean} = \frac{N}{\frac{1}{x_1} + \frac{1}{x_2} + \cdots + \frac{1}{x_N}}. \tag{4-16}$$

Write a Java program that will read in an arbitrary number of positive input values and calculate the harmonic mean of the numbers. Use any method that you desire to read in the input values. Test your program by calculating the harmonic mean of the four numbers 10, 5, 2, and 5.

23. Write a single Java program that calculates the arithmetic mean (average), rms average, geometric mean, and harmonic mean for a set of positive numbers. Use any method that you desire to read in the input values. Compare these values for each of the following sets of numbers:
    a. 4, 4, 4, 4, 4, 4, 4
    b. 4, 3, 4, 5, 4, 3, 5
    c. 4, 1, 4, 7, 4, 1, 7
    d. 1, 2, 3, 4, 5, 6, 7

24. **Mean Time Between Failure Calculations** The reliability of a piece of electronic equipment is usually measured in terms of Mean Time Between Failures (MTBF), where MTBF is the average time that the piece of equipment can operate before a failure occurs in it. For large systems containing many pieces of electronic equipment, it is customary to determine the MTBFs of each component and then to calculate the overall MTBF of the system from the failure rates of the individual components. If the system is structured as is the one shown in Figure 4.13, every component must work in order for the whole system to work, and the overall system MTBF can be calculated as follows:

$$MTBF_{sys} = \frac{1}{\frac{1}{MTBF_1} + \frac{1}{MTBF_2} + \cdots + \frac{1}{MTBF_n}} \, . \qquad (4\text{-}17)$$

Write a program that reads in the number of series components in a system and the MTBFs for each component and then calculates the overall MTBF for the system. To test your program, determine the MTBF for a radar system consisting of an antenna subsystem with an MTBF of 2000 hours, a transmitter with an MTBF of 800 hours, a receiver with an MTBF of 3000 hours, and a computer with an MTBF of 5000 hours.

Figure 4.13. An electronic system containing three subsystems with known MTBFs

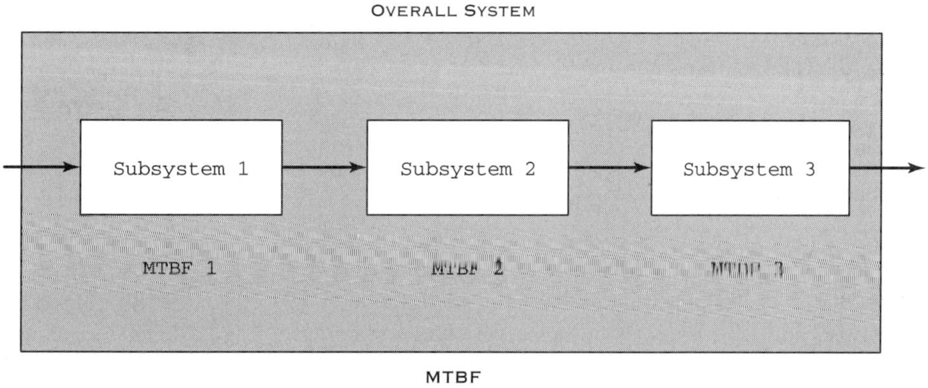

5

Arrays, Strings, File Access, and Plotting

This chapter serves as an introduction to two important data structures: arrays and `Strings`. An **array** is a group of contiguous memory locations that all have the same name and same type. This data structure is implemented as an object in Java.

A **string** is an object containing one or more characters that are treated as a single unit. `String` objects provide convenient methods for creating, comparing, and manipulating groups of characters.

In addition, the chapter introduces convenience classes that allow a programmer to read and write data to files, and to create plots of arrays of data.

SECTIONS

OBJECTIVES

After reading this chapter, you should be able to:

- Understand how to create, initialize, and use arrays
- Be able to read numeric data from, and write numeric data to, disk files using the `FileIn` and `FileOut` convenience classes
- Be able plot arrays of data using the `JPlot2D` class.
- Be able to create, initialize, and use `Strings`.

5.1 INTRODUCTION TO ARRAYS

An **array** is a special object containing (1) a group of contiguous memory locations that all have the same name and same type, and (2) a separate instance variable containing an integer constant equal to the number of elements in the array (see Figure 5.1). An individual value within the array is called an **array element**; it is identified by the name of the array together with a **subscript** in square brackets. The subscript identifies the particular location within the array. Note that *the elements of Java arrays are numbered starting with 0 and working upward.* For example, the first element shown in Figure 5.1 is referred to as `a[0]`, and the fifth variable shown in the figure is referred to as `a[4]`. The subscript of an array must be an integer. Either constants or variables may be used for array subscripts.

The length of any Java array is included as a separate field within the array object itself, and that length can be accessed by appending the field name `.length` to the name of the array. Thus, the length of array `a` in Figure 5.1 can be accessed as `a.length`. The length of a Java array is specified when it is created, and it remains fixed for as long as the array exists.

PROGRAMMING PITFALLS

The elements of an `n`-element Java array `arr` have subscripts numbered 0, 1, 2, . . ., `n` − 1. Note that there is no element `arr[n]`! Novice programmers often make the mistake of trying to use this non-existant element.

As we shall see, arrays can be extremely powerful tools. They permit us to apply the same algorithm over and over again to many different data items with a simple loop. For example, suppose that we need to take the square root of 100 different numbers. If the numbers are stored as elements of an array `a` consisting of 100 real values, then the code

```
for (i = 0; i < 100; i++)
   a[i] = Math.sqrt(a[i]);
```

Figure 5.1. An array object contains a set of elements, all of the same type, occupying successive locations in a computer's memory. It also contains an integer constant set equal to the number of elements in the array.

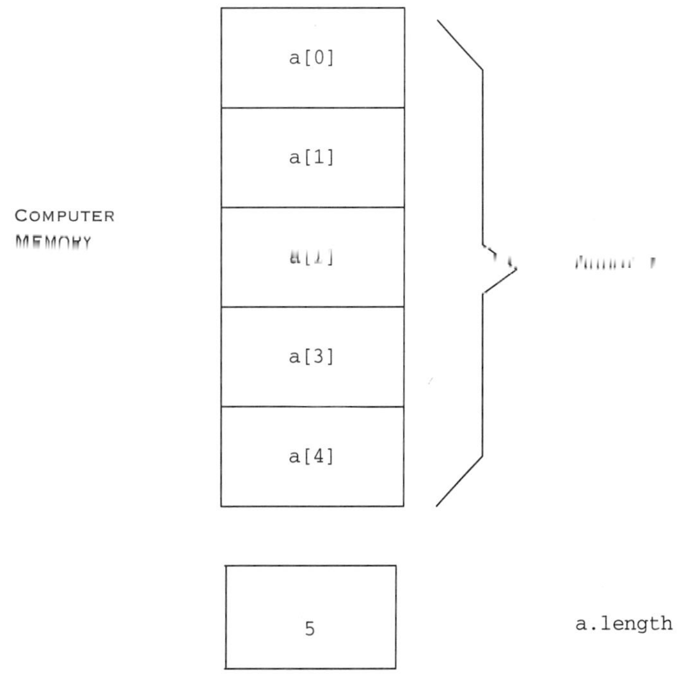

will take the square root of each real number, and store it back into the memory location that it came from. If we wanted to take the square root of 100 numbers without using arrays, we would have to write out

```
a0 = Math.sqrt(a0);
a1 = Math.sqrt(a1);
   . . .
a99 = Math.sqrt(a99);
```

as 100 separate statements! Arrays are obviously a *much* cleaner and shorter way to handle repeated similar operations.

As we shall see, it is possible to manipulate and perform calculations with individual elements of arrays one by one. We will first learn how to declare arrays in Java programs. Then, we will learn how to use arrays in Java statements.

5.2 DECLARING ARRAYS

An array must be created before it can be used. This is a two-step process. First, we must declare a **reference** to an array, and then we actually create the array. A *reference* is a "handle" or "pointer" to an object that permits Java to locate the object in memory when it is needed. A reference to an array is created by listing the array type followed by the name of the reference. It looks just like the declaration of an `int`, `double`, or any other primitive data type, except that the name is followed by square brackets (`[]`). For example, a reference to `double` array is created by the statement

```
double x[];          // Create an array reference
```

When this statement is executed, the array reference `x` is created. However, the value of the reference is initially **`null`**, since it doesn't "point to" an array object yet. Attempting to use a `null` reference in a program will produce a run-time exception.

Once a reference is created, we can create an array object to assign to the reference using the **new** operator. This operator **instantiates** an array object of the specified size. For example, the following statement creates a new 5-element array and sets the reference `x` to refer to that array:

```
x = new double[5];       // Create array object
```

The `new` operator is followed by the data type of the array object to be created and square brackets containing the number of elements to be allocated. When the elements are allocated, they are automatically initialized to zero.

Note that an array reference has a type, and *the reference can only refer to array objects of that type.* Thus, the statements below are illegal, since the attempt to assign a `double` array to an `int` reference:

```
int x[];                // Create array reference
x = new double[5];      // Create array object (illegal)
```

The creation of an array reference and an array object may be written together on a single line as follows:

```
double x[] = new double[5];
```

It is also possible to create multiple reference and array objects in a single declaration. For example, the following declaration creates two `double` arrays `a` and `b`, with 5 and 10 elements respectively.

```
double a[] = new double[5], b[] = new double[10];
```

5.3 USING ARRAY ELEMENTS IN JAVA STATEMENTS

Each element of an array is a variable just like any other variable, and *an array element may be used in any place where an ordinary variable of the same type may be used.* Array elements may be included in arithmetic and logical expressions, and the results of an expression may be assigned to an array element. For example, assume that arrays `index` and `temp` are declared as:

```
int index[] = new int[5];
double temp[] = new double[4];
```

The five elements of array `index` would be addressed as `index[0]`, `index[1]`, `index[2]`, `index[3]`, and `index[4]`, while the four elements of array `temp` would be addressed as `temp[0]`, `temp[1]`, `temp[2]`, and `temp[3]`. With this definition, the following Java statements are perfectly valid:

```
index[0] = 1;
temp[3] = index[0] / 4.;
System.out.println("index[0] = ", index[0]);
```

Arrays are commonly used in loops to allow the same calculation to be applied to many different values stored in the array elements. For example, the simple program shown in Figure 5.2 calculates the squares of the numbers in array `number`, and then prints out the numbers and their squares. The `for` loop applies the same calculations to every element of arrays `number` and `square`.

Note that the `for` loops performing calculations on array `number` use the size of the array `number.length` as their continuation condition. This practice makes the program more flexible, since we can change the sizes of the arrays created in the program,

```
// Calculates the squares of the numbers from 1 to 10
import chapman.io.*;
public class Squares {

   // Define the main method
   public static void main(String[] args) {

      double number[] = new double[10]; // Array of numbers
      double square[] = new double[10]; // Array of squares

      // Calculate squares
      for ( int i = 0; i < number.length; i++ ) {
         number[i] = i + 1;
         square[i] = number[i] * number[i];
      }

      // Write number and square
      for ( int i = 0; i < number.length; i++ ) {
         Fmt.printf("number = %4.1f ", number[i]);
         Fmt.printf("square = %5.1f\n", square[i]);
      }
   }
}
```

Figure 5.2. A program to calculate the squares of the integers from 1 to 10

and the number of passes through the body of the for loops will be updated automatically. Because the elements of the array are labeled `0,1,2,...,number.length-1`, the proper continuation test is `i < number.length`. When this program is executed, the results are:

```
D:\book\java\chap5>java Squares
number =  1.0  square =   1.0
number =  2.0  square =   4.0
number =  3.0  square =   9.0
number =  4.0  square =  16.0
number =  5.0  square =  25.0
number =  6.0  square =  36.0
number =  7.0  square =  49.0
number =  8.0  square =  64.0
number =  9.0  square =  81.0
number = 10.0  square = 100.0
```

GOOD PROGRAMMING PRACTICES

When creating `for` loops to process the elements of an array, use the array object's `length` field in the continuation condition for the loop. This will allow the loop to adjust automatically for different sized arrays. When processing an array `arr` in a `for` loop, use a continuation condition of the form

```
for (j = 0; j < arr.length; j++)
```

The "less than" relational operator is the correct one to use, because the elements of the array are numbered `0,1,...,arr.length-1`.

Although *array elements* may be used freely in Java expressions and statements, the arrays themselves cannot. For a five-element integer array, the following statements are legal:

```
for (int i = 0; i < 5; i++)
   arr[i] += 2;
```

but the statement

```
arr += 2;
```

is illegal and will produce a compilation error.

PROGRAMMING PITFALL

It is illegal to use an unsubscripted array in a Java expression or assignment. Only individual array elements may be used in this manner.

Initializing Array Values

An array object may be created and initialized using an **initializer** when its reference is declared. An *initializer* is a comma-separated list of values enclosed in braces. It may

```
// Calculates the squares of the numbers from 1 to 10
public class Square2 {

   // Define the main method
   public static void main(String[] args) {

      double number[] = {1, 2, 3, 4, 5, 6, 7, 8, 9, 10};
      double square[] = new double[10]; // Array of squares

      // Calculate squares
      for ( int i = 0; i < number.length; i++ )
         square[i] = number[i] * number[i];

      // Write number and square
      for ( int i = 0; i < number.length; i++ ) {
         Fmt.printf("number = %4.1f ", number[i]);
         Fmt.printf("square = %5.1f\n", square[i]);
      }
   }
}
```

Figure 5.3. A program illustrating the use of array initializers

only appear in an array declaration. For example, the following statement declares an array reference `a`, creates and assigns a 5-element array object to the reference, and initializes the array elements to 1, 2, 3, 4, and 5.

```
int a[] = {1, 2, 3, 4, 5};
```

Array initializers only work in declaration statements. They may *not* be used to initialize an object after its reference has been declared. For example, the following statements are illegal and will produce a compile-time error.

```
int a[];
a[] = {1, 2, 3, 4, 5};   // This statement is illegal
```

The simple program shown in Figure 5.3 illustrates the use of an initializer it initialize the elements of array `number`. When this program is executed, the output is identical to the result of the previous program.

Out-of-Bounds Array Subscripts

Each element of an array is addressed using an integer subscript. The range of integers which can be used to address array elements depends on the declared extent of the array. For a `double` array declared as

```
double a[] = new double[5];
```

the integer subscripts 0 through 4 address elements in the array. *Any other integers* (less than 0 or greater than 4) *could not be used as subscripts, since they do not correspond to allocated memory locations.* Such integers subscripts are said to be **out of bounds** for the array. But what would happen if we make a mistake and try to access the out-of-bounds element `a[5]` in a program?

Every Java array "knows" its own length, and Java has automatic **bounds checking** built into the language. If an attempt is made to access an out-of-bounds

```
// Test array bounds checking
public class TestBounds {

   // Define the main method
   public static void main(String[] args) {

      // Declare and initialize array
      int a[] = {1,2,3,4,5};

      // Write array (with an error!)
      for (int i = 0; i <= 5; i++)
         System.out.println("a[" + i + "] = " + a[i]);
   }
}
```

Figure 5.4. A simple program to illustrate the effect of out-of-bounds array references.

array element, a run-time error occurs. Java calls such errors **runtime exceptions**, and the method in which the error occurs is said to **throw an exception**. If an exception occurs and it is not handled, the program containing the exception will abort. The special exception produced by accessing an out-of-bounds array element is called an **`ArrayIndexOutOfBoundsException`**. The program shown in Figure 5.4 illustrates the behavior of a Java program containing incorrect array references. This simple program declares a 5-element `int` array `a`. The array `a` is initialized with the values 1, 2, 3, 4, and 5, and then the program attempts to print out six array elements.

When this program is compiled and executed, the results are

```
C:\book\java\chap5>java TestBounds
a[0] = 1
a[1] = 2
a[2] = 3
a[3] = 4
a[4] = 5
java.lang.ArrayIndexOutOfBoundsException: 5
        at TestBounds.main(TestBounds.java:15)
```

The program checked each array reference, and aborted when an out-of-bounds expression was encountered. Note that the error message tells us what is wrong, and even the line number at which it occurred.

The Use of Named Constants (Final Variables) with Array Declarations

In many Java programs, arrays are used to store large amounts of information. The amount of information that a program can process depends on the size of the arrays it contains. If the arrays are relatively small, the program will be small and will not require much memory to run, but it will only be able to handle a small amount of data. On the other hand, if the arrays are large, the program will be able to handle a lot of information, but it will require a lot of memory to run. The array sizes in such a program are frequently changed to make it run better for different problems or on different processors.

It is good practice to always declare the array sizes using named constants, which are known in Java as final variables. Named constants make it easy to re-size the arrays in a Java program. In the following code, the sizes of all arrays can be changed by simply changing the single named constant ARRAY_SIZE:

```
final int ARRAY_SIZE = 1000;
double array1[] = new double[ARRAY_SIZE];
double array2[] = new double[ARRAY_SIZE];
double array3[] = new double[2*ARRAY_SIZE];
```

Note that by convention, named constants or final variables are written in all capital letters, with underscores separating words.

This may seem like a small point, but it is *very* important to the proper maintenance of large programs. If all related array sizes in a program are declared using named constants, and if the built-in lengths of the arrays are used in any size tests in the program, then it will be much simpler to modify the program later. Imagine what it would be like if you had to locate and change every reference to array sizes within a 50,000 line program! The process could take weeks to complete and debug. By contrast, the size of a well-designed program could be modified in five minutes by changing only one statement.

GOOD PROGRAMMING PRACTICE

Declare the sizes of arrays in a Java program using named constants (final variables) to make them easy to change.

EXAMPLE 5-1: FINDING THE LARGEST AND SMALLEST VALUES IN A DATA SET

To illustrate the use of arrays, we will write a simple program that reads in data values, and finds the largest and smallest numbers in the data set. The program will then write out the values, with the word "LARGEST" printed by the largest value and the word "SMALLEST" printed by the smallest value in the data set.

SOLUTION

This program must ask the user for the number of values to read, create an array large enough to hold those values, and then read the input values into the array. Once the values are all read, it must go through the data to find the largest and smallest values in the data set. Finally, it must print out the values, with the appropriate annotations beside the largest and smallest values in the data set.

1. **State the problem.** We have not yet specified the type of data to be processed. If we are processing integer data, then the problem may be stated as follows:

   > Develop a program to read a user-specified number of integer values from the standard input device, locate the largest and smallest values in the data set, and write out all of the values with the words "LARGEST" and "SMALLEST" printed by the largest and smallest values in the data set.

2. **Define the inputs and outputs**. There are two types of inputs to this program:
   (a) An integer containing the number of integer values to read. This value will come from the standard input device.
   (b) The integer values in the data set. These values will also come from the standard input device.

   The outputs from this program are the values in the data set, with the word "LARGEST" printed by the largest value, and the word "SMALLEST" printed by the smallest value.
3. **Decompose the program into classes and their associated methods**. We will not be creating multiple objects until Chapter 7, so for now, there will be a single class, and only the `main` method within that class. We will call the class `Extremes`, and make it a subclass of the root class `Object`.
4. **Design the algorithm that you intend to implement for each method**. The `main` method can be broken down into four major steps:

```
Get the number of values to read
Read the input values into an array
Find the largest and smallest values in the array
Write out the data with the words 'LARGEST' and
       'SMALLEST' at the appropriate places
```

   The first two major steps of the program are to get the number of values to read in and to read the values into an input array. We must prompt the user for the number of values to read, and create an array of that size. Then we should read in the data values. The detailed pseudocode for these steps is:

```
Prompt user for the number of input values nvals
Read in nvals
Create an integer array of size nvals
for (j = 0; j < nvals; j++) {
   Read in input value
}
...
...(Further processing here)
...
```

   Next we must locate the largest and smallest values in the data set. We will use variables `large` and `small` as pointers to the array elements having the largest and smallest values. The pseudocode to find the largest and smallest values is:

```
// Find largest value
temp ← input[0];
large ← 0;
for ( j = 1; j < nvals; j++ ) {
   if (input[j] > temp) {
      temp ← input[j];
      large ← j;
   }
}

// Find smallest value
temp ← input[0];
small ← 0;
for ( j = 1; j < nvals; j++ ) {
   if (input[j] < temp) {
```

```
            temp ← input[j];
            small ← j;
        }
    }
```

The final step is writing out the values with the largest and smallest numbers labeled:

```
for ( j = 0; j < nvals; j++ ) {
   if (small == j)
      Write input[j] and "SMALLEST"
   else if (large == j)
      Write input[j] and "LARGEST"
   else
      Write input[j]
}
```

5. **Turn the algorithm into Java statements**. The resulting Java program is shown in Figure 5.5.

```
/*
   Purpose:
      To find the largest and smallest values in a data set,
      and to print out the data set with the largest and
      smallest values labeled.

   Record of revisions:
       Date       Programmer           Description of change
       ====       ==========           =====================
     4/02/98    S. J. Chapman          Original code
*/
import chapman.io.*;
public class Extremes {

   // Define the main method
   public static void main(String[] args) {

      // Declare variables, and define each variable
      int j;                  // loop index
      int large;              // Index of largest value
      int nvals;              // Number of vals in data set
      int small;              // Index of smallest value
      int temp;               // Temporary variable

      // Create a StdIn object
      StdIn in = new StdIn();

      // Get the number of points to input
      System.out.print("Enter number of elements in array: ");
      nvals = in.readInt();

      // Create array of proper size
      int input[] = new int[nvals];

      // Get values
      for (j = 0; j < nvals; j++) {
         System.out.print("Enter value " + (j+1) + ": ");
         input[j] = in.readInt();
      }
```

Figure 5.5. *(cont.)*

```
        // Find largest value
        temp = input[0];
        large = 0;
        for ( j = 1; j < nvals; j++ ) {
           if (input[j] > temp) {
              temp = input[j];
              large = j;
           }
        }

       // Find smallest value
        temp = input[0];
        small = 0;
        for ( j = 1; j < nvals; j++ ) {
           if (input[j] < temp) {
              temp = input[j];
              small = j;
           }
        }
  // Write out results
        System.out.print("\n The values are:\n");
        for ( j = 0; j < nvals; j++ ) {
           if (small == j)
              Fmt.printf("%6d SMALLEST\n", input[j]);
           else if (large == j)
              Fmt.printf("%6d LARGEST\n", input[j]);
           else
              Fmt.printf("%6d\n", input[j]);
        }
     }
  }
```

Figure 5.5. A program to read in a data set from the standard input device, find the largest and smallest values, and print the values with the largest and smallest values labeled

6. **Test the program**. To test this program, we will a data set with 6 values: –6, 5, –11, 16, 9, and 0.

```
D:\book\java\chap5>java Extremes
Enter number of elements in array: 6
Enter value 1: -6
Enter value 2: 5
Enter value 3: -11
Enter value 4: 16
Enter value 5: 9
Enter value 6: 0

The values are:
    -6
     5
   -11 SMALLEST
    16 LARGEST
     9
     0
```

The program correctly labeled the largest and smallest values in the data set. Thus, the program gives the correct answer for our test data set.

PRACTICE!

This quiz provides a quick check to see if you have understood the concepts introduced in the first three sections. If you have trouble with the quiz, reread the sections, ask your instructor, or discuss the material with a fellow student. The answers to this quiz are found in the back of the book.

1. What is an array? What components are found with a Java array?
2. What is a reference? How do you declare a reference to an array?
3. How is an array object created?
4. How may an array be initialized?
5. Suppose you have created an 100-element array. What range of subscripts may be used to address the elements of this array?

Determine which of the following Java statements are valid. For each valid statement, specify what will happen in the program.

6.
```
double arr[];
arr = new double[10];
```
7.
```
double aaa[];
aaa = new int[100];
```
8.
```
double bbb[];
bbb[] = {1., 2., 3., 4., 5., 6.};
```
9.
```
double aaa[] = {1., 2., 3., 4., 5.};
for (i = 1; i < aaa.length; i++)
   System.out.println("i = " + aaa[i]);
```
10.
```
double aaa[] = {1., 2., 3., 4., 5.};
for (i = aaa.length-1; i >= 0; i--)
   System.out.println("i = " + aaa[i]);
```

5.4 READING DATA FROM AND WRITING DATA TO FILES

Arrays are designed to hold and manipulate large amounts of data, so that we can apply the same basic calculations to many different values in just a few statements. Unfortunately, it is not quite so easy to read in and write out the large quantities of data. For example, suppose that we were working with a 10,000-element input array. Can you imagine entering all 10,000 elements by typing them one-by-one to the standard input stream?

What we really need is a convenient way to read and write data to disk files. There are two possible ways to accomplish this. One possible approach is to read data from the standard input stream and to write data to the standard output stream, but to *redirect* the standard input stream and standard output stream to disk files using **command-line redirection**. The standard input stream is redirected by using a less-than sign (<) followed by the file name on the command line, and the standard output stream is redirected by using a greater-than sign (>) followed by the file name on the command line.[1] If all of the inputs needed for a program are placed in the input file in the proper order, the program will be able to execute without further operator input. For example, the following command line starts program `Example`, which reads data from file `infile` and writes data to file `outfile`.

```
D:\book\java\chap5>java Example < infile > outfile
```

[1]Command line redirection works for PC and Unix systems, but is not available on the Macintosh.

The other, more flexible approach is to open files and to read or write to them directly from inside a Java program. Unfortunately, the Java I/O system is very complex, and it is not easy to simply open a file, read the data you want, and close it again. We will not study the details of of the Java I/O system in this abbreviated text, but we will use a couple of convenience classes that simplify the reading and writing of formatted files.[2]

The `chapman.io` package includes two classes called `FileIn` and `FileOut` that allow a program to easily read and write formatted files.

Reading Files with Class `FileIn`

Class `FileIn` is designed to read numeric data stored in formatted input files. The numbers can be arranged freely within the input file. This class contains methods `readDouble()`, `readFloat()`, `readInt()`, and `readLong()` to read `double`, `float`, `int`, and `long` values from the file, as well as a status variable `readStatus` to indicate the success or failure of each operation. The methods in class `FileIn` are summarized in Table 5-1, and the possible values of the status variable `readStatus` are summarized in Table 5-2.

To open a file with this class, we simply create a new `FileIn` object with the name of the input file. For example, the statement

```
FileIn in = new FileIn("infile");
```

creates a new `FileIn` object, sets reference `in` to refer to it, and simultaneously opens file `infile` for reading. If the file was found and opened successfully, variable `in.readStatus` will be set to value `in.READ_OK`. If the file was not found, variable `in.readStatus` will be set to value `in.FILE_NOT_FOUND`, and so forth for other types of errors. The program in Figure 5.6 shows how to use class `FileIn` to read `double` values into an array. Note that the program checks to see if the file was opened successfully, and writes a warning message if not. Then, the program reads the input file until the end of file is reached, storing the data into array `a`.

TABLE 5-1 Methods in Class `FileIn`

METHOD	DESCRIPTION
`close()`	Closes the file, making it available for other programs to use.
`readDouble()`	Reads a `double` value from the file.
`readFloat()`	Reads a `float` value from the file.
`readInt()`	Reads an `int` value from the file.
`readLong()`	Reads a `long` value from the file.

TABLE 5-2 Possible Values of `readStatus` for Class `FileIn`

VALUE	DESCRIPTION
`READ_OK`	Value read successfully from the file.
`INVALID_FORMAT`	The value in the file had an invalid format. For example, a string was found where numeric data was expected.
`FILE_NOT_FOUND`	The specified file was not found when the `FileIn` object was created.
`EOF`	End-of-file reached.
`IO_EXCEPTION`	An unspecified I/O error occurred.

[2]A brief discussion of the Java I/O system is available for download at the books Web site, which is http://www.prenhall.com/chapman_java.

```
// Test FileIn for double values
import chapman.io.*;
public class TestFileIn {

   // Define the main method
   public static void main(String arg[]) {

      // Declare variables and arrays
      double a[] = new double[100];   // Input array
      int i = 0;                      // Loop index
      int nvals = 0;                  // Number of values read

      // Open file
      FileIn in = new FileIn("infile");

      // Check for valid open
      if ( in.readStatus != in.FILE_NOT_FOUND ) {

         // Read numbers into array
         while ( in.readStatus != in.EOF ) {
            a[i++] = in.readDouble();
         }
         nvals = i;

         // Close file
         in.close();

         // Display the numbers read
         for ( i = 0; i < nvals; i++ ) {
            Fmt.printf("a[%3d] = ", i);
            Fmt.printf("%10.4f\n", a[i]);
         }
      }

      // Get here if file not found. Tell user
      else {
         System.out.println("File not found: infile");
      }
   }
}
```

Figure 5.6. A program to read a data set from a file into an array, and then display the data

To test this program, we will create a file `infile` containing the following data:

```
12                    34.56
3.14159265358979      45.5
1                     -7
0
```

When this program is executed, the results are:

```
D:\book\java\chap5>java TestFileIn
a[ 0] =    12.0000
a[ 1] =    34.5600
a[ 2] =     3.1416
a[ 3] =    45.5000
a[ 4] =     1.0000
a[ 5] =    -7.0000
a[ 6] =     0.0000
```

As you can see, the program read the input file successfully.

TABLE 5-3 Methods in Class `FileOut`

METHOD	DESCRIPTION
`close()`	Closes the file, making it available for other programs to use.
`printf(fmt,value)`	Writes data to the file using `printf` conventions.

TABLE 5-4 Possible Values of `writeStatus` for Class `FileOut`

VALUE	DESCRIPTION
`WRITE_OK`	Value written successfully to the file.
`IO_EXCEPTION`	An unspecified I/O error occurred.

Writing Files with Class `FileOut`

Class `FileOut` is designed to write data to formatted output files. Its principal method is `printf`, which works identically to the `printf` to the standard output stream in the `Fmt` class. The methods in class `FileOut` are summarized in Table 5-3, and the possible values of the status variable `writeStatus` are summarized in Table 5-4.

There are two ways open a file with this class.

```
FileOut out = new FileOut("outfile");      // Open and replace
FileOut out = new FileOut("outfile",true); // Open and append
```

The first way to open the file creates a new `FileOut` object, sets reference `out` to refer to it, and simultaneously opens file `outfile` for writing. If the file already exists, it will be deleted and an empty file will be opened. The second way to open the file behaves the same, except that it *appends* data to an existing file instead of deleting the file.

The program in Figure 5.7 shows how to use class `FileOut` to write data to a file.

```
// Test FileOut
import chapman.io.*;
public class TestFileOut {

   // Define the main method
   public static void main(String arg[]) {

      double a[] = new double[10];

      // Test open without append
      FileOut out = new FileOut("outfile");

      // Check for valid open
      if ( out.writeStatus != out.IO_EXCEPTION ) {

         // Write values
         out.printf("double = %20.14f\n",Math.PI);
         out.printf("long   = %20d\n",12345678901234L);
         out.printf("char   = %20c\n",'A');
         out.printf("String = %20s\n","This is a test.");
      }

      // Close file
      out.close();

      // Now test the append option
      out = new FileOut("outfile",true);

      // Check for valid open
      if ( out.writeStatus != out.IO_EXCEPTION ) {
```

Figure 5.7. *(cont.)*

```
            // Write values
            out.printf("double = %20.14f\n",Math.PI);
            out.printf("long   = %20d\n",12345678901234L);
            out.printf("char   = %20c\n",'A');
            out.printf("String = %20s\n","This is a test.");
         }

         // Close file
         out.close();
      }
   }
```

Figure 5.7. A program to write data to an output file. Note that this program test both the replace and append options.

When this program is executed, the results are:

```
double =       3.14159265358979
long =           12345678901234
char =                        A
String =        This is a test.
double =       3.14159265358979
long =           12345678901234
char =                        A
String =        This is a test.
```

As you can see, the program wrote the output successfully, appending to the existing data the second time that it opened the file.

5.5 EXAMPLE PROBLEMS

Now we will examine two example problems that illustrate the use of arrays.

EXAMPLE 5-2: SORTING DATA

In many scientific and engineering applications, it is necessary to take a random input data set and to sort it so that the numbers in the data set are either all in *ascending order* (lowest-to-highest) or all in *descending order* (highest-to-lowest). For example, suppose that you were a zoologist studying a large population of animals, and that you wanted to identify the largest 5% of the animals in the population. The most straightforward way to approach this problem would be to sort the sizes of all of the animals in the population into ascending order, and take the top 5% of the values.

Sorting data into ascending or descending order seems to be an easy job. After all, we do it all the time. It is simple matter for us to sort the data (10, 3, 6, 4, 9) into the order (3, 4, 6, 9, 10). How do we do it? We first scan the input data list (10, 3, 6, 4, 9) to find the smallest value in the list (3), and then scan the remaining input data (10, 6, 4, 9) to find the next smallest value (4), *etc.* until the complete list is sorted.

In reality, however, sorting can be a very difficult job. As the number of values to be sorted increases, the time required to perform the simple sort described above increases rapidly, since we must scan the input data set once for each value sorted. For very large data sets, this technique just takes too long to be practical. Even worse, how would we sort the data if there were too many numbers to fit into the main memory of the computer? The development of efficient sorting techniques for large data sets is an active area of research, and is the subject of whole courses all by itself.

In this example, we will confine ourselves to the simplest possible algorithm to illustrate the concept of sorting. This simplest algorithm is called the **selection sort**. It is just a computer implementation of the mental math previously described. The basic algorithm for the selection sort is as follows:

1. Scan the list of numbers to be sorted to locate the smallest value in the list. Place that value at the front of the list by swapping it with the value currently at the front of the list. If the value at the front of the list is already the smallest value, then do nothing.
2. Scan the list of numbers from position 2 to the end to locate the next smallest value in the list. Place that value in position 2 of the list by swapping it with the value currently at that position. If the value in position 2 is already the next smallest value, then do nothing.
3. Scan the list of numbers from position 3 to the end to locate the third smallest value in the list. Place that value in position 3 of the list by swapping it with the value currently at that position. If the value in position 3 is already the third smallest value, then do nothing.
4. Repeat this process until the next-to-last position in the list is reached. After the next-to-last position in the list has been processed, the sort is complete.

Note that if we are sorting `n` values, this sorting algorithm requires `n-1` scans through the data to accomplish the sort. This process is illustrated in Figure 5.8. Since there are 5 values in the data set to be sorted, we will make 4 scans through the data. During the first pass through the entire data set, the minimum value is 3, so the 3 is swapped with the 10 that was in position 1. Pass 2 searches for the minimum value in positions 2 through 5. That minimum is 4, so the 4 is swapped with the 10 in position 2. Pass 3 searches for the minimum value in positions 3 through 5. That minimum is 6, which is already in position 3, so no swapping is required. Finally, pass 4 searches for the minimum value in positions 4 through 5. That minimum is 9, so the 9 is swapped with the 10 in position 4, and the sort is completed.

Figure 5.8. An example problem demonstrating the selection sort algorithm

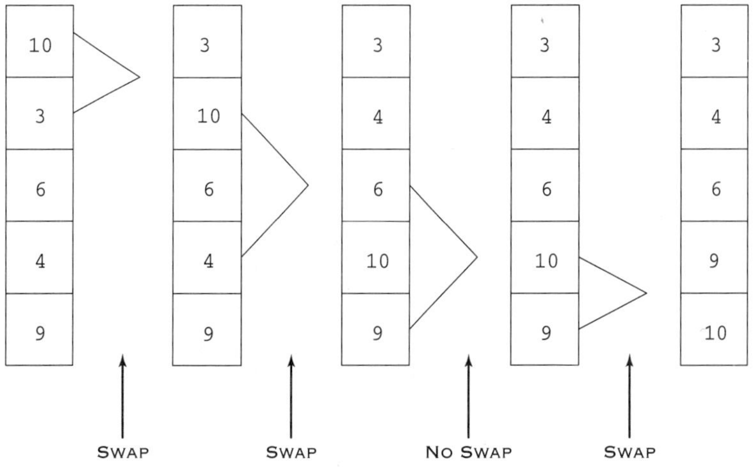

GOOD PROGRAMMING PRACTICE

The selection sort algorithm is the easiest sorting algorithm to understand, but it is computationally inefficient. *It should never be applied to sort really large data sets* (say, sets with more than 1000 elements). Over the years, computer scientists have developed much more efficient sorting algorithms. We will encounter one such algorithm in Chapter 6.

We will now develop a program to read a data set from the standard input stream into an array, sort it into ascending order, and display the sorted data set.

SOLUTION

This program must be able to read input data from the disk file, sort the data, and write out the sorted data. The design process for this problem is given below.

1. **State the problem**. We have not yet specified the type of data to be sorted. If the data is `double`, then the problem may be stated as follows:

   Develop a program to read an arbitrary number of `double` input data values from an input file, sort the data into ascending order, and write the sorted data to the standard output device.

2. **Define the inputs and outputs**. There are two types of inputs to this program:
   (a) The name of the file to open;
   (b) The values in the file, which will be the `double` values to sort. The outputs from this program are the sorted data values written to the standard output device.
3. **Decompose the program into classes and their associated methods**. There will be a single class, and only the `main` method within that class. We will call the class `SelectionSort`.
4. **Design the algorithm that you intend to implement for each method**. The `main` method can be broken down into four major steps:

```
Get file name and open input file
Read the input data into the array
Sort the data in ascending order
Write the sorted data
```

The first major steps of the program are to read the input file name, open the file, and read in the data. The pseudocode for these steps is as follows:

```
Read fileName from standard input device
Open input file
while ( in.readStatus != in.EOF ) {
   arr[i++] = in.readDouble();
}
nvals ← i;
```

Next we have to sort the data. We will need to make `nvals-1` passes through the data, finding the smallest remaining value each time. We will use a pointer to locate the smallest value in each pass. Once the smallest value is found, it will be swapped to the top of the list of it is not already there. (Note that in order to make `nvals-1` passes, the `for` loop must run from 0 to `nvals-2`.) The pseudocode for this step is as follows:

```
for ( i = 0; i <= nvals-2; i++ ) {

   // Find the minimum value in arr[i] through arr[nvals-1]
   iptr ← i;
   for ( j = i+1; j <= nvals-1; j++ ) {
      if (arr[j] < arr[iptr])
         iptr ← j;
   }

   // iptr now points to the min value, so swap arr[iptr] with
   // with arr[i] if iptr != i.
   if (i != iptr) {
      temp ← arr[i];
      arr[i] ← arr[iptr];
      arr[iptr] ← temp;
   }
}
```

The final step is writing out the sorted values. No refinement of the pseudocode is required for that step. The final pseudocode is the combination of the reading, sorting and writing steps.

5. **Turn the algorithm into Java statements**. The resulting Java program is shown in Figure 5.9.

```
/*
   Purpose:
     To read in a set of double values from an input file, sort it
     into ascending order using the selection sort algorithm, and to
     write the sorted data to the standard output stream.

   Record of revisions:
       Date        Programmer      Description of change
       ====        ==========      =====================
     4/02/98     S. J. Chapman     Original code
*/
import chapman.io.*;
public class SelectionSort {

   // Define the main method
   public static void main(String[] args) {

      // Define maximum array size
      final int MAXVAL = 1000;

      // Declare variables, and define each variable
      double arr[] = new double[MAXVAL];
                            // Array of input measurements
      String fileName;      // Input file name
      int i = 0, j;         // Loop index
      int iptr;             // Pointer to smallest value
      int nvals;            // Number of data values to sort
      double temp;          // Temporary variable for swapping
```

Figure 5.9. *(cont.)*

```
         // Create StdIn object
         StdIn in = new StdIn();

         // Get input file name
         System.out.print("Enter file name: ");
         fileName = in.readString();

         // Open file
         FileIn in1 = new FileIn(fileName);

         // Check for valid open
         if ( in1.readStatus != in1.FILE_NOT_FOUND ) {

            // Read numbers into array
            while ( in1.readStatus != in1.EOF ) {
               arr[i++] = in1.readDouble();
            }
            nvals = i;

            // Close file
            in1.close();

            // Sort values
            for ( i = 0; i <= nvals-2; i++ ) {

               // Find the minimum value in arr[i] through arr[nvals-1]
               iptr = i;
               for ( j = i+1; j <= nvals-1; j++ ) {
                  if (arr[j] < arr[iptr])
                     iptr = j;
               }

               // iptr now points to the min value, so swap
               // arr[iptr] with arr[i] if iptr != i.
               if (i != iptr) {
                  temp = arr[i];
                  arr[i] = arr[iptr];
                  arr[iptr] = temp;
               }
            }

            // Write out sorted values
            for ( i = 0; i < nvals; i++ )
               System.out.println(arr[i]);
         }

         // Come here if file not found
         else {
            System.out.println("File " + fileName + " not found!");
         }
      }
   }
```

Figure 5.9. A program to read values from an input file, and to sort them into ascending order.

6. **Test the program.** To test this program, we will create an input data file containing the data to sort. The data set will contain a mixture of positive and negative numbers as well as at least one duplicated value to see if the program works properly under those conditions. The following data set will be placed in file `input1`:

```
13.3
12.
-3.0
 0.
 4.0
 6.6
 4.
-6.
```

The first value in the file is the number of values to read, and the remaining values are the data set to sort. Running this file values through the program yields the following result (user-input values are displayed in boldface):

```
D:\book\java\chap5>java SelectionSort
Enter file name: input1
-6.0
-3.0
 0.0
 4.0
 4.0
 6.6
12.0
13.3
```

The program gives the correct answers for our test data set. Note that it works for both positive and negative numbers as well as for repeated numbers.

EXAMPLE 5-3: THE MEDIAN

In Chapter 4, we examined two common statistical measures of data: averages (or means) and standard deviations. Another common statistical measure of data is the median. The median of a data set is the value such that half of the numbers in the data set are larger than the value and half of the numbers in the data set are smaller than the value. If there are an even number of values in the data set, then there cannot be a value exactly in the middle. In that case, the median is usually defined as the average of the two elements in the middle. The median value of a data set is often close to the average value of the data set, but not always. For example, consider the following data set:

```
  1
  2
  3
  4
100
```

The average or mean of this data set is 22, while the median of this data set is 3!

An easy way to compute the median of a data set is to sort it into ascending order, and then to select the value in the middle of the data set as the median. If there are an even number of values in the data set, then average the two middle values to get the median.

In a language such as Java, where the array subscripts run from `0` to `nvals-1`, the median of a sorted data set is defined as:

$$\text{median} = \begin{cases} \texttt{a[nvals/2]} & \texttt{if nvals is odd} \\ \texttt{(a[nvals/2 - 1] + a[nvals/2])/2} & \texttt{if nvals is even} \end{cases} \tag{5-1}$$

For example, in the sorted 5-element array shown below, the middle element is `a[nvals/2]`, which is `a[2]`, or 3.

```
a[0]          1
a[1]          2
a[2]          3
a[3]          4
a[4]        100
```

Similarly, in the sorted four-element array shown below, the median is `(a[nvals/2-1] + a[nvals/2])/2`, which is `(a[1] + a[2])/2`, or 2.5.

```
a[0]          1
a[1]          2
a[2]          3
a[3]          4
```

Write a program to calculate the mean, median, and standard deviation of an input data set that is read from a the standard input stream.

SOLUTION

This program must be able to read input measurements from a disk file and calculate the mean, median, and standard deviation of the data set. Note that the data will have to be sorted in order to calculate the median. The design process for this problem is given below.

1. **State the problem**. Calculate the average, median, and standard deviation of a set of measurements which are read from an input file, and write those values out on the standard output device.
2. **Define the inputs and outputs**. (a) The name of the input file. (b) The input data values in the file, which will be measurements of type `double`. The outputs from this program are the average, median, and standard deviation of the input measurements. They are written to the standard output device.
3. **Decompose the program into classes and their associated methods**. There will be a single class, and only the `main` method within that class. We will call the class `Stats3`.
4. **Design the algorithm that you intend to implement for each method**. The `main` method can be broken down into five major steps:

```
Get file name and open input file
Read the input data into the array
Sort the measurements in ascending order
Calculate the average, mean, and standard deviation
Write average, median, and standard deviation
```

The detailed pseudocode for the first three steps is similar to that of the previous example:

```
Read fileName from standard input device
Open input file
while ( in.readStatus != in.EOF ) {
   arr[i++] = in.readDouble();
}
for ( i = 0; i <= nvals-2; i++ ) {

   // Find the minimum value in arr[i] through arr[nvals-1]
   iptr ← i;
   for ( j = i+1; j <= nvals-1; j++ ) {
      if (arr[j] < arr[iptr])
         iptr ← j;
   }
```

```
         // iptr now points to the min value, so swap
         // arr[iptr] with arr[i] if iptr != i.
         if (i != iptr) {
            temp ← arr[i];
            arr[i] ← arr[iptr];
            arr[iptr] ← temp;
         }
      }
```

The fourth step is to calculate the required average, median, and standard deviation. To do this, we must first accumulate some statistics on the data (Σx and Σx^2), and then apply the definitions of average, median, and standard deviation given previously. The pseudocode for this step is

```
for (i = 0; i < nvals; i++) {
   sumX ← sumX + arr[i];
   sumX2 ← sumX2 + arr[i]*arr[i];
}
if (nvals >= 2) {
   xBar ← sumX / nvals;
   stdDev ← Math.sqrt((nvals*sumX2 - sumX*sumX) /(nvals*(nvals-1)));
   if nvals is an even number
      median ← (arr[nvals/2-1] + arr[nvals/2]) / 2;
   else
      median ← arr[nvals/2];
}
else {
   Tell user about insufficient data.
}
```

We will decide if `nvals` is an even number by using the modulo operator `nvals%2`. If `nvals` is even, this operation will return a 0; if `nvals` is odd, it will return a 1. Finally, we must write out the results.

```
Write out average, median, standard deviation, and no. of
points
```

5. **Turn the algorithm into Java statements**. The resulting Java program is shown in Figure 5.10.

```
/*
   Purpose:
      To read in a set of double values from an input file, and calcu-
      late the mean, median, and standard deviation of the input data.
      The mean, median, and standard deviation are written to the stan-
      dard output stream.

   Record of revisions:
      Date        Programmer     Description of change
      ====        ==========     =====================
     4/02/98     S. J. Chapman   Original code
*/
import chapman.io.*;
public class Stats3 {
```

Figure 5.10. *(cont.)*

```
// Define the main method
public static void main(String[] args) {

   // Define maximum array size
   final int MAXVAL = 1000;

   // Declare variables, and define each variable
   double arr[] = new double[MAXVAL];
                           // Array of input measurements
   String fileName;        // Input file name
   int i = 0, j;           // Loop index
   int iptr;               // Pointer to smallest value
   double median;          // Median of the input measurements
   int nvals;              // Number of data values to sort
   double stdDev = 0;      // The standard deviation of the input samples.
   double sumX = 0;        // The sum of the input values.
   double sumX2 = 0;       // The sum of the squares of the input values.
   double x = 0;           // An input data value.
   double xBar = 0;        // Average of the input measurements
   double temp;            // Temporary variable for swapping

   // Create a StdIn object
   StdIn in = new StdIn();

   // Get input file name
   System.out.print("Enter file name: ");
   fileName = in.readString();

   // Open file
   FileIn in1 = new FileIn(fileName);

   // Check for valid open
   if (in1.readStatus != in1.FILE_NOT_FOUND) {

      // Read numbers into array
      while (in1.readStatus != in1.EOF) {
         arr[i++] = in1.readDouble();
      }
      nvals = i;

      // Close file
      in1.close();

      // Sort values
      for (i = 0; i <= nvals-2; i++) {

         // Find the minimum value in arr[i] through arr[nvals-1]
         iptr = i;
         for (j = i+1; j <= nvals-1; j++) {
            if (arr[j] < arr[iptr])
         }

         // iptr now points to the min value, so swap a[iptr]
         // with a[i] if iptr != i.
         if (i != iptr) {
            temp = arr[i];
            arr[i] = arr[iptr];
            arr[iptr] = temp;
         }
      }
```

Figure 5.10. *(cont.)*

```
         // Calculate sums
         for (i = 0; i < nvals; i++) {
            sumX += arr[i];
            sumX2 += arr[i]*arr[i];
         }

         // Check to see if we have enough input data.
         if (nvals >= 2) {

            // There is enough information, so calculate the
            // mean, median, and standard deviation
            xBar = sumX / nvals;
            stdDev = Math.sqrt((nvals*sumX2-sumX*sumX)/(nvals*(nvals-1)));
            if (nvals%2 == 0)
               median = (arr[nvals/2-1] + arr[nvals/2]) / 2.;
            else
               median = arr[nvals/2];

            // Tell user.
            Fmt.printf("mean               = %8.3f\n",xBar);
            Fmt.printf("median             = %8.3f\n",median);
            Fmt.printf("standard deviation = %8.3f\n",stdDev);
            Fmt.printf("No. of data points = %8d\n",nvals);
         }
         else {
            System.out.println("At least 2 values must be entered!");
         }
      }

      // Come here if file not found
      else {
         System.out.println("File " + fileName + " not found!");
      }
   }
}
```

Figure 5.10. A program to read values from an input file, and to calculate their mean, median, and standard deviation

6. **Test the program**. To test this program, we will calculate the answers by hand for a simple data set, and then compare the answers to the results of the program. If we use five input values: 5, 3, 4, 1, and 9, then the mean, standard deviation, and median are

$$\bar{x} = \frac{1}{N}\sum_{i=1}^{N} x_i = \frac{1}{5}\,22 = 4.4,$$

$$s = \sqrt{\frac{N\sum_{i=1}^{N} x_i^2 - \left(\sum_{i=1}^{N} x_i\right)^2}{N(N-1)}} = 2.966,$$

$$\text{median} = 4.$$

If these values are placed in the file `input2` and the program is run with that file as an input, the results are

```
C:\book\java\chap5>java Stats3
Enter file name: input2
```

```
mean                = 4.400
median              = 4.000
standard deviation  = 2.966
No. of data points  =     5
```

The program gives the correct answers for our test data set.

5.6 INTRODUCTION TO PLOTTING

One of the great advantages of Java compared to other languages such as C and Fortran is that *device-and platform-independent graphics is built directly into the standard language.* While we will not learn the details of Java graphics until a later chapter, we will begin using Java's graphics capabilities here. The `chapman.graphics` package includes classes suitable for plotting arrays of data, and we will now learn how to use a class called **`JPlot2D`** to create two-dimensional plots of our data.

Java graphics are displayed in windows known as **frames**. A frame is a rectangular portion of a graphics display device that can be used to display Java graphics, including buttons, text boxes, plots, etc. A Java program can open as many frames as desired, and each frame can display different types of graphical information. The support for frames is included in Java's Abstract Windowing Toolkit, and in the Swing package. Any program that uses graphics must import packages `java.awt.*`, `javax.swing.*`, and `java.awt.event.*` from the toolkit. For the time being, we will introduce a template that creates a frame without explaining the details of its operation, and we will use that template to display our plots. Figure 5.11 shows a sample program that creates and displays a plot. The program creates arrays generated from the equations $y(x) = \sin x$ and $y(x) = 1.2 \cos x$ over the range $0 \le x \le 2\pi$, and plots the contents of those arrays.

```
import java.awt.*;
import java.awt.event.*;
import javax.swing.*;
import chapman.graphics.JPlot2D;
public class TestPlot2D {

   /**
   * This method is an example of how to use Class JPlot2D. The
   * method creates a frame, and then places a new JPlot2D
   * object within the frame. It draws two curves on the
   * object, and also illustrates the use of line styles,
   * titles, labels, and grid
   */
   public static void main(String s[]) {

      //*********************************************************
      //
      //  Create data to plot
      //
      //*********************************************************

      // Define arrays to hold the two curves to plot
      double x[] = new double[81];
      double y[] = new double[81];
      double z[] = new double[81];
```

Figure 5.11. *(cont.)*

```
            // Calculate a sine and a cosine wave
            for ( int i = 0; i < x.length; i++ ) {
               x[i] = (i+1) * 2 * Math.PI / 40;
               y[i] = Math.sin(x[i]);
               z[i] = 1.2 * Math.cos(x[i]);
            }
            //*****************************************************
            //
            // Create plot object and set plot information.
            //
            //*****************************************************
            JPlot2D pl = new JPlot2D( x, y );
            pl.setPlotType ( JPlot2D.LINEAR );
            pl.setLineColor( Color.blue );
            pl.setLineWidth( 2.0f );
            pl.setLineStyle( JPlot2D.LINESTYLE_SOLID );
            pl.setMarkerState( JPlot2D.MARKER_ON );
            pl.setTitle( "Plot of sin(x) and 1.2*cos(x) vs. x" );
            pl.setXLabel( "x" );
            pl.setYLabel( "sin(x) and 1.2 cos(x)" );
            pl.setGridState( JPlot2D.GRID_ON );

            // Add a second curve to the plot.
            pl.addCurve( x, z );
            pl.setLineWidth( 3.0f );
            //*****************************************************
            //
            // Create a frame and place the plot in the center of the
            // frame. Note that the plot will occupy all of the
            // available space.
            //
            //*****************************************************
            JFrame fr = new JFrame("Plot2D ...");

            // Create a Window Listener to handle "close" events
            WindowHandler l = new WindowHandler();
            fr.addWindowListener(l);

            fr.getContentPane().add(pl, BorderLayout.CENTER);
            fr.setSize(500,500);
            fr.setVisible(true);
         }
      }
      //*********************************************************
      //
      // Create a window listener to close the program.
      //
      //*********************************************************
      class WindowHandler extends WindowAdapter {

         // This method implements a simple listener that detects
         // the "window closing event" and stops the program.
         public void windowClosing(WindowEvent e) {
            System.exit(0);
         };
      }
```

Figure 5.11. Program to create and plot a data set. The portions of the program in bold face must be modified to create differrent types of plots; the remainder of the program is Java "boilerplate"

TABLE 5-5 Common Methods in Class `JPlot2D`

METHOD	DESCRIPTION
`addCurve(double[] x, double[] y)`	Add an additional curve to the plot.
`setGridState(int gs)`	Set the grid state. Legal values are as follows: `JPlot2D.GRID_OFF` (default), `JPlot2D.GRID_ON`
`setLineColor(Color c)`	Set the line color. Legal values are any color defined in class `java.awt.Color`.
`setLineState(boolean b)`	Determines whether or not lines are to be plotted between data points on a curve. By default, lines *are* plotted. Legal values are: `JPlot2D.LINE_ON`, and `JPlot2D.LINE_OFF`.
`setLineStyle(int ls)`	Set the line style. Legal values are as follows: `JPlot2D.LINESTYLE_SOLID`, `JPlot2D.LINESTYLE_DOT`, `JPlot2D.LINESTYLE_LONGDASH`, and `JPlot2D.LINESTYLE_SHORTDASH`.
`setLineWidth(float w)`	Set the line width, in pixels.
`setMarkerColor(Color c)`	Set the marker type. Legal values are any color defined in class `java.awt.Color`.
`setMarkerState(boolean b)`	Determines whether or not markers are to be plotted at data points on a curve. By default, markers are *not* plotted. Legal values are: `Plot2D.MARKER_ON`, and `JPlot2D.MARKER_OFF`.
`setMarkerStyle(int ms)`	Set the marker style. Legal values are as follows: `JPlot2D.MARKER_CIRCLE`, and `JPlot2D.MARKER_SQUARE`.
`setPlotType(int t)`	Set the plot type. Legal values are as follows: `JPlot2D.LINEAR (default)`, `JPlot2D.SEMILOGX`, `JPlot2D.SEMILOGY`, `JPlot2D.LOGLOG`, `JPlot2D.BAR`, and `JPlot2D.POLAR`.
`setTitle(String s)`	Set the plot title.
`setXLabel(String s)`	Set the *x* label.
`setYLabel(String s)`	Set the *y* label.

The results of this program are shown in Figure 5.12 (see the top of the next page)
A new plot is created by instantiating a `JPlot2D` object, as follows:

```
JPlot2D pl = new JPlot2D(x,y);
```

This plot object will now automatically plot x versus y in any frame to which the object is added. Once the plot object has been created, a large number of methods may be called to modify and enhance the basic plot. Some of those methods are summarized in Table 5-5, and all of them are included in the on-line Java documentation for class `JPlot2D`.

Figure 5.12. Java plot of the equations $y(x) = \sin x$ and $y(x) = 1.2 \cos x$ over the range $0 \leq x \leq 2\pi$

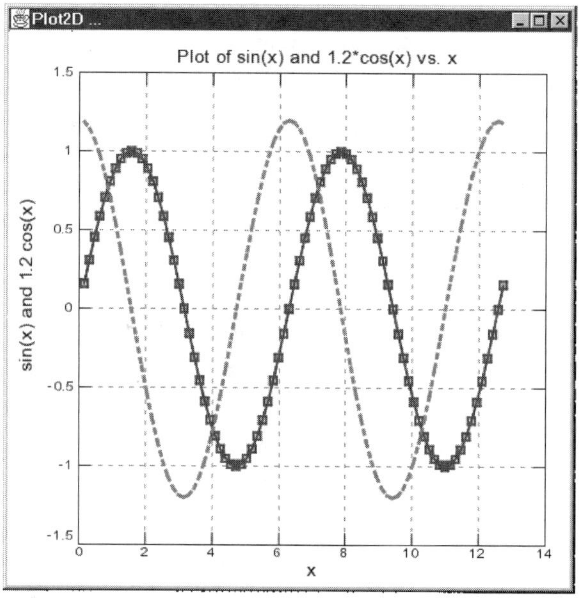

5.7 STRINGS

A **string** is a group of one or more characters treated as a single unit. For example, in the statement

```
System.out.println("Value = " + value);
```

the term `"Value = "` is a string. The principal importance of strings in our programs is that *they represent information in a form easily readable by humans*, and so are regularly used to enter information into a program and to display the results of the program.

It is important to understand the difference between a string representing a value and the value itself as it is stored inside a computer. For example, let's compare the integer value `123` and the string `"123"`. The integer value `123` is stored as the binary pattern 01111011 in a memory location inside the computer. It is in the form that the computer can use directly in its calculations, but it *cannot* be displayed on a computer screen in that form. In contrast, the string `"123"` consists of three Unicode characters encoding the characters 1, 2, and 3 respectively. A computer cannot use the string `"123"` directly in numerical calculations, but humans can easily recognize the number that it represents.

Because humans enter data for computer programs and read the results of the programs, most input and output from programs is in the form of character strings. On the other hand, most internal calculations are performed with the binary representations of numbers. Thus, we must be able to convert from strings to numbers when data is read into a program, and from numbers to strings when data is printed out of a program. The ability to work with strings is vital if we humans are to actually use computers. This section teaches us how to manipulate character strings in Java.

There are two types strings in Java: `Strings` and `StringBuffers`. Both types are implemented as objects. The fundamental difference between the two classes is that the `String` class consists of strings that *never change* once they are created, while the `StringBuffer` class consists of modifiable strings. Only `Strings` will be discussed in this chapter[3].

Creating and Initializing `Strings`

A reference to a `String` object is created by a statement such as

```
String str1;            // Create a String reference
```

When this statement is executed, reference `str1` is created and set to `null`, since it does not yet point to a `String` object. Attempting to use a null reference in a program will produce a run-time exception.

`String` objects may be created and assigned to a `String` reference in several ways. The easiest way is with a **`String` literal.** `String` literals (which are also known as `String` constants or **anonymous `String` objects**) are written as a string of characters between double quotation marks. For example,

```
"This is a test."
```

is a `String` object. A `String` reference can be created and a `String` literal can be assigned to that reference in a single statement:

```
String str1 = "This is a test.";
```

`String` objects can also be created using `String` constructors. `String` constructors can create new strings from other strings, from arrays of bytes, and from arrays of characters. Some examples of `String` constructors are shown below:

```
String str1 = new String("This is a test.");  // From another string
String str2 = new String(charArray);          // From a char array
String str3 = new String(byteArray);          // From a byte array
String str4 = new String();                   // New empty string
```

Once it has been created, each `String` object contains a fixed number of characters. The number of characters in any string can always be found using the `length` method. For example, the expression `str1.length()` will return a value of 15 for the string `str1` defined above.

`String` Methods

There are many methods available for manipulating `Strings` in Java programs. Some of the more important `String` methods are summarized in Table 5-6, shown at the top of the next page. All of the `String` methods are described in the Java API documentation in class `java.lang.String`, but we will discuss only a few of the most important ones here.

Substrings

A **substring** is a portion of a string. The method `substring` allows a substring to be extracted from a Java string. There are two forms of this method. The first form is

```
s.substring(int st)
```

[3]For a discussion of `StringBuffers`, see Chapman, Stephen J., *Introduction to Java for Scientists and Engineers,* 1st ed., Prentice-Hall, 1999.

TABLE 5-6 Selected `String` Methods

METHOD	DESCRIPTION
`char charAt(int index)`	Returns the character at a specified index in the string.
`int compareTo(String s)`	Compares the `String` object to another string lexicographically. Returns the following value: 0 if the string is equal to `s` < 0 if the string less than `s` > 0 if the string greater than `s`
`String concat(String s)`	Concatenates the `string s` to the end of this string.
`boolean endsWith(String suffix)`	Returns `true` if this string ends with the specified suffix.
`boolean equals(Object o)`	Returns `true` if `o` is a `String`, and `o` contains exactly the same characters as the string.
`boolean equalsIgnoreCase(String s)`	Returns `true` if `s` contains exactly the same characters as the string, disregarding case.
`void getChars((int i1, int i2, char[] dst, int i3))`	Copies the characters in the `String` from position `i1` to position `i2` into character array `dst`, starting at index `i3`.
`int indexOf(char ch)`	Returns the index of the *first* location of `ch` in the string.
`int indexOf(char ch, int start)`	Returns the index of the *first* location of `ch` at or after position `start` in the string.
`int indexOf(String s)`	Returns the index of the *first* location of substring `s` in the string.
`int indexOf(String s, int start)`	Returns the index of the *first* location of substring `s` at or after position `start` in the string.
`int lastIndexOf(char ch)`	Returns the index of the *last* location of `ch` in the string.
`int lastIndexOf(char ch, int start)`	Returns the index of the *last* location of `ch` at or before position `start` in the string.
`int lastIndexOf(String s)`	Returns the index of the *last* location of substring `s` in the string.
`int lastIndexOf(String s, int start)`	Returns the index of the *last* location of substring `s` at or before position `start` in the string.
`String replace(char old, char new)`	Returns a new string with every occurrence of character `old` replaced by character `new`.
`boolean startsWith(String p)`	Returns true if the beginning of the string exactly matches the string `p`.
`boolean startsWith(String p, int off1)`	Returns true if the beginning of the substring starting at index `off1` exactly matches the string `p`.
`String substring(int st)`	Returns the substring starting at index `st` and going to the end of the string.
`String substring(int st, int en)`	Returns the substring starting at index `st` and going to index `en-1`.
`String toLowerCase()`	Converts the string to lower case.
`String toUpperCase()`	Converts the string to upper case.
`String trim()`	Removes white space from either end of the string.

It takes one integer argument that specifies the starting index of the substring only, and returns a new string containing the characters from that index to the end of the original string. The second form is

```
s.substring(int st, int en)
```

It takes two integer arguments and specifies both the starting and ending indexes of the substring. (The actual string returned consists of the characters from position `st` to position `en-1`.) If the indexes specified in either case are less than zero or greater than the number of characters in the string, then a `StringIndexOutOfBounds` exception will occur.

The following class illustrates both types of `substring` methods.

```
// Substrings
public class Substring {

   public static void main(String[] args) {
      String s = "abcdefghijABCDEFGHIJabcdefghij";

      // Test substring methods
      System.out.println("String = \"" + s + "\"");
      System.out.println("Substring starting at 18 = "
           + "\" + s.substring(18) + "\"");

      System.out.println("Substring from 18 to 24 = "
           + "\"" + s.substring(18,24) + "\"");
   }
}
```

When this program is executed, the results are:

```
D:\book\java\chap5>java Substring
String = "abcdefghijABCDEFGHIJabcdefghij"
Substring starting at 18 = "IJabcdefghij"
Substring from 18 to 24 = "IJabcd"
```

Concatenating Strings

Two strings may be **concatenated** (joined end-to-end) using the `concat` method. The expression

```
s1.concat(s2)
```

creates a new string by appending the characters in `s2` to the end of `s1`. The original `String` objects `s1` and `s2` are not affected. The class shown below illustrates the use of `concat`.

```
// Concatenation
public class Concatenate {

   public static void main(String[] args) {
      String s1 = "abc";
      String s2 = "def";

      // Test concatenation method
      System.out.println("Test concatenation:");
      System.out.println("s1 = " + s1);
      System.out.println("s2 = " + s2);
      System.out.println("s1.concat(s2) ="
          + s1.concat(s2));

      // Watch what happens here!
      System.out.println("\nBefore assignment:");
      System.out.println("s1 = " + s1);
      System.out.println("s2 = " + s2);
      s1 = s1.concat(s2);
      System.out.println("\nAfter assignment:");
      System.out.println("s1 = " + s1);
      System.out.println("s2 = " + s2);
   }
}
```

When this program is executed, the results are:

```
D:\book\java\chap5>java Concatenate
Test concatenation:
s1 = abc
s2 = def
s1.concat(s2) = abcdef

Before assignment:
s1 = abc
s2 = def

After assignment:
s1 = abcdef
s2 = def
```

Note that the `concat` method concatenated string `s1` with string `s2`.

The second part of the program illustrates a common source of confusion of novice Java programmers. We stated at the beginning of this section that `Strings` never change once they are created, and yet `s1` changes value between lines 17 and 22 in the above program! What is happening here?

The answer is very simple. Remember that `s1` and `s2` are *references to `Strings`,* not strings themselves. Lines 5 and 6 of the program create two `String` objects containing the strings `"abc"` and `"def"`, and set references `s1` and `s2` to refer to them. Therefore, lines 17 and 18 print out `s1 = "abc"` and `s2 = "def"`. The `concat` method in line 19 creates a *new* `String` object containing the string `"abcdef"`, and sets the reference `s1` to point to the new object. Therefore, line 21 prints out `s1 = "abcdef"`. This process is illustrated in Figure 5.13. What happened to the original `String` object pointed to by `s1`? It is still present, but there is no longer any reference to it. This object can never be used again.

Comparing Strings

Two strings may be compared using methods `equals`, `equalsIgnoreCase`, `compareTo`, and `regionMatches`. When two strings are compared, they are compared according to the **lexicographic sequence** of the characters in the strings. This is the sequence in which the character appear within the character set used by the computer. For example, Appendix A shows the order of the first 127 letters according to the Unicode character set. In this set, the letter `'A'` is character 64 and the letter `'a'` is 96, so `'A'` is lexicographically less than `'a'`. *Note that capital letters are not the same as lower-case letters* when they are compared lexicographically.

If the strings being compared are more than one character long, then the comparison starts with the first character in each string. If the two characters are equal, then the comparison moves to the second character in the strings, etc., until the first difference is found. If there is no difference between the two strings, then they are considered equal. The method `equals` compares two strings and returns `true` if they contain identically the same value. For example, if

```
s1 = new String("Hello");
s2 = new String("Hello");
s3 = new String("hello");
```

then the expression `s1.equals(s2)` will be `true` because the two strings are identically the same, while the expression `s1.equals(s3)` will be `false` because the two strings are differ in the first position.

Note that *comparing two strings with the method `equals` is not the same as comparing two strings with the == operator.* When the == operator compares two object ref-

Figure 5.13. (a) Initially reference `s1` points to the `String` object containing `"abc"`, and reference `s2` points to the `String` object containing `"def"`. (b) After line 19 is executed, reference `s1` points to the `String` object containing `"abcdef"`, and reference `s2` points to the `String` object containing `"def"`. The object containing `"abc"` no longer has a reference, and cannot be used.

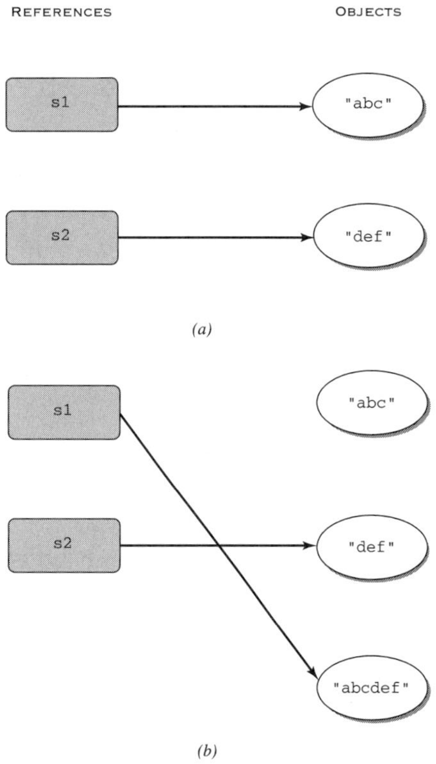

erences, *it checks to see if the two references point to the same object.* If you are comparing two objects with the == operator, they will not be equal even if they have identically the same contents. For example, if `s1` and `s2` are as previously defined, the expression

```
s1.equals(s2)
```

will return `true`, because the contents of the two objects are equal, but the expression

```
s1 == s2
```

will return `false`, because `s1` and `s2` are two different objects.

PROGRAMMING PITFALLS

Never attempt to compare two objects for equality with the == operator. This operator checks to see if two references point to the same object, and not to see if two objects have equal values.

The method `equalsIgnoreCase` is similar to `equals`, except that it ignores the case of letters when comparing the two strings. Thus the expression `s1.equalsIgnoreCase(s3)` will be `true` because the two strings are only differ in the case of the letter `'H'`.

The method `s1.compareTo(s2)` compares two strings `s1` and `s2`, and returns an integer equal to the *difference in lexicographic position* between the corresponding letters at the first differing location in the strings. This difference will be negative if `s1 < s2` and positive if `s1 > s2`. If the strings are equal, then the method will return a zero. For example, if

```
s1 = new String("Good");
s2 = new String("Help");
s3 = new String("HELP");
s4 = new String("HELP");
```

then the expression `s1.compareTo(s2)` will be –1 because the letter `'G'` appears one character before `'H'` in the Unicode character set. Similarly, the expression `s2.compareTo(s3)` will be 32 because the strings first differ in the second position, and the letter `'e'` appears 32 characters after `'E'` in the Unicode character set. Finally, the expression `s3.compareTo(s4)` will be zero, since the strings are equal.

The methods `startsWith` and `endsWith` can be used to determine whether or not a string starts with or ends with a particular substring. For example, if

```
s1 = new String("started");
```

then the method call

```
s1.startsWith("st")
```

would return `true`, because "`started`" does start with "`st`". Similarly, the method call

```
s1.endsWith("ed")
```

would return `true`, because "`started`" does end with "`ed`".

All of these comparison methods are illustrated in the Java program shown below.

```
// Comparisons
public class Compare {

   public static void main(String[] args) {

      String s1, s2, s3, s4; // Declare references

      // Test equality and inequality
      System.out.println("Test equality:");
      s1 = new String("Hello");
      s2 = new String("Hello");
      s3 = new String("hello");

      System.out.println("s1 = \"" + s1 + "\"");
      System.out.println("s2 = \"" + s2 + "\"");
      System.out.println("s3 = \"" + s3 + "\"");
      System.out.println("s1.equals(s2) = " + s1.equals(s2));
      System.out.println("s1.equals(s3) = " + s1.equals(s3));
      System.out.println("s1.equalsIgnoreCase(s3) = "
          + s1.equalsIgnoreCase(s3));
      System.out.println("s1 == s2 = " + (s1 == s2));

      // Test comparison
      System.out.println("\nTest compare:");
      s1 = new String("Good");
      s2 = new String("Help");
      s3 = new String("HELP");
      s4 = new String("HELP");
```

```
            System.out.println("s1 = \"" + s1 + "\"");
            System.out.println("s2 = \"" + s2 + "\"");
            System.out.println("s3 = \"" + s3 + "\"");
            System.out.println("s4 = \"" + s4 + "\"");
            System.out.println("s1.compareTo(s2) = " + s1.compareTo(s2));
            System.out.println("s2.compareTo(s3) = " + s2.compareTo(s3));
            System.out.println("s3.compareTo(s4) = " + s3.compareTo(s4));
        }
    }
```

When this program is executed, the results agree with our previous discussions:

```
D:\book\java\chap5>java Compare
Test equality:
s1 = "Hello"
s2 = "Hello"
s3 = "hello"
s1.equals(s2) = true
s1.equals(s3) = false
s1.equalsIgnoreCase(s3) = true
s1 == s2 = false

Test compare:
s1 = "Good"
s2 = "Help"
s3 = "HELP"
s4 = "HELP"
s1.compareTo(s2) = -1
s2.compareTo(s3) = 32
s3.compareTo(s4) = 0
```

PRACTICE!

This quiz provides a quick check to see if you have understood the concepts introduced in Section 5.7. If you have trouble with the quiz, reread the section, ask your instructor, or discuss the material with a fellow student. The answers to this quiz are found in the back of the book.

Determine which of the following Java statements are valid. For each valid statement, specify what will happen in the program. For invalid statements, explain why the are invalid.

1.
```
String s1 = new String("abcdefg");
String s2 = s1.substring(1,8);
```

2.
```
String s1, s2, s3;
s1 = new String("abcdefg");
s2 = new String("123");
s3 = s1.substring(1,3);
s3 = s3.concat(s2);
System.out.println(s3);
```

3.
```
String s1 = new String("Hello");
String s2 = new String("hello");
System.out.println(s1.equals(s2));
System.out.println(s1.equalsIgnoreCase(s2));
```

4.
```
String s1 = new String("Hello");
String s2 = new String("Hello");
System.out.println(s1 == s2);
```

SUMMARY

- An **array** is a special object containing (1) a group of contiguous memory locations that all have the same name and same type, and (2) a separate instance variable containing an integer constant equal to the number of elements in the array.
- The length of a Java array is specified when it is created, and it remains fixed for as long as the array exists.
- A *reference* is a "handle" or "pointer" to an object that permits Java to locate the object in memory when it is needed. A reference is declared by naming the object type followed by the reference name.
- Once an array reference exists, new arrays are created using the `new` operator.
- Arrays may be initialized with an array initializer, which is just a comma-separated list of valued enclosed in braces.
- Java programs always include automatic bounds checking, and any attempt to address an out-of-bounds array element will produce an exception.
- Reading from and writing to disk files can be rather complicated in Java, but the task is simplified by the use of convenience classes such as the classes `FileIn` and `FileOut` supplied with this book.
- The contents of arrays may be plotted using the `chapman.graphics.JPlot2D` class supplied with this book.
- Strings are groups of one or more characters treated as a single unit. Objects of type `String` never change once they are created.
- `String` method `length()` returns the number of characters in a `String`.
- `String` method `s1.concat(s2)` returns a new `String` that is the concatenation of strings `s1` and `s2`.
- `String` method `s1.equals(s2)` returns `true` if the contents of strings `s1` and `s2` are identical.
- `String` method `s1.equalsIgnoreCase(s2)` returns `true` if the contents of strings `s1` and `s2` are identical ignoring case.
- The operator `s1 == s2` returns `true` if references `s1` and `s2` point to identically the same object in memory.
- `String` method `s1.compareTo(s2)` returns 0 if `s1` and `s2` are equal, a negative number if `s1` is lexicographically less than `s2`, and a positive number if `s1` is lexicographically greater than `s2`.

Other common `String` methods are summarized in Table 5-6.

APPLICATIONS: SUMMARY OF GOOD PROGRAMMING PRACTICES

The following guidelines introduced in this chapter will help you to develop good programs:

1. When creating `for` loops to process the elements of an array, use the array object's `length` field in the continuation condition for the loop. This will allow the loop to adjust automatically for different sized arrays.
2. When processing an array `arr` in a `for` loop, use a continuation condition of the form

   ```
   for (j = 0; j < arr.length; j++).
   ```

   The "less than" relational operator is the correct one to use, because the elements of the array are numbered `0,1,..., arr.length-1`.
3. Declare the sizes of arrays in a Java program using named constants to make them easy to change.

KEY TERMS

anonymous `String` object
array
array element
`ArrayIndexOutOfBoundsException`
array initializer
bounds checking
command-line redirection
concatenate
frame
instantiate
lexciographic sequence
`null`
out of bounds
reference
`String` literal
subscript
substring

Problems

1. How may arrays be declared?
2. What is the difference between an array and an array element?
3. Assume that `values` is a 101-element array containing a list of measurements from a scientific experiment, which has been declared by the statement

```
double values[] = new double[101];
```

Write the Java statements that would count the number of positive values, negative values, and zero values in the array, and write out a message summarizing how many values of each type were found.

4. Write Java statements that would print out every fifth value in the array `values` described in Exercise 5.3. The output should take the form

```
values[0] = x.xx
values[5] = x.xx
. . .
values[100] = x.xx
```

5. Modify the selection sort program in Example 5-2 so that it writes the sorted data set to a user-specified file.
6. **Polar to Rectangular Conversion** A *scalar quantity* is a quantity that can be represented by a single number. For example, the temperature at a given location is a scalar. In contrast, a *vector* is a quantity that has both a magnitude and a direction associated with it. For example, the velocity of an automobile is a vector, since it has both a magnitude and a direction.

   Vectors can be defined either by a magnitude and a direction, or by the components of the vector projected along the axes of a rectangular coordinate system. The two representations are equivalent. For two-dimensional vectors, we can convert back and forth between the representations using the following equations

$$\mathbf{V} = V\angle\theta = V_x\mathbf{i} + V_y\mathbf{j}$$
$$V_x = V\cos\theta$$
$$V_y = V\sin\theta$$
$$V = \sqrt{V_x^2 + V_y^2}$$
$$\theta = \tan^{-1}\frac{V_y}{V_x}$$

where **i** and **j** are the unit vectors in the x- and y-directions respectively. The representation of the vector in terms of magnitude and angle is known as *polar coordinates,* and the

Figure 5.14. Representations of a vector

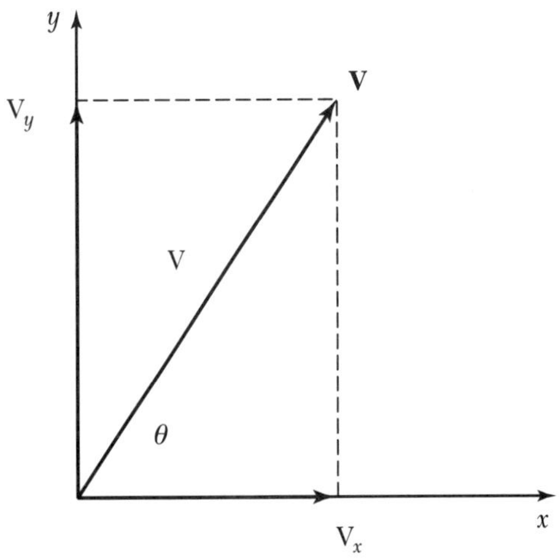

representation of the vector in terms of components along the axes is know as *rectangular coordinates.* Write a program that reads the polar coordinates (magnitude and angle) of a two-dimensional vector into a two-element array `polar` (`polar[0]` will contain the magnitude V and `polar[1]` will contain the angle θ in degrees), and converts the vector from polar to rectangular form, storing the result in a two-element array `rect`. The first element of `rect` should contain the x-component of the vector, and the second element should contain the y-component of the vector. After the conversion, display the contents of array `rect`. Test your program by converting the following polar vectors to rectangular form. The number before then angle sign ($\angle$) is the magnitude, while the number after the angle sign is the direction.

a. $5\angle-36.87°$

b. $10\angle45°$

c. $25\angle233.13°$

7. **Rectangular to Polar Conversion** Write a program that reads the rectangular components of a two-dimensional vector into a two-element array `rect` (`rect[0]` will contain the component V_x, and `rect[1]` will contain the component V_y) and converts the vector from rectangular to polar form, storing the result in a two-element array `polar`. The first element of `polar` should contain the magnitude of the vector, and the second element should contain the angle of the vector in degrees. After the conversion, display the contents of array `polar`. (*Hint:* Look up method `Math.atan2`.) Test your program by converting the following rectangular vectors to polar form:

   a. $3\mathbf{i} - 4\mathbf{j}$

   b. $5\mathbf{i} + 5\mathbf{j}$

   c. $-5\mathbf{i} + 12\mathbf{j}$

8. **Dot Product** A three-dimensional vector can be represented in rectangular coordinates as

$$\mathbf{V} = V_x\mathbf{i} + V_y\mathbf{j} + V_z\mathbf{k},$$

where V_x is the component of vector $\mathbf{V}$ in the x-direction, V_y is the component of vector $\mathbf{V}$ in the y-direction, and V_z is the component of vector $\mathbf{V}$ in the z-direction. Such a vector can be stored in a three-element array, since there are 3 dimensions in the coordinate system. The same idea applies to an n-dimensional vector. An n-dimensional vector can be stored in a rank-1 array containing n elements.

Figure 5.15. A force **F** applied to an object moving with velocity **v**

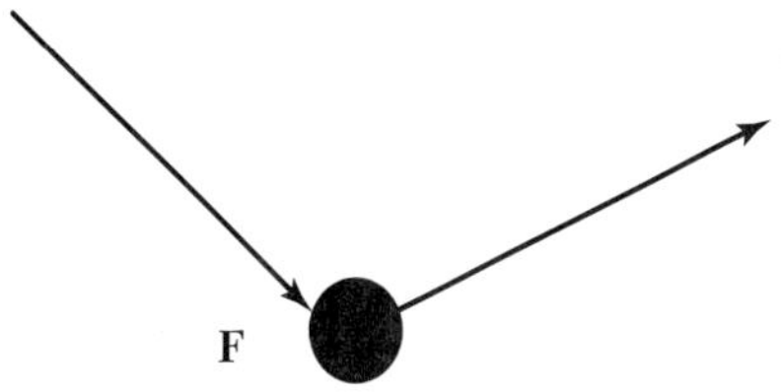

One common mathematical operation between two vectors is the *dot product*. The dot product of two vectors $\mathbf{V}_1 = V_{x1}\mathbf{i} + V_{y1}\mathbf{j} - V_{z1}\mathbf{k}$ and $\mathbf{V}_2 = V_{x2}\mathbf{i} + V_{y2}\mathbf{j} + V_{z2}\mathbf{k}$ is a scalar quantity defined by the equation

$$V_1 \bullet V_2 = V_{x1}V_{x2}V + V_{y1}V_{y2} + V_{z1}V_{z2}.$$

Write a Java program that will read two vectors $\mathbf{V}_1$ and $\mathbf{V}_2$ into two one-dimensional arrays in computer memory, and then calculate their dot product according to the equation given above. Test your program by calculating the dot product of vectors $\mathbf{V}_1 = 5\,\mathbf{i} - 3\,\mathbf{j} + 2\,\mathbf{k}$ and $\mathbf{V}_2 = 2\,\mathbf{i} + 3\,\mathbf{j} + 4\,\mathbf{k}$.

9. **Power Supplied to an Object** If an object is being pushed by a force **F** at a velocity **v**, then the power supplied to the object by the force is given by the equation

$$\mathbf{P} = \mathbf{F} \bullet \mathbf{v}$$

where the force **F** is measured in newtons, the velocity **v** is measured in meters per second, and the power P is measured in watts. Use the Java program written in the Exercise 5.8 to calculate the power supplied by a force of $\mathbf{F} = 4\,\mathbf{i} + 3\,\mathbf{j} - 2\,\mathbf{k}$ newtons to an object moving with a velocity of $\mathbf{v} = 4\,\mathbf{i} - 2\,\mathbf{j} + 1\,\mathbf{k}$ meters per second.

10. **Cross Product** Another common mathematical operation between two vectors is the *cross product*. The cross product of two vectors $\mathbf{V}_1 = V_{x1}\mathbf{i} + V_{y1}\mathbf{j} + V_{z1}\mathbf{k}$ and $\mathbf{V}_2 = V_{x2}\mathbf{i} + V_{y2}\mathbf{j} + V_{z2}\mathbf{k}$ is a vector quantity defined by the equation

$$\mathbf{V}_1 \times \mathbf{V}_2 = (V_{y1}V_{z2} - V_{y2}V_{z1})\mathbf{i} + (V_{z1}V_{x2} - V_{z2}V_{x1})\mathbf{j} + (V_{x1}V_{y2} - V_{x2}V_{y1})\mathbf{k}$$

Write a Java program that will read two vectors $\mathbf{V}_1$ and $\mathbf{V}_2$ into arrays in computer memory, and then calculate their cross product according to the equation given above. Test your program by calculating the cross product of vectors $\mathbf{V}_1 = 5\,\mathbf{i} - 3\,\mathbf{j} + 2\,\mathbf{k}$ and $\mathbf{V}_2 = 2\,\mathbf{i} + 3\,\mathbf{j} + 4\,\mathbf{k}$.

11. **Velocity of an Orbiting Object** The vector angular velocity ω of an object moving with a velocity **v** at a distance **r** from the origin of the coordinate system is given by the equation

$$\mathbf{v} = \mathbf{r} \times \omega,$$

where **r** is the distance in meters, ω is the angular velocity in radians per second, and **v** is the velocity in meters per second. If the distance from the center of the earth to an orbiting satellite is $\mathbf{r} = 300000\,\mathbf{i} + 400000\,\mathbf{j} + 50000\,\mathbf{k}$ meters, and the angular velocity of the satellite is $\omega = -6 \times 10^{-3}\,\mathbf{i} + 2 \times 10^{-3}\,\mathbf{j} - 9 \times 10^{-4}\,\mathbf{k}$ radians per second, what is the velocity of the satellite in meters per second (see Figure 5.16)? Use the program written in the previous exercise to calculate the answer.

12. Plot the function $y(t) = e^{-0.8t} \cos 4\pi t$ from $t = 0$ to $t = 3$. For what fraction of the interval is $y(t) > 0.5$?

13. Plot the function $y(x) = 2e^{-0.2x}$ from $x = 0$ to $x = 10$ on both a linear and a semilog scale. What does this function look like on each scale?

14. Figure 5.17 shows an electrical load with a voltage applied to it and a current flowing into it. Assume that the voltage applied to this load is given by the equation $v(t) = 17 \sin 377t$ V and

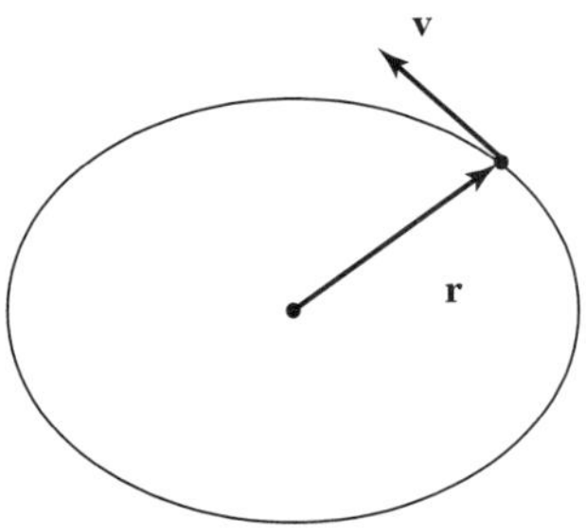

Figure 5.16. The velocity of an orbiting object

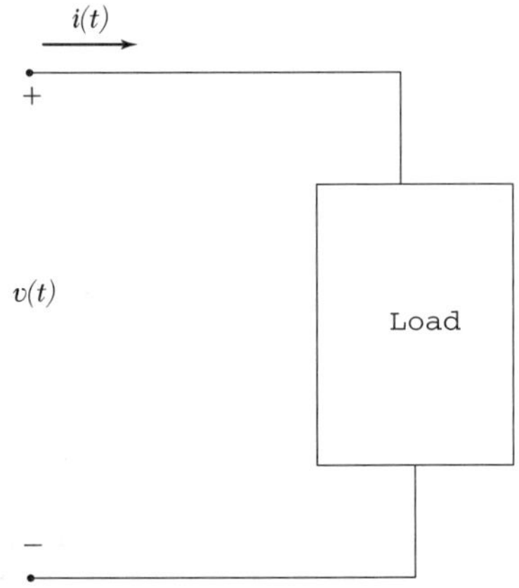

Figure 5.17. AC voltage and current supplied to an electrical load

the current flowing into the load is given by the equation $i(t) = 1.414\sin(377t - \pi/3)$ A. Write a Java program the performs the following actions:

a. Create two arrays storing the voltage and current applied to the load as a function of time. There should be 1000 samples in each array, spaced at intervals of 0.1 ms.

b. Calculate the instantaneous power supplied to the load at each time from the equation $P(t) = v(t)i(t)$, and store the power in an array.

c. Plot the voltage, current, and power as a function of time from $t = 0$ s to $t = 0.1$ s.

d. Calculate the average power supplied to the load over this interval.

15. The location of any point P in a three-dimensional space can be represented by a set of three values (x, y, z), where x is the distance along the x-axis to the point, y is the distance along the y-axis to the point, and z is the distance along the z-axis to the point. If two points P_1 and P_2 are represented by the values (x_1, y_1, z_1) and (x_2, y_2, z_2), respectively, then the distance between the points P_1 and P_2 can be calculated from the equation

$$distance = \sqrt{(x_1 - x_2)^2 + (y_1 + y_2)^2 + (z_1 - z_2)^2}\ .$$

Write a Java program to read in two points (x_1, y_1, z_1) and (x_2, y_2, z_2) and to calculate the distance between them. Test your program by calculating the distance between the points (–1, 4, 6) and (1, 5, –2).

16. Assume that `s1` and `s2` are `Strings`, and that

```
String s = "abcdefghijABCDEFGHIJabcdefghij";
```

What will be the contents of `s1` and `s2` after the following statements are executed?

a.
```
s1 = s.substring(10);
s2 = s.substring(10,12);
```

b.
```
s1 = s.substring(1,3);
s2 = s.substring(7,9);
s1 = s1.concat(s2);
```

17. Assume the definitions

```
String s1 = "Test1";
String s2 = "test1";
String s3 = "Test1";
String s4 = "Test2";
String s5 = s1;
```

What will be the results of the following expressions?

a. `s1.equals(s2);`

b. `s1.equals(s3);`

c. `s1.equalsIgnoreCase(s2);`

d. `s1 == s3;`

e. `s1 == s5;`

f. `s1.compareTo(s2);`

g. `s1.compareTo(s4);`

h. `(s1.toLowerCase()).compareTo(s2);`

18. Modify the selection sort program in Example 5-2 to sort `Strings` into alphabetical order. Use the `compareTo()` method to determine the order in which the strings should be sorted. Test your program by sorting file `input4`, which contains the following strings:

```
deBrincat, Charles
Chapman, Stephen
Johnson, James
Chapman, Rosa
Anderson, William
Johnston, Susan
Johns, Joe
```

(*Hint:* Read the strings into the program using the `readLine()` method in class `chapman.io.FileInLines`. You can find the descriptions of this class in the on-line documentation of the `chapman.io` package.)

19. Note that the results of Exercise 5.18 were not in strict alphabetical order, since lower case letters appear after upper case letters in the Unicode character set. Modify the program of the pervious example to sort in strict alphabetical order by converting both strings to upper case before performing the comparison. Use the `toUpperCase()` method for this purpose.

6

Methods

A **method** is separate piece of code that can be called by a main program or another method to perform some specific function. Each method must be defined within the body of some class, and many methods can be defined within a single class.

A method may be **called** or **invoked** by a main program or another method. The method is called by naming both the method and the object that it is defined within in an expression, together with a list of **parameters** in parentheses. (Parameters are values passed to the method to use in its calculations.) When the method finishes its calculations, it usually returns a *single* value to the calling program, and the calling program uses that value in its own calculations.

Figure 6.1 shows a simple program consisting of a single class containing two methods. One of these methods is the `main` method that we have seen before, and the second one is method `square`. Method `square` is defined on lines 17-21 of the program. This method accepts a single integer parameter, and calculates and returns the square of that parameter to the calling program.

To use method `square`, the main program must first create an object of the type that the method is defined

SECTIONS

OBJECTIVES

After reading this chapter, you should be able to:

- Learn how to create and use methods
- Learn about the pass-by-value scheme used to pass parameters to methods
- Learn about variable duration and scope
- Learn about recursive methods
- Learn how to use methods in the `java.util.Arrays` class
- Learn how to use methods in the `chapman.math` package

```
1  // Calculates the squares of the numbers from 1 to 10
2  public class SquareInt {
3
4     // Define the main method
5     public static void main(String[] args) {
6
7        // Instantiate a SquareInt object
8        SquareInt sq = new SquareInt();
9
10       // Write number and square
11       for ( int i = 1; i <= 10; i++ ) {
12          System.out.print("number = " + i);
13          System.out.println(" square = " + sq.square(i));
14       }
15    }
16
17    // Definition of square method
18    public int square (int x) {
19
20       return x * x;
21    }
22 }
```

Figure 6.1. A program to calculate the squares of integers from 1 to 10. This program uses method `square` to calculate the squares of the integers.

within. This is done on line 8, which creates a new `SquareInt` object and assigns it to reference `sq`. Method `square` is actually called in line 13 with the expression `sq.square(i)`. Note that the program names the object containing the method, followed a period and the method name, followed by the calling parameters in parentheses. The value of integer `i` is passed to the method and stored in parameter `x`, and the method calculates and returns x^2. The `main` program then uses the returned value in the `System.out.println` function on line 13 (see Figure 6.2).

When this program is executed, the results are

```
D:\book\java\chap6>java Squares
number = 0 square = 0
number = 1 square = 1
number = 2 square = 4
number = 3 square = 9
number = 4 square = 16
number = 5 square = 25
number = 6 square = 36
number = 7 square = 49
number = 8 square = 64
number = 9 square = 81
```

Figure 6.2. When the method named `square` appears in the expression, the method is called with the specified parameter `i`. The method then calculates i^2 and returns the result to the calling method. The calling method uses that result in its `println` statement.

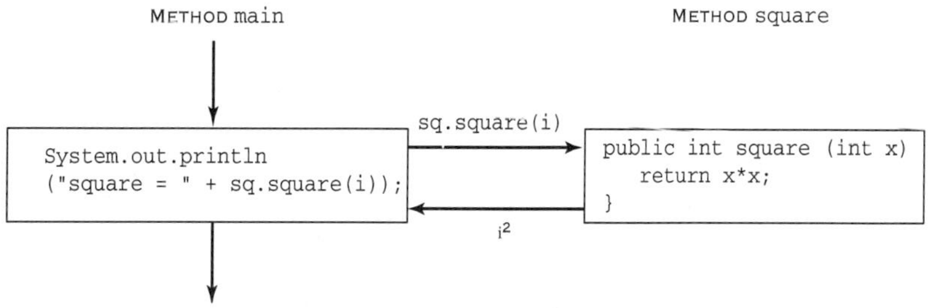

Note that method `square` was used 10 times to calculate 10 different squares. This same method could also have been used anywhere else within the program that we wanted to calculate the square of an integer.

6.1 WHY USE METHODS?

In Chapter 3, we learned the importance of good program design. The basic technique that we employed was **top-down design**. In top-down design, the programmer starts with a statement of the problem to be solved and the required inputs and outputs. Next, he or she describes the algorithm to be implemented by the program in broad outline, and applies *decomposition* to break the algorithm down into logical subdivisions called sub-tasks. Then, the programmer breaks down each sub-task until he or she winds up with many small pieces, each of which does a simple, clearly understandable job. Finally, the individual pieces are turned into Java code.

Although we have followed this design process in our examples, the results have been somewhat restricted, because we have had to combine the final Java code generated for each sub-task into a single large program. There has been no way to code, verify, and test each sub-task independently before combining them into the final program.

Fortunately, Java has two special features designed to make sub-tasks easy to develop and debug independently before building the final program: **classes** and **methods**. Classes are separately-compiled units containing data, together with the methods to manipulate that data. Methods are procedures that describe how to modify or manipulate the data contained in the class. Both classes and methods allow us to chop large programs up into smaller pieces which can be independently coded, tested, and verified. They also allow us to create *reusable* software, since a class created for one program can often be used intact in another program that needs the same features. We are studying methods in this chapter, and followed by classes in Chapter 7.

Well-designed classes and methods enormously reduce the effort required on a large programming project. Their benefits include:

1. **Independent testing of sub-tasks**. Each sub-task can be coded and compiled as an independent unit. The sub-task can be tested separately to ensure that it performs properly by itself before combining it into the larger program. This step is know as **unit testing**. It eliminates a major source of problems before the final program is even built.
2. **Reusable code**. In many cases, the same basic sub-task is needed in many parts of a program. For example, it may be necessary to sort a list of values into ascending order many different times within a program, or even in other programs. It is possible to design, code, test, and debug a *single* method to do the sorting, and then to reuse that method whenever sorting is required. This reusable code has two major advantages: it reduces the total programming effort required, and it simplifies debugging, since the sorting method only needs to be debugged once.
3. **Isolation from unintended side effects**. A caller communicates with the methods that it invokes through a list of values called an **argument list**. *The only values in the caller that are visible to the method are those in the argument list*. This is very important, since accidental programming mistakes can only affect the variables in the method in which the mistake occurred.

Once a large program is written and released, it has to be *maintained*. Program maintenance involves fixing bugs and modifying the program to handle new and unforeseen circumstances. The programmer who modifies a program during maintenance is often not the person who originally wrote it. In poorly written programs, it is common for the programmer modifying the program to make a change in one region of the code, and to have that change cause unintended side effects in a totally different part of the program. This happens because variable names are re-used in different portions of the program. When the programmer changes the values left behind in some of the variables, those values are accidentally picked up and used in other portions of the code.

The use of well-designed methods and classes minimizes this problem by **information hiding**. Except for the values in its argument list, *all of the variables within a called method are completely invisible from its calling method*, and *all of the variables in the calling method are completely invisible from the called method*. Since the variables in one method are invisible to the other method, one method cannot accidentally change the other method's variables. Thus, modifications in one of the methods will not cause unintended side-effects in the other method.

GOOD PROGRAMMING PRACTICE

Break large programming tasks into classes and methods whenever practical to achieve the important benefits of independent component testing, reusability, and isolation from unintended side effects.

6.2 METHOD DEFINITIONS

The general form of a method definition is as follows:

```
[keywords] return-value-type method-name( parameter-list ) {

   declarations and statements
   (return statement)
}
```

The *method-name* is any valid Java identifier. The *return-value-type* is the data type of the result returned from the method to the caller. If no value is returned from the method, then the *return-value-type* is **void**. A *return-value-type* must be specified for every method. If the *return-value-type* is not **void**, then a **return** statement must appear in the body of the method to specify the value to be passed back to the calling method.

The *parameter-list* is a comma-separated list containing the declarations of the parameters received by the method whenever it is called. If the method does not receive any parameters, then the *parameter-list* is empty, but the parentheses are still required. A type must be declared for every parameter in the list.

The *declarations and statements* within the braces form the **method body**. The method body is a compound statement defining the local variables and actions performed by the method.

When a method is called, execution begins at the first statement in the method body, and continues until either a **return** statement is executed or the end of the body is reached. There are two forms of the `return` statement. The simple statement

```
return;
```

stops execution of the method and returns control to the point at which the method was invoked *without returning a value to the calling method*. The statement

```
return expression;
```

stops execution of the method and returns control to the point at which the method was invoked *returning the value of expression to the calling method*. The first form of `return` is used with `void` methods, and the second form is used with methods that return a value to the caller.

Another example method is shown in bold face in Figure 6-3. This method calculates the hypotenuse of a right triangle from the lengths of the other two sides.

This method has two arguments in its parameter list. These arguments are placeholders for the values that will be passed to the method when it is executed. The variable `hypot` is actually defined within the method. It is used in the method, but it is not accessible to any calling program. Variables that are used within a method and that are not accessible by calling methods are called **local variables**.

After `hypot` is declared, the method calculates the hypotenuse of the right triangle from the information about the two sides. Finally, the `return` statement returns the value of `hypot` to the calling method.

To test a method, it is necessary to write a program called a **test driver**. The test driver is a small method that instantiates an object containing the method, and then

```
// Calculates the hypotenuse of a right triangle
import chapman.io.*;
public class Triangle {

   // Define the main method
   public static void main(String[] args) {

      // Declare variables
      double side1;          // Side 1 of right triangle
      double side2;          // Side 2 of right triangle

      // Create a StdIn object
      StdIn in = new StdIn();

      // Instantiate a Triangle object
      Triangle tri = new Triangle();

      // Get the two sides
      System.out.print("Enter side 1: ");
      side1 = in.readDouble();
      System.out.print("Enter side 2: ");
      side2 = in.readDouble();

      // Calculate hypotenuse and display result
      System.out.println("Hypotenuse = " + tri.hypotenuse (side1,side2));

   }
   // Definition of method hypotenuse
   public double hypotenuse (double side1, double side2) {

      double hypot;
      hypot = Math.sqrt( side1*side1 + side2*side2 );
      return hypot;
   }
}
```

Figure 6.3. A method to calculate the hypotenuse of a right triangle

invokes the new method with a sample data set for the specific purpose of testing it. The test driver for method `hypotenuse` is the `main` method of the class. When this program is executed, the results are as follows:

```
D:\book\java\chap6>java Triangle
Enter side 1: 3
Enter side 2: 4
Hypotenuse = 5.0
```

6.3 VARIABLE PASSING IN JAVA: THE PASS-BY-VALUE SCHEME

Java programs communicate with their methods using a **pass-by-value** scheme. When a method call occurs, Java *makes a copy* of each calling argument and places that copy in the corresponding parameter of the method. This scheme is called pass-by-value, because the *value* of the calling argument is placed in the corresponding parameter. Similarly, the value returned from a method is returned by value.

Note that the method has a *copy* of value of the original argument, not the argument itself. This fact means that the method cannot accidentally modify the original argument even if it modifies the parameter during its calculations. Thus the program is protected against an error caused by the method accidentally modifying data in the method that called it.

A program illustrating the pass-by-value scheme is shown in Figure 6.4. The `main` method of this program declares a variable `i`, initializes it to 5, and prints out the 5. It then calls method `test(i)`, passing the *value* of `i` to the method. Method `test` increments the value of the parameter `i` that it receives, and prints out a 6. However, it does not affect the original value in the `main` method. After execution returns from method `test`, the main method prints `i` again, and its value is still 5.

```
// Tests the pass-by-value scheme
public class TestPassByValue {

   // Define the main method
   public static void main(String[] args) {

      // Instantiate a TestPassByValue object
      TestPassByValue t = new TestPassByValue();

      // Initialize a value and print it out
      int i = 5;
      System.out.println("Before test: i = " + i);

      // Now call method test()
      int j = t.test(i);

      // Print out value after call
      System.out.println("After test: i = " + i);
   }

   // Definition of test()
   public int test (int i) {

      int j = ++i;
      System.out.println("In test: i = " + i);
      return j;
   }
}
```

Figure 6.4. A program illustrating the pass-by-value scheme

When this program is executed, the results are as follows:

```
D:\book\java\chap6>java TestPassByValue
Before test: i = 5
In test:     i = 6
After test:  i = 5
```

GOOD PROGRAMMING PRACTICE

The pass-by-value scheme prevents a method from accidentally modifying its calling arguments.

The pass-by-value scheme also applies to objects as well, but the results are different for that case. If an object such as an array is passed as a calling argument, the program *makes a copy of the reference* to the array and places the copy of the reference in the method parameter. Because the reference was copied, the called method cannot modify the original reference in the calling method. However, it *can* use the reference that was passed to it to read or modify the actual array or other object being referred to.

A program illustrating this effect is shown in Figure 6.5. The `main` method of this program declares an array a, initializes it, and prints out the values. It then calls method `test(a)`, passing the *value of the reference to* `a`. Method `test` uses that reference to modify `x[3]`, prints out the results, and returns to the calling method. After execution returns from method `test`, the `main` method prints `a` again, and we can see that its value has been modified (see Figure 6.6).

```
// Tests the pass-by-value scheme
public class TestPassByValue2 {

   // Define the main method
   public static void main(String[] args) {

      // Instantiate a TestPassByValue object
      TestPassByValue2 t = new TestPassByValue2();

      // Initialize values and print them out
      int a[] = {1, 2, 3, 4, 5};
      System.out.print("Before test: a = ");
      for ( int i = 0; i < a.length; i++ )
         System.out.print(a[i] + " ");
      System.out.println();

      // Now call method test()
      t.test(a);

      // Print out values after call
      System.out.print("After test: a = ");
      for ( int i = 0; i < a.length; i++ )
         System.out.print(a[i] + " ");
      System.out.println();
   }

   // Definition of test()
   public static void test (int x[]) {

      x[3] = -10;
      System.out.print("In test: x = ");
      for ( int i = 0; i < x.length; i++ )
      System.out.print(x[i] + " ");
      System.out.println();
   }
}
```

Figure 6.5. A program showing that a method can modify an array or other object whiose reference is passed to it by value.

Figure 6.6. Reference `a` in `main` refers to a 5-element array. When method `test` is called, reference `a` is copied into parameter `x`, so now method `test` can use and modify the same 5-element array. Note that method `test` cannot modify reference `a` in the main program, but it can modify the object that `a` refers to.

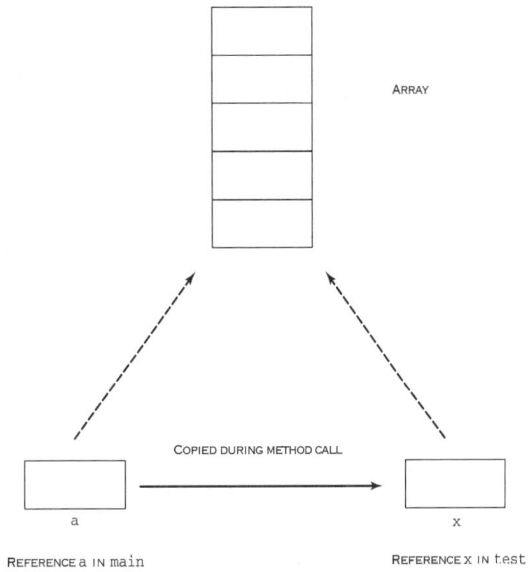

When this program is executed, the results are as follows:

```
D:\book\java\chap6>java TestPassByValue2
Before test: a = 1 2 3 4 5
In test: x = 1 2 3 -10 5
After test: a = 1 2 3 -10 5
```

There are advantages and disadvantages to passing references to arrays and objects by reference instead of copying the objects themselves. On the one hand, passing by reference weakens the security of the program, because the method can modify the calling method's data. On the other hand, the program is *much* more efficient. Imagine how much work would be involved if we attempted to pass a 10,000-element array by value! Every time that the array occurred in a method call, the program would have to make a new copy of all 10,000 elements.

A programmer must always be careful when working with arrays or other objects that are passed to methods. He or she must be certain that the method's calculations do not cause unintended modifications to the data in the arrays passed to the method. Such unintentional modifications are called **unintended side effects**, and they can create quite subtle and hard-to-find bugs.

PROGRAMMING PITFALLS

Beware of unintended side effects caused by accidentally modifying a calling method's data with references passed as parameters.

PRACTICE!

This quiz provides a quick check to see if you have understood the concepts introduced in the first three sections. If you have trouble with the quiz, reread the section, ask your instructor, or discuss the material with a fellow student. The answers to this quiz are found in the back of the book.

For questions 1 through 3, determine whether the method invocations are correct or not. If they are in error, specify what is wrong with them. If they are correct, explain what the methods do.

1.
```
public class Test1 {
   public static void main(String[] args) {
      Test1 t = new Test1();
      int i = 5; int j[] = {1, -2, 3, -4, 5};
      t.method1(i,j);
   }
   public int method1(int i[], int j) {
      for ( int k = 0; k < j; k++ )
         System.out.println(i[k] );
   }
}
```

2.
```
public class Test2 {
   public static void main(String[] args) {
      Test2 t = new Test2();
      double i = 5; int j = 5;
      System.out.println("Result = " + t.method2(i,j));
   }
   public void method2(double x, int y) {
      return (x / y);
   }
}
```

3.
```
public class Test3 {
   public static void main(String[] args) {
      Test3 t = new Test3();
      int x[] = {1, -2, 3, -4, 5, -6};
      System.out.println(t.method3(x));
   }
   public double method3(int i[]) {
      int sum = 0;
      for ( int j = 0; j < i.length; j++)
         sum += i[j];
      return ( (double) sum / i.length );
   }
}
```

6.4 EXAMPLE PROBLEM

Let us now re-examine the sorting problem of Example 5-2, using methods where appropriate.

EXAMPLE 6-1: SORTING DATA

Develop a program to read in a data set from a file, sort it into ascending order, and write the sorted data to an output file. Use methods where appropriate.

SOLUTION

The program in Example 5-2 read an arbitrary number of real input data values from an input file, sorted the data into ascending order, and wrote the sorted data to the standard output stream. The sorting process would make a good candidate for a separate method, since only the array `arr` and its length `nvals` are in common between the sorting process and the rest of the program. The other change from Example 5-2 is that we must write the sorted output to a file instead of the standard output stream.

The rewritten program using a sorting method and an output file is shown in Figure 6.7.

```
/*
   Purpose:
     To read in a set of double values from an input file, sort it into
     ascending order using the selection sort algorithm, and to write
     the sorted data to the standard output stream. The main method
     calls method  "sort" to do the actual sorting.

 Record of revisions:
    Date        Programmer          Description of change
    ====        ==========          =====================
  4/06/98       S. J. Chapman       Original code
*/
import chapman.io.*;
public class SelectionSort2 {

   // Define the main method
   public static void main(String[] args) {

      // Define maximum array size
      final int MAXVAL = 1000;

      // Declare variables, and define each variable
      double arr[] = new double[MAXVAL];
                              // Array of input measurements
      String inFileName;      // Input file name
      int i = 0, j;           // Loop index
      int iptr;               // Pointer to smallest value
      int nvals;              // Number of data values to sort
      String outFileName;     // Output file name
      double temp;            // Temporary variable for swapping

      // Create StdIn object
      StdIn in = new StdIn();

      // Create a SelectionSort2 object
      SelectionSort2 s = new SelectionSort2();

      // Get input file name
      System.out.print("Enter input file name: ");
      inFileName = in.readString();

      // Get output file name
      System.out.print("Enter output file name: ");
      outFileName = in.readString();

      // Open input file
      FileIn in1 = new FileIn(inFileName);

      // Check for valid open
      if ( in1.readStatus != in1.FILE_NOT_FOUND ) {

         // Read numbers into array
         while ( in1.readStatus != in1.EOF ) {
            arr[i++] = in1.readDouble();
         }
         nvals = i;
```

Figure 6.7. *(cont.)*

```
         // Close file
         in1.close();

         // Sort values
         s.sort( arr, nvals );

         // Open output file
         FileOut out = new FileOut(outFileName);

         // Check for valid open
         if ( out.writeStatus != out.IO_EXCEPTION ) {

            // Write sorted output values
            for ( i = 0; i < nvals; i++ )
               out.printf("%10.4f\n",arr[i]);
         }

         // Close file
         out.close();
      }

      // Come here if file not found
      else {
         System.out.println("File " + inFileName + " not found!");
      }
   }

   // Define the sort method
   public void sort(double arr[], int nvals) {

      // Declare variables, and define each variable
      int i, j;              // Loop index
      int iptr;              // Pointer to smallest value
      double temp;           // Temporary variable for swapping

      // Sort values
      for ( i = 0; i <= nvals-2; i++ ) {

         // Find the minimum value in arr[i] through arr [nvals-1]
         iptr = i;
         for ( j = i+1; j <= nvals-1; j++ ) {
            if (arr[j] < arr[iptr])
               iptr = j;
         }

         // iptr now points to the min value, so swap a[iptr]
         // with a[i] if iptr != i.
         if (i != iptr) {
            temp = arr[i];
            arr[i] = arr[iptr];
            arr[iptr] = temp;
         }
      }
   }
}
```

Figure 6.7. Program to sort real data values into ascending order using a sort method

Note that the sort method is declared to be `void`, meaning that it does not return a value to the calling method. Instead, it uses the reference to array `arr` passed by the calling method to directly manipulate the values in the array.

This new program can be tested just as the original program was, with identical results. The following data set will be placed in file `input1`:

```
13.3
12.
-3.0
 0.
 4.0
 6.6
 4.
-6.
```

Running this file values through the program yields the following result:

```
D:\book\java\chap6>java SelectionSort2
Enter input file name: input1
Enter output file name: output1

D:\book\java\chap6>type output1
    -6.0000
   -3.0000
    0.0000
    4.0000
    4.0000
    6.6000
   12.0000
   13.3000
```

The program gives the correct answers for our test data set, as before.

Method `sort` performs the same function as the sorting code in the original example, but now `sort` is an independent method that we can re-use unchanged whenever we need to sort any array of `double` numbers.

Note that the array was declared in the `sort` method as

```
public static void sort(double arr[], int nvals)
```

The statement tells the Java compiler that parameter `arr` is a one-dimensional array of `double` values. It does *not* specify the length of the array, since that will not be known until the program is actually executed.

6.5 AUTOMATIC VARIABLES

Every variable in Java is characterized by a duration and a scope. The *duration* of a variable is the time during which it exists, and the *scope* of a variable is the portion of the program from which the variable can be addressed.

Variables defined with the body of a method are sometimes called **local variables**. These variables may be defined anywhere within the method, and they are automatically created and initialized when program execution reaches that point in the method. Once created, such local variables continue to exist until the program exits the block in which they were defined. When that happens, the variables are automatically destroyed.

Such variables are said to have **automatic duration**, because they are automatically created when they are needed, and automatically destroyed when they are no longer needed. Variables with automatic duration are known as **automatic variables**.

For example, consider the following example method:

```
int sum ( int array[] ) {

   int i = 0, total = 0;
   for ( int i = 0; i < array.length; i++ ) {
      total += array[i];
   return (total);
}
```

Figure 6.8. An example method declaring two automatic variables

In this method, integer variables `i` and `total` are automatic variables. They are automatically created and initialized to zero. The `for` loop sums up all of the elements in the input array, and the `total` is returned to the calling method. When the method stops executing, variables `i` and `total` are automatically destroyed. Note that if the method is called again, new variables will be created and initialized to zero, and the process will start over again.

Automatic variables are very useful, because they conserve memory in a program. If they remained in memory, the local variables of all methods not currently being executed would just be taking up space without serving a useful purpose.

GOOD PROGRAMMING PRACTICE

Automatic variables conserve memory in a program by automatically removing unused variables when they are no longer needed in memory.

It is also possible to declare variables that persist from the moment that the class defining them is first loaded into memory until the program stops executing. These variables are said to have **static duration**, and they are known as **static variables**. They can be used to preserve information between invocations of a method, or to share data between methods, as we will see in Chapter 7. We will learn how to declare static variables in Chapter 7, where we will also learn how to declare static methods.

6.6 SCOPE

The **scope** of a variable is the portion of the program from which the variable can be addressed. There are two possible scopes for Java variables: **class scope** and **block scope**.

Methods and the instance variables of a class have *class scope*. Class scope begins at the opening left brace (`{`) of a class definition and ends at the closing right brace (`}`) of the class definition. Class scope allows any method in the class to directly invoke any other method of the class (or any method inherited from a superclass), and to directly access any instance variable of the class. Instance variables are effectively *global* within a class, since they can be seen from within any method in the class. Thus, instance variables can be used to communicate between methods in a class, or to retain information between invocations of a given method in the class.

Variables defined inside a block have *block scope*. A **block** is defined as a *compound statement*, which consists of all the statements between an open brace (`{`) and the corresponding closing brace (`}`). Variables defined within a block are visible within the block and within any blocks contained within the block, but they are not visible after the block finishes executing. For example, the local variables defined at the beginning of a method have block scope. They are visible between the open brace (`{`) and the corresponding closing brace (`}`) of the method. They are also visible within any blocks contained within the block, such as within the `for` loop of Figure 6.8.

Any block may contain variable declarations which will only be valid for the duration of the block. For example, it is possible to declare a loop index within a `for` loop itself:

```
for (int j = 0; j < n; j++ ) {
   ...
   ...
}
```

The variable `j` in this loop is an automatic variable that is only defined while the `for` loop is executing. It is automatically destroyed when loop execution ends.

However, it is illegal to declare a variable in an inner block that has the same name as a variable in an outer block. Thus, the following code would produce a compile-time error:

```
int sum ( int array[] ) {
   int i = 0, total = 0;
   i += array[0];
   for ( int i = 0; i < array.length; i++ ) {
      total += array[i];
   return (total);
}
```

PROGRAMMING PITFALLS

It is illegal to declare a variable in an inner block that has the same name as a variable in an outer block.

It *is* legal to declare a variable within a block that has the same name as an instance variable within a class. If this is done, the instance variable is "hidden" until he block terminates execution. We will learn how to access such "hidden" instance variables in Chapter 7.

6.7 RECURSIVE METHODS

For some classes of problems, it is convenient for a Java method to invoke itself. A **recursive method** is a method that invokes itself either directly or indirectly through another method. Java methods are inherently recursive, since all local variables are automatic, and new copies of each variable are created each time that the method is invoked.

There are certain classes of problems that are easily solved recursively. For example, the factorial function can be defined as follows:

$$N! = \begin{cases} N(N-1)! & \text{for} \quad N \geq 1 \\ 1 & \text{for} \quad N = 0 \end{cases} \tag{6-1}$$

This definition can easily be implemented recursively, with the procedure that calculates $N!$ calling itself to calculate $(N - 1)!$, and that procedure calling itself to calculate $(N - 2)!$, etc. until finally the procedure is called to calculate $0!$.

A recursive method to calculate the factorial function is shown in Figure 6-9. Note that if `n > 1`, method `factorial` calls itself with the argument `n-1`. Method `factorial` in this class includes two `println` statements, so that we can see happens as the method calls itself recursively.

```
/*
   Purpose:
      To calculate the factorial function N! through a
      recursive method.

   Record of revisions:
      Date         Programmer        Description of change
      ====         ==========        =====================
     4/07/98      S. J. Chapman     Original code
*/
import chapman.io.*;
public class Factorial {

   // Define method factorial
   public int factorial ( int n ) {
      int answer;     // Result of calculation
      System.out.println("In factorial: n = " + n);
      if ( n >= 1 )
         answer = n * factorial(n-1);
      else
         answer = 1;
      System.out.println("In factorial: n = "
                   + n + " answer = " + answer);
      return answer;
   }

   // Define the main method
   public static void main(String[] args) {

      int n; // Integer to calculate factorial of

      // Create a StdIn object
      StdIn in = new StdIn();

      // Instantiate a Factorial object
      Factorial f = new Factorial();

      // Get number to calculate factorial of
      System.out.print("Enter int to calculate factorial of: ");
      n = in.readInt();

      // Output factorial
      System.out.print(n + "! = " + f.factorial(n));
   }
}
```

Figure 6.9. A method to recursively implement the factorial function

When this program is used to calculate the value of 5!, the results are as follows:

```
D:\book\java\chap6>java Factorial
Enter integer to calculate factorial of: 5
In factorial: n = 5
In factorial: n = 4
In factorial: n = 3
In factorial: n = 2
In factorial: n = 1
In factorial: n = 0
In factorial: n = 0 answer = 1
```

```
In factorial: n = 1 answer = 1
In factorial: n = 2 answer = 2
In factorial: n = 3 answer = 6
In factorial: n = 4 answer = 24
In factorial: n = 5 answer = 120
5! = 120
```

Note that the method was called by the main program with `n = 5`, and then the method called itself with `n = 4`, etc. It is easy to verify by hand calculation that this method produced the correct answer.

6.8 METHOD OVERLOADING

Java allows several methods to be defined with the same name, as long as the methods have different sets of parameters (based on the number, types, and order of the parameters). This is called **method overloading**. When an overloaded method is called, the Java compiler selects the proper method by examining the number, type, and order of the calling arguments. Method overload is commonly used to create several methods that perform the same function, but with different data types. For example, the program in Figure 6.10 includes two methods to calculate the square of a number, one for `ints` and the other for `doubles`. Java determines which of the two methods to call based on the argument of a particular method call.

```
// This program illustrates method overloading
public class Overload1 {

   // Define the main method
   public static void main(String[] args) {

      int i; double x;

      // Instantiate an Overload1 object
      Overload1 ov = new Overload1();

      // Write number and square using integers
      System.out.println("Using integers:");
      for ( i = 1; i <= 5; i++ ) {
         System.out.print("number = " + i);
         System.out.println(" square = " + ov.square(i));
      }

      // Write number and square using doubles
      System.out.println("Using doubles:");
      for ( i = 1; i <= 5; i++ ) {
         x = i;
         System.out.print("number = " + x);
         System.out.println(" square = " + ov.square(x));
      }
   }

   // Definition of square method 1
   public int square (int x) {
      return x * x;
   }

   // Definition of square method 2
   public double square (double x) {
      return x * x;
   }
}
```

Figure 6.10. A Java program illustrating method overloading

When this program is executed, the results are as follows:

```
D:\book\java\chap6>java Overload1
Using integers:
number = 1 square = 1
number = 2 square = 4
number = 3 square = 9
number = 4 square = 16
number = 5 square = 25
Using doubles:
number = 1.0 square = 1.0
number = 2.0 square = 4.0
number = 3.0 square = 9.0
number = 4.0 square = 16.0
number = 5.0 square = 25.0
```

Note that Java used the first method when the method was invoked with an `int` argument, and the second method when the method was invoked with a `double` argument.

Overloaded methods are distinguished by their **signature**, which is a combination of the method name and the number, type, and order of parameters. If two methods have the same signature, then the Java compiler cannot distinguish between them. Defining two methods with the same signature produces a compile-time error. This is true *even if the two methods return different data types*. For example, the program shown in Figure 6.11 contains two methods with an identical signature but different return data types.

```
// This program contains two methods with identical
// signatures, which is an error.
class Overload2 {

   // Define the main method
   public static void main(String[] args) {

      // Instantiate an Overload2 object
      Overload2 ov = new Overload2();

      // Write number and square using integers
      for ( int i = 1; i <= 10; i++ ) {
         System.out.print("number = " + i);
         System.out.println(" square = " + ov.square(i));
      }
   }

   // Definition of square method 1
   public int square (int x) {
      return x * x;
   }

   // Definition of square method 2
   public double square (int x) {
      return (double) x * x;
   }
}
```

Figure 6.11. A Java program containing two different methods with the same signature in the same class.

When this program is compiled, the results are as follows:

```
D:\book\java\chap6>javac Overload2.java
Overload2.java:24:
Methods can't be redefined with a different return
type: double square(int) was int square(int)
   public double square (int x) {
                 ^
1 error
```

PROGRAMMING PITFALLS

It is an error to define two different methods with the same signature in a single class.

EXAMPLE 6-2: STATISTICS METHODS

Develop a set of reusable methods capable of determining the average and standard deviation of a data set consisting of numbers in an array. The method should work for data input arrays of `int`, `float`, and `double` types.

SOLUTION

To solve this problem, we will need to create six different methods, one each to determine the average and the standard deviation of `int`, `float`, and `double` arrays. Note that we will be using method overloading so that the same method name may be called to get the average of a data set, whether it is an `int`, a `float`, or a `double`. In addition, we will need to define a `main` method to test our six other methods.

1. **State the problem**. The problem is clearly stated above. We will write six different methods: three forms of `average` to calculate the average of the three types of data sets, and three forms of `stdDev` to calculate the standard deviation of the three types of data sets.
2. **Define the inputs and outputs**. The input to each method will be array of `int`, `float`, or `double` values. The output will be a `double` value containing either the average or the standard deviation of the data in the array. Note that we are using a `double` value in all cases, because the average and standard deviations of even the integer arrays will not in general be integer values.
3. **Decompose the program into classes and their associated methods**. We have not yet studied classes in detail, so for now there will be only one class. The class will contain six methods to calculate the average and standard deviations, plus the `main` method to contain text driver code. We will call the class `Stats`, and make it a subclass of the root class `Object`.
4. **Design the algorithm that you intend to implement for each method**. The pseudocode for the `average` methods is as follows:

```
Initialize sumX to zero
if (arr.length >= 1) {
   for ( int i = 0; i < arr.length; i++ ) {
      sumX ← sumX + arr[i];
   }
   xBar ← sumX / arr.length;

else {
   Insufficient data: set xBar to 0.
}
return xBar;
```

The pseudocode for the `stdDev` methods is:

```
Initialize sumX and sumX2 to zero.
if (arr.length >= 2) {
   for ( int i = 0; i < arr.length; i++ ) {
      sumX ← sumX + arr[i];
```

```
      sumX2 ← sumX2 + arr[i]*arr[i];
   }
   stdDev ← Math.sqrt((nvals*sumX2 - sumX*sumX)
          / (arr.length*(arr.length-1)));
else {
   Insufficient data: set stdDev to 0.
}
return stdDev;
```

The `main` method will have to test all six calculational methods, so it must define three arrays containing `int`, `float`, or `double` values, and call both `average` and `stdDev` with each array. The pseudocode for the `main` method is:

```
int arr1 = { 8, 9, 10, 11, 12 };
float arr2 = { 8.F, 9.F, 10.F, 11.F, 12.F };
double arr3 = { 8., 9., 10., 11., 12. };
System.out.println("Integer array:");
System.out.println("Average = " + average(arr1));
System.out.println("Standard Deviation = " + stdDev(arr1));
System.out.println("Float array:");
System.out.println("Average = " + average(arr2));
System.out.println("Standard Deviation = " + stdDev(arr2));
System.out.println("Double array:");
System.out.println("Average = " + average(arr3));
System.out.println("Standard Deviation = " + stdDev(arr3));
```

5. **Turn the algorithm into Java statements**. The resulting Java program is shown in Figure 6.12.

```
/*
   Purpose:
     To calculate the average and standard deviation of
     integer, float, or double arrays.  This class
     illustrates method overloading.

   Record of revisions:
      Date       Programmer          Description of change
      ====       ==========          =====================
     4/10/98    S. J. Chapman        Original code
*/
public class Stats {

   // Define the average method for integer arrays
   public double average( int arr[] ) {

      // Declare variables, and define each variable
      double sumX = 0;      // Sum of the input samples
      double xBar = 0;      // Average of input samples

      if (arr.length >= 1) {
         for ( int i = 0; i < arr.length; i++ ) {
            sumX += arr[i];
         }
         xBar = sumX / arr.length;
      }
```

Figure 6.12. *(cont.)*

```
      else {
         // Insufficient data
         xBar = 0;
      }
      return xBar;
   }

   // Define the average method for float arrays
   public double average( float arr[] ) {

      // Declare variables, and define each variable
      double sumX = 0;      // Sum of the input samples
      double xBar = 0;      // Average of input samples

      if (arr.length >= 1) {
         for ( int i = 0; i < arr.length; i++ ) {
            sumX += arr[i];
         }
         xBar = sumX / arr.length;
      }
      else {
         // Insufficient data
         xBar = 0;
      }
      return xBar;
   }

   // Define the average method for double arrays
   public double average( double arr[] ) {

      // Declare variables, and define each variable
      double sumX = 0;      // Sum of the input samples
      double xBar = 0;      // Average of input samples

      if (arr.length >= 1) {
         for ( int i = 0; i < arr.length; i++ ) {
            sumX += arr[i];
        }
        xBar = sumX / arr.length;
      }
      else {
         // Insufficient data
         xBar = 0;
      }
      return xBar;
   }

   // Define the stdDev method for integer arrays
   public double stdDev( int arr[] ) {

      // Declare variables, and define each variable
      double stdDev = 0;    // Std deviation of input samples
      double sumX = 0;      // Sum of the input samples
      double sumX2 = 0;     // Sum of squares of input samples
      double xBar = 0;      // Average of input samples

      if (arr.length >= 2) {
         for ( int i = 0; i < arr.length; i++ ) {
            sumX += arr[i];
```

Figure 6.12. *(cont.)*

```
            sumX2 += arr[i] * arr[i];
         }
         stdDev = Math.sqrt((arr.length*sumX2 - sumX*sumX)
               / (arr.length*(arr.length-1)));
      }
      else {
         // Insufficient data
         stdDev = 0;
      }
      return stdDev;
   }

   // Define the stdDev method for float arrays
   public double stdDev( float arr[] ) {

      // Declare variables, and define each variable
      double stdDev = 0;   // Std deviation of input samples
      double sumX = 0;     // Sum of the input samples
      double sumX2 = 0;    // Sum of squares of input samples
      double xBar = 0;     // Average of input samples

      if (arr.length >= 2) {
         for ( int i = 0; i < arr.length; i++ ) {
            sumX += arr[i];
            sumX2 += arr[i] * arr[i];
         }
         stdDev = Math.sqrt((arr.length*sumX2 - sumX*sumX)
               / (arr.length*(arr.length-1)));
      }
      else {
         // Insufficient data
         stdDev = 0;
      }
      return stdDev;
   }

   // Define the stdDev method for double arrays
   public double stdDev( double arr[] ) {

      // Declare variables, and define each variable
      double stdDev = 0;   // Std deviation of input samples
      double sumX = 0;     // Sum of the input samples
      double sumX2 = 0;    // Sum of squares of input samples
      double xBar = 0;     // Average of input samples

      if (arr.length >= 2) {
        for ( int i = 0; i < arr.length; i++ ) {
           sumX += arr[i];
           sumX2 += arr[i] * arr[i];
        }
        stdDev = Math.sqrt((arr.length*sumX2 - sumX*sumX)
               / (arr.length*(arr.length-1)));
      }
      else {
         // Insufficient data
         stdDev = 0;
      }
      return stdDev;
     }
```

Figure 6.12. *(cont.)*

```
   // Define the main method
   public static void main(String[] args) {

      // Instantiate a Stats object
      Stats st = new Stats();

      // Declare test arrays
      int arr1[] = { 8, 9, 10, 11, 12 };
      float arr2[] = { 8.F, 9.F, 10.F, 11.F, 12.F };
      double arr3[] = { 8., 9., 10., 11., 12. };

      // Calculate average and standard dev of each array
      System.out.println("Integer array:");
      System.out.println("Average       = " + st.average(arr1));
      System.out.println("Std Deviation = " + st.stdDev(arr1));
      System.out.println("Float array:");
      System.out.println("Average       = " + st.average(arr2));
      System.out.println("Std Deviation = " + st.stdDev(arr2));
      System.out.println("Double array:");
      System.out.println("Average       = " + st.average(arr3));
      System.out.println("Std Deviation = " + st.stdDev(arr3));
   }
}
```

Figure 6.12. Program to calculate the average and standard deviation of `int`, `float`, or `double` arrays. This program illustrates method overloading.

6. **Test the resulting Java programs.** To test this program, we can calculate the average and standard deviation of the values 8, 9, 10, 11, 12 from the definitions of average and standard deviation:

$$\bar{x} = \frac{1}{N}\sum_{i=1}^{N} x_i = \frac{1}{5}\,50 = 10.0, \tag{6-1}$$

$$s = \sqrt{\frac{N\sum_{i=1}^{N} x_i^2 - \left(\sum_{i=1}^{N} x_i\right)^2}{N(N-1)}} = 1.5811. \tag{6-2}$$

When the program is executed, the results are as follows:

```
D:\book\java\chap6>java Stats
Integer array:
Average            = 10.0
Standard Deviation = 1.5811388300841898
Float array:
Average            = 10.0
Standard Deviation = 1.5811388300841898
Double array:
Average            = 10.0
Standard Deviation = 1.5811388300841898
```

The results of the program agree with our hand calculations to the number of significant digits that we performed the calculation.

6.9 CLASS `java.util.Arrays`

Beginning with the Java Development Kit version 1.2, Java includes a special class that contains some methods designed to make manipulating arrays easier. This class is named `Arrays`, and it is located in the `java.util` package.

The `Arrays` class contains several overloaded methods designed to make working with arrays easier. Some of these methods are summarized in Table 6-1.

The quicksort method implemented in this class is much more efficient than the selection sort method that we developed in Example 5-2. You will be asked to compare the sorting speeds of these two methods in an end-of chapter exercise.

The `Arrays` class must be imported into a Java program before it can be used. The easiest way to do this is to include the statement

```
import java.util.*;        // Import Java utils package
```

before the class definition in which the methods will be used. Once the `Arrays` class has been imported, the methods can be invoked as `Arrays.sort()`, `Arrays.fill()`, and so forth.

TABLE 6-1 Selected Methods in `java.util.Arrays`

METHOD NAME	DESCRIPTION

`int binarySearch(long[] a, long key)`
`int binarySearch(int[] a, int key)`
`int binarySearch(short[] a, short key)`
`int binarySearch(char[] a, char key)`
`int binarySearch(byte[] a, byte key)`
`int binarySearch(double[] a, double key)`
`int binarySearch(float[] a, float key)`
`int binarySearch(Object[] a, Object key)`
	These overloaded methods search array `a` for the specified value `key` using the binary search algorithm. The array *must* be sorted prior to making this call. The method returns the index of the search key, if it is contained in the array, and a negative number, if it is not contained in the array.

`boolean equals(long[] a, Object o)`
`boolean equals(int[] a, Object o)`
`boolean equals(short[] a, Object o)`
`boolean equals(char[] a, Object o)`
`boolean equals(byte[] a, Object o)`
`boolean equals(double[] a, Object o)`
`boolean equals(float[] a, Object o)`
`boolean equals(Object[] a, Object o)`
	These overloaded methods test to see if array `a` is equal to the given object `o`. The array and the object are considered equal if the object is an array of the same type, both arrays contain the same number of elements, and all corresponding pairs of elements in the two arrays are equal. In other words, the two arrays are equal if they contain the same elements in the same order.

`void fill(long[] a, long val)`
`void fill(void[] a, void val)`
`void fill(short[] a, short val)`
`void fill(char[] a, char val)`
`void fill(byte[] a, byte val)`
`void fill(double[] a, double val)`
`void fill(float[] a, float val)`
`void fill(Object[] a, Object val)`
	These overloaded methods fill each element of array `a` with the value `val`.

`void sort(long[] a)`
`void sort(int[] a)`
`void sort(short[] a)`
`void sort(char[] a)`
`void sort(byte[] a)`
`void sort(double[] a)`
`void sort(float[] a)`
`void sort(Object[] a)`
	These overloaded methods sort the array `a` into ascending numerical order. The sorting algorithm is a tuned quicksort. Note that is possible to sort arrays of `Objects` as well as primitive data types, provided that the objects are mutually comparable.

Note that these methods are all `static`, like the ones in the `Math` class. We will explain `static` methtods in Chapter 7. For now, the importance of the `static` declaration is that the methods can be invoked directly without first instantiating an object of type `Arrays`.

GOOD PROGRAMMING PRACTICE

Use the `sort()` method in the `java.util.Arrays` class to sort arrays in practical programs. This method is very efficient, already debugged, and built right into the basic Java environment.

6.10 ADDITIONAL METHODS SUPPLIED WITH THIS BOOK

Because Java was created as a computer scientist's language, it lacks some of the functions that engineers are accustomed to using in other languages such as Fortran, Gnu C, MATLAB, etc. A number of classes have been supplied with this book to partially fill in the gap. Two of these classes (containing extended math methods and array manipulations) are described in this section, while other classes for complex numbers, signal processing, etc. are described later in the book. You should check the book's Web site (given in the Chapter 2) for the latest versions of these classes, as enhancements and bug fixes are expected throughout the life of the book.

Class `Math1` contains a collection of static methods to supplement the methods included in Java's `Math` class. These methods may be used in the same way as Java's standard math methods, but you must import package `chapman.math` into your program with the statement

```
import chapman.math.*;
```

The methods in class `Math1` are summarized in Table 6-2.

Class `Array` contains set of `static` methods designed to manipulate one- and two-dimensional arrays in a convenient fashion. These methods permit a programmer to treat an array as an entity, and to perform operations on an entire array with a single method invocation. It is also in package `chapman.math`.

Various methods in this class support data of types `int`, `long`, `float`, and `double`, although not all data types are supported for all functions. Consult the detailed on-line documentation accompanying the package to determine which data types are supported by each method. The methods in this class fall into several categories, as follows:

1. **Calculational Methods**. These methods perform calculations on the elements of an array. Examples include `add`, `sub`, `mul`, `div`, `sum`, `product`, `dotProduct`, `sin`, `sinc`, `sind`, `sinh`, `cos`, `cosd`, `cosh`, `tan`, `tand`, `tanh`, `asin`, `asind`, `asinh`, `acos`, `acosd`, `acosh`, `atan`, `atand`, `atanh`, `abs`, `exp`, `log`, `log10`, and `pow`.
2. **Inquiry Methods**. These methods extract information from an array. Examples include `maxAbs`, `maxAbsLoc`, `maxVal`, `maxLoc`, `minVal`, and `minLoc`.
3. **Relational Methods**. These methods perform relational comparisons on the data in an array, producing a `boolean` result. Examples include

TABLE 6-2 Mathematical Methods in Class `chapman.math.Math1`

METHOD NAME AND PARAMETERS	METHOD VALUE	PARAMETER TYPE	RESULT TYPE	COMMENTS
`Math1.acosd(x)`	$\cos^{-1} x$	`Double`	`Double`	Returns inverse cosine of x for $-1 \le x \le 1$ (results in *degrees*)
`Math1.acosh(x)`	$\cosh^{-1} x$	`Double`	`Double`	Returns inverse hyperbolic cosine of x
`Math1.asind(x)`	$\sin^{-1} x$	`Double`	`Double`	Returns inverse sine of x for $-1 \le x \le 1$ (results in *degrees*)
`Math1.asinh(x)`	$\sinh^{-1} x$	`Double`	`Double`	Returns inverse hyperbolic sine of x
`Math1.atand(x)`	$\tan^{-1} x$	`Double`	`Double`	Returns inverse tangent of x (results in *degrees* in the range $-90° \le x \le 90°$)
`Math1.atan2d(y,x)`	$\tan^{-1} y/x$	`Double`	`Double`	Returns inverse tangent of x (results in *degrees* in the range $-180 \le x \le 180$)
`Math1.atanh(x)`	$\tanh^{-1} x$	`Double`	`Double`	Returns inverse hyperbolic tangent of x
`Math1.cosd(x)`	$\cos x$	`Double`	`Double`	Returns cosine of x, where x is in degrees
`Math1.cosh(x)`	$\cosh x$	`Double`	`Double`	Returns hyperbolic cosine of x
`Math1.log10(x)`	$\log_{10} x$	`Double`	`Double`	Returns logarithm to the base 10 of x, for $x > 0$
`Math1.randomGaussian()`		`N/A`	`Double`	Returns a normally-distributed Gaussian random number
`Math1.randomRayleigh()`		`N/A`	`Double`	Returns a Rayleigh-distributed random number
`Math1.sinc(x)`	$\text{sinc}\, x$	`Double`	`Double`	Returns the value of sinc x, where sinc $x = \sin x / x$
`Math1.sind(x)`	$\sin x$	`Double`	`Double`	Returns sine of x, where x is in degrees
`Math1.sinh(x)`	$\sinh x$	`Double`	`Double`	Returns hyperbolic sine of x
`Math1.tand(x)`	$\tan x$	`Double`	`Double`	Returns tangent of x, where x is in degrees
`Math1.tanh(x)`	$\tanh x$	`Double`	`Double`	Returns hyperbolic tangent of x

isGreaterThan, isGreaterThanOrEqual, isLessThan, isLessThanOrEqual, isEqual, and isNotEqual.

4. **Logical Methods**. These methods manipulate boolean arrays, which can be produced by the relational methods described above. Examples include all, any, count, and, or, xor, and not. The resulting boolean arrays can be used as a mask to control certain operations.
5. **Random Number Methods**. These methods return arrays of random values. Examples include random, randomGaussian, and randomRayleigh.

EXAMPLE 6-3: USING ARRAY METHODS

The methods provided in class chapman.math.Array allow certain types of engineering problems to be solved with a relatively few statements. As a quick example, suppose that we want to determine the percentage of samples in a Rayleigh-distributed random data set that are greater than 3.0. Figure 6.13 shows a Rayleigh probability distribution. Clearly, some samples will exceed the threshold, but how many?

SOLUTION

One approach to solving this problem is by Monte Carlo simulation, which is the process of making many random trials, and determining for each trial whether the result exceeds 3.0. If we do enough trials, the average number of times that the random number exceeds 3.0 will approximate the theoretical answer.

Figure 6.13. A Rayleigh probability distribution

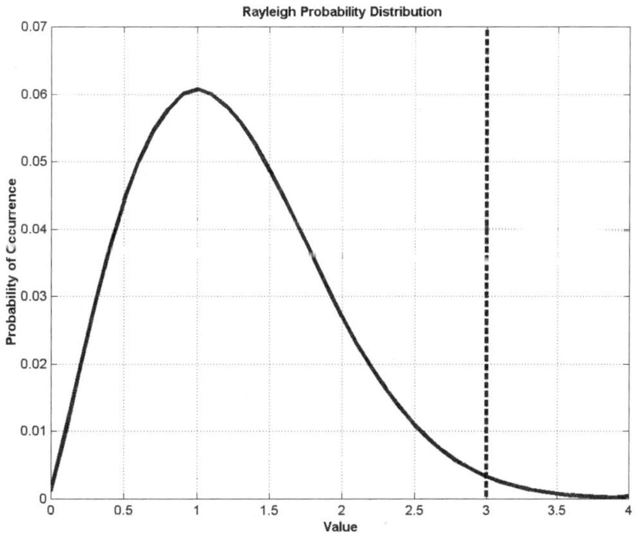

```
/*
   Purpose:
      To calculate the percentage of random samples from a
      Rayleigh distribution that exceed 3.0. This class
      demonstrates the use of methods from chapman.math.Array.

   Record of revisions:
      Date        Programmer             Description of change
      ====        ==========             =====================
     9/10/98     S. J. Chapman          Original code
*/
import chapman.io.*;
import chapman.math.*;
public class ThresholdCrossing {

   public static void main(String[] args) {

      double arr[];       // Array of random samples
      int count;          // Number of samples > 3.0

      // Get an array of 50000 random values
      arr = Array.randomRayleigh(50000);

      // Find out how many samples are > 3.0.
      count = Array.count(Array.isGreaterThan(arr, 3.));

      // Print result
      Fmt.printf("Percentage = %6.2f %", 100.*count/arr.length);
   }
```

Figure 6.14. Program to calculate the percentage of samples from a Rayleigh distribution that exceed 3.0

Figure 6.14 shows a Java program that solves this problem by Monte Carlo simulation. First, it uses method `Array.randomRayleigh` to generate an array of 50,000 samples from a Rayleigh distribution. Next the program uses method `Array.isGreaterThan` to generate a `boolean` array whose elements are `true` at any location where the sample was greater than 3, and method `Array.count` to count the number of `true` values in the `boolean` array.

When the program is executed, the results are as follows:

```
D:\book\java\chap6>java ThresholdCrossing
Percentage = 1.08 %
```

About 1% of the samples will exceed 3.0.

PRACTICE!

This quiz provides a quick check to see if you have understood the concepts introduced in Sections 6.4 through 6.10. If you have trouble with the quiz, reread the sections, ask your instructor, or discuss the material with a fellow student. The answers to this quiz are found in the back of the book.

1. What is the duration of a variable? What types of duration exist in Java?
2. What is the scope of a variable? What types of scope exist in Java?
3. What is the duration and scope of local variables defined within a Java method?
4. What is a recursive method?
5. What is method overloading?

For questions 6 and 7, determine whether there are any errors in these programs. If so, tell what the errors are. If not, tell what the output from each program will be.

6.
```
public class Test {

   public static void main(String[] args) {

      int i = 2; int j = 3;

      while (j > 0) {
         int i = j;
         System.out.println(i*j);
         j--;
      }
   }
}
```

7.
```
public class Test {

   public static void main(String[] args) {

      Test t = new Test();

      int a = 5, b = 10;
      System.out.println(t.m1(a,b));
   }

   public static double m1 (int i, int j) {
      return (i + 2*j);
      }

   public static double m1 (int x, int y) {
      return (x - 2*y);
   }
}
```

SUMMARY

- A **method** performs a specific function and returns (at most) a *single* value to the calling method. All methods must be defined within a Java class definition.
- A method definition includes four components: (1) optional keywords, (2) the data type of the returned value, (3) the method name, and (4) the list of parameters that the method expects to receive. This definition is followed by the body of the method within braces ({}).
- A method calls or invokes another method by including its name, together with appropriate calling arguments in parentheses, in an expression.
- A method executes until the end of the body is reached, or until a `return` statement is executed. The value returned by the method is the value of the expression in the `return` statement.

- It is possible for a method to return no value to a calling methods. In that case, the method is declared with a `void` data type.
- Local variables are variables defined within a method and not accessible to calling methods. These variables are automatic variables, meaning that they are automatically created when the methods starts executing, and automatically destroyed when the method stops executing.
- Java uses a *pass-by-value* scheme to pass the value of calling arguments to method parameters. A *copy* of the value of each calling argument is placed in the corresponding parameters. As a result, any modifications of the parameters within the method have no effect on the calling arguments.
- When an array or object is passed to a method, Java copies the value of the *reference* to the array or object and places it in the corresponding parameter. The method can then use that reference to modify the original array or object.
- Automatic variables are created when a body is executed, and destroyed when the execution of the body is completed.
- Static variables are created when the class defining them is first loaded into memory, and persist until the program stops executing.
- The scope of a variable is the portion of the program from which the variable can be addressed.
- Variables and methods with class scope can be addressed from anywhere within the class in which they are defined. For example, instance variables have class scope.
- Variables with block scope are defined within a block, and are visible within that block and within any blocks contained within that block.
- A block is a compound statement, consisting of all the statements between an open brace (`{`) and the corresponding close brace (`}`).
- A recursive method is a method that calls itself, either directly or indirectly.
- Method overloading is the definition of two or more methods with the same name, distinguishable by the type, number, and order of their calling parameters.
- Overload methods are distinguished by their **signature**, which is a combination of the method name and the number, type, and order of parameters. No two methods in a single class may have the same signature.

APPLICATIONS: SUMMARY OF GOOD PROGRAMMING PRACTICES

The following guidelines introduced in this chapter will help you to develop good programs:

1. Break large program tasks into classes and methods whenever practical to achieve the important benefits of independent component testing, reusability, and isolation from unintended side effects.
2. The pass-by-value scheme prevents a method from accidentally modifying its calling arguments.
3. Automatic variables conserve memory in a program by automatically removing unused variables when they are no longer needed in memory.
4. Use the `sort()` method in the `java.util.Arrays` class to sort arrays in practical programs. This method is very efficient, already debugged, and built right into the basic Java environment.

KEY TERMS

argument list
automatic duration
automatic variables
block
block scope
`chapman.math.Array` class
`chapman.math.Math1` class
class scope
duration
information hiding
`java.util.Arrays` class
local variables
method
method body
method call
method invocation
method overloading
parameters
pass-by-value
recursive method
`return` statement
scope
signature
static duration
static variables
test driver
top-down design
unit testing
`void`

Problems

1. When a method is called, how is data passed from the calling method to the called method, and how are the results of the method returned to the calling program?
2. What are the advantages and disadvantages of the pass-by-value scheme used in Java?
3. Suppose that a fifteen-element array `a` is passed to a method as a calling argument. What will happen if the method attempts to access element `a[15]`?
4. Determine whether the following method calls are correct or not. If they are in error, specify what is wrong with them. If they are correct, describe what the program does.

a.
```
public class Test {
   public static void main(String[] args) {
      Test t = new Test();

      int arr[] = { 1, 2, 3, 4, 5};
      System.out.println(t.sum(arr));
   }

   public int sum (int a) {
      int sum = 0;
      for ( int i = 0; i < a.length; i++ )
         sum += a[i];
      return (sum);
   }
}
```

b.
```
public class Test {
   public static void main(String[] args) {
      Test t = new Test();

      int arr[] = {1, 2, 3, 4, 5};
      int i;

      // Print array
      System.out.print("Before: ");
      for (i = 0; i < arr.length; i++)
         System.out.print(arr[i] + " ");
      System.out.println();

      t.calc(arr,6);

      System.out.print("After: ");
      for (i = 0; i < arr.length; i++)
         System.out.print(arr[i] + " ");
      System.out.println();
   }

   public static void calc (int a[], int b) {
      int sum = 0;
      for ( int i = 0; i < a.length; i++ ) {
         a[i] *= b;
         sum += a[i];
      }
      return (sum);
   }
}
```

5. Modify the selection sort method developed in this chapter so that it sorts `double` values in *descending* order.

6. The mathematical method `Math.random()` returns a sample value from a uniform distribution in the range [0,1). Each time that the method is called, a random value in the range 0 ≤ `value` < 1 is returned, with every possible value having an equal probability of occurrence. A method like `Math.random()` can be used to introduce an element of chance into a program.

   Every possible number between 0 and 1 should have an equal probability of being returned as a result from this method. Test the distribution of values returned from this method by calling it 10,000 times and calculating the number of values falling between 0 and 0.1, 0.1 and 0.2, etc. Are the values evenly distributed between 0 and 1? Plot the number of values falling in each interval using the plotting classes provided in package `chapman.graphics`.

7. Use method `Math.random()` to generate arrays containing 1000, 10,000, and 100,000 random values between 0.0 and 1.0. Then, use the statistical methods developed in this chapter to calculate the average and standard deviation of values in the arrays. The theoretical average of a uniform random distribution in the range [0,1) is 0.5, and thetheoretical standard deviation of the uniform random distribution is $1/(\sqrt{12})$. How close do the random arrays generated by `Math.random()` come to behaving like the theoretical distribution?

8. Write a method that uses method `Math.random()` to generate a random value in the range [–1.0, 1.0).

9. **Dice Simulation** It is often useful to be able to simulate the throw of a fair die. Write a Java method `dice()` that simulates the throw of a fair die by returning some random integer between 1 and 6 every time that it is called. (*Hint:* Call `Math.random()` to generate a random number. Divide the possible values out of `Math.random()` into six equal intervals, and return the number of the interval that a given random number falls into.)

10. **Road Traffic Density** Method `Math.random()` produces a number with a *uniform* probability distribution in the range [0.0, 1.0). This method is suitable for simulating random events if each outcome has an equal probability of occurring. However, in many events, the probability of occurrence is *not* equal for every event, and a uniform probability distribution is not suitable for simulating such events.

    For example, when traffic engineers studied the number of cars passing a given location in a time interval of length t, they discovered that the probability of k cars passing during the interval is given by the equation

$$P(k,t) = e^{-\lambda t}\frac{(\lambda t)^k}{k!} \text{ for } t \geq 0, \lambda > 0, \text{ and } k = 0, 1, 2, \ldots \qquad (6\text{-}2)$$

    This probability distribution is known as the *Poisson distribution;* it occurs in many applications in science and engineering. For example, the number of calls k to a telephone switchboard in time interval t, the number of bacteria k in a specified volume t of liquid, and the number of failures k of a complicated system in time interval t all have Poisson distributions.

    Write a method to evaluate the Poisson distribution for any k, t, and λ. Test your method by calculating the probability of 0, 1, 2, . . ., 5 cars passing a particular point on a highway in 1 minute, given that λ is 1.6 per minute for that highway.

11. Write three Java methods to calculate the hyperbolic sine, cosine, and tangent functions:

$$\sinh(x) = \frac{e^x - e^{-x}}{2}, \qquad \cosh(x) = \frac{e^x + e^{-x}}{2}, \qquad \tanh(x) = \frac{e^x - e^{-x}}{e^x + e^{-x}}$$

Use your methods to calculate the hyperbolic sines, cosines, and tangents of the values between –2.0 and 2.0 in steps of 0.25. Compare your answers with the ones produced by the methods `Math1.sinh()`, `Math1.cosh()`, and `Math1.tanh()`. Create plots of the shapes of the hyperbolic sine, cosine, and tangent functions.

12. **Cross Product** Write a method to calculate the cross product of two `double` vectors $\mathbf{V}_1$ and $\mathbf{V}_2$. The cross product is defined as

$$\mathbf{V}_1 \times \mathbf{V}_2 = (V_{y1}V_{z2} - V_{y2}V_{z1})\mathbf{i} + (V_{z1}V_{x2} - V_{z2}V_{x1})\mathbf{j} + (V_{x1}V_{y2} - V_{x2}V_{y1})\mathbf{k},$$

where $\mathbf{V}_1 = V_{x1}\,\mathbf{i} + V_{y1}\,\mathbf{j} + V_{z1}\,\mathbf{k}$ and $\mathbf{V}_2 = V_{x2}\,\mathbf{i} + V_{y2}\,\mathbf{j} + V_{z2}\,\mathbf{k}$. Note that this method will return a `double` array as its result. Use the method to calculate the cross product of the two vectors $\mathbf{V}_1 = [-2, 4, 0.5]$ and $\mathbf{V}_2 = [0.5, 3, 2]$.

13. **Sort with Carry** It is often useful to sort an array `arr1` into ascending order, while simultaneously carrying along a second array `arr2`. In such a sort, each time an element of array `arr1` is exchanged with another element of `arr1`, the corresponding elements of array `arr2` are also swapped. When the sort is over, the elements of array `arr1` are in ascending order, while the elements of array `arr2` that were associated with particular elements of array `arr1` are still associated with them. For example, suppose we have the following two arrays:

Element	arr1	arr2
1.	6.	1.
2.	1.	0.
3.	2.	10.

After sorting array `arr1` while carrying along array `arr2`, the contents of the two arrays will be:

Element	arr1	arr2
1.	1.	0.
2.	2.	10.
3.	6.	1.

Write a method to sort one `double` array into ascending order while carrying along a second one. Test the method with the following two 9-element arrays:

```
double a[] = {  1., 11., -6., 17.,-23.,   0.,   5.,   1., -1. };
double b[] = { 31.,101., 36.,-17.,   0., 10., -8., -1., -1. };
```

14. **Comparing Sort Algorithms** Write a program to compare the sorting speed of the selection sort method developed in Example 6-1 with the quicksort sorting method included in the `java.util.Arrays` class. Use method `Math.random()` to generate two arrays containing 1000 and 10,000 random values between 0.0 and 1.0. Then, use both sorting methods to sort copies of these arrays. How does the sorting time compare for these two methods? (*Note:* The method `System.currentTimeMillis()` returns the current system time in milliseconds as a `double` value. You can determine the elapsed time required by a sorting algorithm by calling this method before and after the call to each sorting algorithm.)

15. **Linear Least Squares Fit** Develop a method that will calculate slope m and intercept b of the least-squares line that "best fits" an input data set. The input data points (x,y) will be passed to the method in two input arrays, `x` and `y`. The equations describing the slope and intercept of the least-squares line are

$$y = m\,x + b \tag{6-3}$$

$$m = \frac{(\Sigma xy) - (\Sigma x)\bar{y}}{(\Sigma x^2) - (\Sigma x)\bar{x}} \tag{6-4}$$

and

$$b = \bar{y} - m\,\bar{x} \tag{6-5}$$

where

Σx is the sum of the x values
Σx^2 is the sum of the squares of the x values
Σxy is the sum of the products of the corresponding x and y values
$\bar{x}$ is the mean (average) of the x values
$\bar{y}$ is the mean (average) of the y values

Test your method using a test driver program that calculates the least-squares fit to the following 20-point input data set, and plots both the original input data and the resulting least-squares fit line:

Sample Data to Test Least Squares Fit Method

NO.	x	y	NO.	X	Y
1	–4.91	–8.18	11	–0.94	0.21
2	–3.84	–7.49	12	0.59	1.73
3	–2.41	–7.11	13	0.69	3.96
4	–2.62	–6.15	14	3.04	4.26
5	–3.78	–5.62	15	1.01	5.75
6	–0.52	–3.30	16	3.60	6.67
7	–1.83	–2.05	17	4.53	7.70
8	–2.01	–2.83	18	5.13	7.31
9	0.28	–1.16	19	4.43	9.05
10	1.08	0.52	20	4.12	10.95

16. **Correlation Coefficient of Least Squares Fit** Develop a method that will calculate both the slope m and intercept b of the least-squares line that best fits an input data set, and also the correlation coefficient of the fit. The input data points (x,y) will be passed to the method in two input arrays, x and y. The equations describing the slope and intercept of the least-squares line are given in the previous problem, and the equation for the correlation coefficient is

$$r = \frac{n(\Sigma xy) - (\Sigma x)(\Sigma y)}{\sqrt{[(n\Sigma x^2) - (\Sigma x)^2][(n\Sigma y^2) - (\Sigma y)^2]}}, \tag{6-6}$$

where

Σx is the sum of the x values
Σy is the sum of the y values
Σx^2 is the sum of the squares of the x values
Σy^2 is the sum of the squares of the y values
Σxy is the sum of the products of the corresponding x and y values
n is the number of points included in the fit

Test your method using a test driver program and the 20-point input data set given in the previous problem.

17. **The Birthday Problem** The Birthday Problem is: if there are a group of n people in a room, what is the probability that two or more of them have the same birthday? It is possible to determine the answer to this question by simulation. Write a method that calculates the probability that two or more of n people will have the same birthday, where n is a calling argument. (*Hint:* To do this, the method should create an array of size n and generate n birthdays in the range 1 to 365 randomly. It should then check to see if any of the n birthdays are identical. The method should perform this experiment at least 5000 times, and calculate the fraction of those times in which two or more people had the same birth-

day.) Write a program that calculates and prints out the probability that 2 or more of n people will have the same birthday for $n = 2, 3, \ldots, 40$. Then, plot the probability as a function of the number of people in the room. At what size group does the probability of having two people with the same birthday exceed 80%?

18. **Evaluating Infinite Series** The value of the exponential function can be calculated by evaluating the following infinite series:

$$e^x = \sum_{n=0}^{\infty} \frac{x^n}{n!}$$

Write a Java method that calculates using the first 12 terms of the infinite series. Compare the result of your method with the result of the intrinsic method `Math.exp(x)` for $x = -10, -5, -1, 0, 1, 5$, and 10.

19. **Gaussian (Normal) Distribution** Method `Math.random()` returns a uniformly-distributed random variable in the range [0,1), which means that there is an equal probability of any given number in the range occurring on a given call to the method. Another type of random distribution is the Gaussian Distribution, in which the random value takes on the classic bell-shaped curve shown in Figure 6.15. A Gaussian Distribution with an average of 0.0 and a standard deviation of 1.0 is called a *standardized normal distribution*, and the probability of any given value occurring in the standardized normal distribution is given by the equation

$$p(x) = \frac{1}{\sqrt{2\pi}} e^{-x^2/2}. \qquad (6\text{-}7)$$

It is possible to generate a random variable with a standardized normal distribution starting from a random variable with a uniform distribution in the range [−1, 1) as follows:

1. Select two uniform random variables x_1 and x_2 from the range [−1, 1) such that $x_1^2 + x_2^2 < 1$. To do this, generate two uniform random variables in the range [−1, 1), and see if the sum of their squares happens to be less than 1. If so, use them. If not, try again.
2. Then each of the values y_1 and y_2 in the equations below will be a normally-distributed random variable (note that "ln" is the symbol for the natural logarithm to the base e).

$$y_1 = \sqrt{\frac{-2\ln r}{r}}\, x_1, \qquad (6\text{-}8)$$

Figure 6.15. A Normal probability distribution

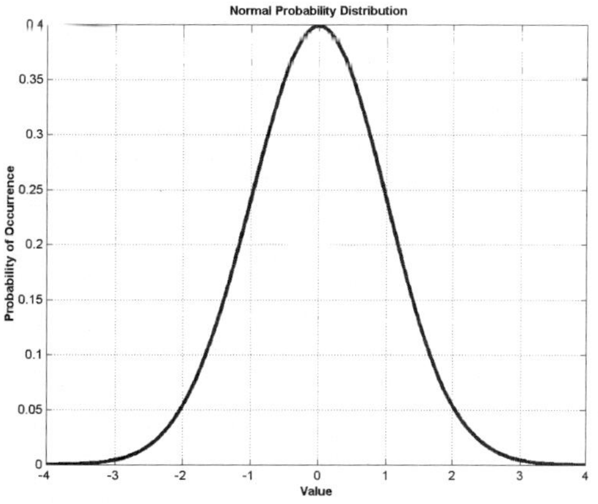

$$y_2 = \sqrt{\frac{-2\ln r}{r}}\,x_2, \qquad (6\text{-}9)$$

where
$$r = x_1^2 + x_2^2. \qquad (6\text{-}10)$$

Write a method that returns a normally-distributed random value each time that it is called. Test your method by getting 1000 random values and calculating the standard deviation. How close to 1.0 was the result?

20. Use the method developed in the previous problem to generate an array of 50,000 random samples. Test the distribution of the values returned from this method by calculating the number of values falling within the ranges –4.0 to –3.9, –3.9 to –3.8, etc., up to 3.9 to 4.0. Divide the number of values in each bin by the total number of samples (50,000), and plot the resulting distribution.

21. **Gravitational Force** The gravitational force F between two bodies of masses m_1 and m_2 is given by the equation

$$F = \frac{Gm_1m_2}{r^2}, \qquad (6\text{-}11)$$

where G is the gravitational constant (6.672×10^{-11} N m^2 / kg^2), m_1 and m_2 are the masses of the bodies in kilograms, and r is the distance between the two bodies. Write a method to calculate the gravitational force between two bodies given their masses and the distance between them. Test you method by determining the force on an 800 kg satellite in orbit 38,000 km above the Earth. (The mass of the Earth is 5.98×10^{24} kg.)

7

Classes and Object-Oriented Programming

This chapter expands on the definition and use of classes and objects. It is the most important chapter of this book, since it explains how to properly create and use objects.

As we said in Chapter 1, Java is an *object-oriented language*. Everything in Java (except for variables of primitive data types) is an object. Objects encapsulate data (*properties*) and methods (*behaviors*), hiding the details of their internal manipulation from the outside world. Thus objects inherently support **information hiding**. Since code outside of the object cannot directly "see" the data in an object, it cannot accidentally modify that data and introduce unintended bugs into the program. Objects interact with the outside world through well-defined *interfaces* defined and implemented by their associated methods.

In C, Fortran, and similar *procedural programming languages*, programmers tend to concentrate on writing procedures, which describe the actions to be performed by the program. The basic unit of programming in C is called a *function* (like a Java method); it describes how data is manipulated. The structure of C program is oriented towards the *actions* to be performed, with the result that

SECTIONS

OBJECTIVES

After reading this chapter, you should be able to:

- Understand the structure of a class.
- Understand class scope and the use of references to access instance variables and methods.
- Learn how to create and use packages.
- Understand how and why to use member access modifiers.
- Understand finalizers and the garbage collection process.
- Understand `static` variables and methods.

the various components tend to be interdependent. This interdependence makes it harder to modify the program later without introducing unintentional bugs.

By contrast, the basic units in Java are **classes**, which are descriptions of the *data* in a program. Each class contains some type of data together with the specific instructions (methods) for manipulating that data. A class isolates its data from the outside world through the specific interface formed by the class's methods. This approach to programming has three critical advantages:

- **Modularity**: The source code for a class can be written and maintained independently of the source code for other classes. As long as the interface between a class and the outside world (the class's method definitions) doesn't change, the rest of the program won't care about the details of the class's implementation. Therefore, individual classes can be modified and upgraded without "breaking" the rest of the program.
- **Reusability**: Since a class is only connected to the rest of the program through well-defined interfaces, it is easy to re-use the class in other programs. Once a class has been written once, it never needs to be re-written. This leads to a "snowball effect". The first object-oriented project that you do will take a long time, since you will have to write every object from scratch. The second and subsequent projects will become quicker and easier, because you will have an ever-expanding collection of pre-defined and debugged classes to apply to your new projects.
- **Information Hiding**: A class has a public interface that other classes can use to communicate with it. But the class can maintain private information, and code outside the class cannot accidentally modify this information. This reduces the occurrence of bugs caused by unintended side-effects.

Classes are effectively new **programmer-defined types**, with each class defining data (called **fields**) and a set of methods to manipulate the data. Recall from Chapter 1 that classes are the "blueprint" or template from which objects are created. The fields in the class serve as a template for the **instance variables** that will be created when objects are instantiated (created) from that class. Each time that an object is instantiated from a class, *a new set of instance variables is created and initialized for that object*. This fact means that if two objects are created from the same class, they will have independent copies of all the instance variables defined in the class. The instance variables in a object can be modified and manipulated through calls to the object's methods.

It is important to remember that all classes from a part of a **class hierarchy**. Every class is a **subclass** of some other class, and the class inherits both instance variables and methods from its parent class. The class can add additional instance variables and methods, and can also override the behavior of methods inherited from its parent class.

Any class above a specific class in the class hierarchy is known as a **superclass** of that class. The class just above a specific class in the hierarchy is known as the **immediate superclass** of the class. Any class below a specific class in the class hierarchy is known as a subclass of that class.

Inheritance is another major advantage of object-oriented programming; once a behavior (method) is defined in a superclass, that behavior is automatically inherited by all subclasses unless it is explicitly overridden with a modified method. Thus behaviors only need to be coded *once*, and they can be used by all subclasses. A subclass need only provide methods to implement the *differences* between itself and its parent.

The highest class in the hierarchy is class `Object`; all classes inherit behaviors from this class.

7.1 THE STRUCTURE OF A CLASS

The major components (class members) of any class are:

1. **Fields**: Fields define the instance variables that will be created when an object is instantiated from a class. Instance variables are the data encapsulated inside an object. A new set of instance variables is created each time that an object is instantiated from the class.
2. **Constructors**: **Constructors** are special methods that specify how to initialize the instance variables in an object when it is created. Constructors are easy to identify, because the have the same name as the class that they are initializing, and they do not have a return data type. Constructors can be overloaded as long as the different constructors can be distinguished by their signatures.
3. **Methods**: Methods implement the behaviors of a class. Some methods may be explicitly defined in a class, while other methods may be inherited from superclasses of the class. As we learned in Chapter 6, methods may be overloaded as long as the different methods with the same name can be distinguished by their signatures.
4. **Finalizer**: Just before an object is destroyed, it makes a call to a special method called a **finalizer**. The method performs any necessary clean-up (releasing resources, etc.) before the object is destroyed. This special method is always named `finalize`. There can be at most one finalizer in a class, and many classes do not need a finalizer at all.

The members of a class, whether variables or methods, are accessed by referring to an object created from the class using the **member access operator**, also known as the **dot operator**. For example, suppose that a class `MyClass` contains an instance variable `a` and a method `processA`. If an object of this class is named `obj`, then the instance variable in `obj` would be accessed as `obj.a`, and the method would be accessed as `obj.processA()`.

7.2 IMPLEMENTING A `Timer` CLASS

When developing software, it is often useful to be able to determine how long a particular part of a program takes to execute. This measurement can help us locate the "hot spots" in the code, the places where the program is spending most of its time, so that we can try to optimize them. This is usually done with an *elapsed time calculator*.

Figure 7.1. A class consists of fields (data), one or more constructors to initialize the data, one or more methods to modify and manipulate the data, and up to one finalizer to clean up before the object is destroyed. Note that both fields and methods may be inherited from a superclass.

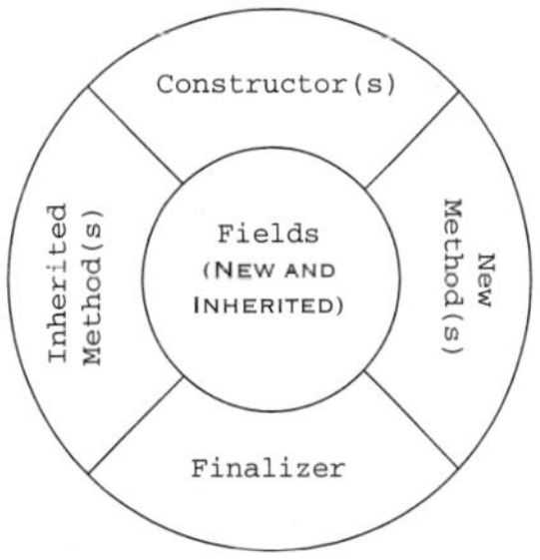

An elapsed time calculator needs to contain the following components:

1. A method to reset the timer to zero.
2. A method to return the elapsed time since the last reset.
3. An instance variable to store the time that the timer started running, for use by the elapsed time method.

This method must be able to determine the current time whenever one of its methods is called. Fortunately, the `System` class of the standard Java core API includes a method to read the current system time in milliseconds from the computer's system clock: **`System.currentTimeMillis()`**. This method will provide the current time information needed by the class.

The `Timer` class is shown in Figure 7.2.

```
/*
   Purpose:
     Object to measure the elapsed time between the most
     recent call to method resetTimer() and the call to
     method elapsedTime(). This class creates and starts
     a timer when a Timer object is instantiated, and
     returns the elapsed time in seconds whenever
     elapsedTime() is called.

   Record of revisions:
      Date        Programmer          Description of change
      ====        ==========          =====================
     4/13/98      S. J. Chapman       Original code
*/
public class Timer {

   // Define instance variables
   private double savedTime;    // Saved start time in ms

   // Define class constructor
   public Timer() {
      resetTimer();
   }

   // ResetTimer() method
   public void resetTimer() {
      savedTime = System.currentTimeMillis();
   }

   // elapsedTime() method returns elapsed time in seconds
   public double elapsedTime() {
      double eTime;
      eTime = (System.currentTimeMillis() - savedTime) / 1000;
      return eTime;
   }
}
```

Figure 7.2. The `Timer` class

This class contains a single instance variable `savedTime`, a single constructor `Timer()`, and two methods `resetTimer()` and `elapsedTime()`.

To use this class in a program, the programmer must first instantiate a `Timer` object with a statement like

```
Timer t = new Timer();
```

Figure 7.3. The statement `Timer t = new Timer();` instantiates (or creates) a new `Timer` object from the template provided by the class definition, and makes reference `t` point to that object. This object has its own unique copy of the instance variable `savedTime`

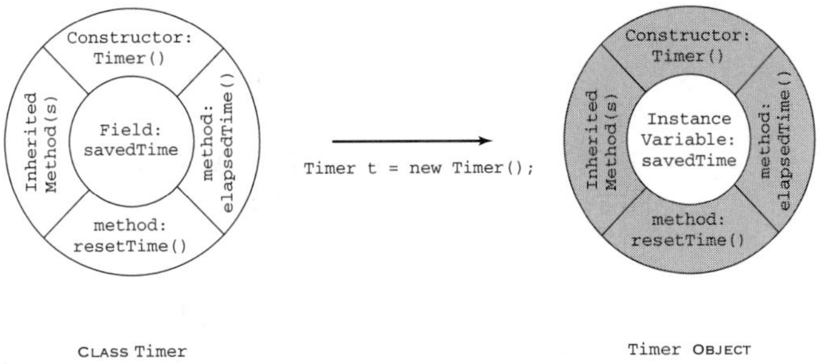

This statement defines a reference `t` to refer to a `Timer` object, and then instantiates a new `Timer` object (See Figure 7.3). When the `Timer` object is instantiated, Java automatically calls the class constructor `Timer` to initialize the object. The constructor for this class begins on line 21; it resets the elapsed time counter so that by default it measures the elapsed time since the creation of the `Timer` object.

A program can reset the elapsed timer to zero at any time by calling method `resetTimer()`, and can get the elapsed time by calling method `elapsedTime()`. An example program that uses the `Timer` object is shown in Figure 7.4. The program tests this class by measuring the time required to perform 25,000,000 iterations of a pair of nested `for` loops.

```
/*
   Purpose:
      Class to test the operation of Timer.

   Record of revisions:
      Date         Programmer          Description of change
      ====         ==========          =====================
     4/13/98       S. J. Chapman       Original code
*/
public class TestTimer {

   // Define the main method to test this object
   public static void main(String[] args) {

      // Declare variables, and define each variable
      double arr[];           // Data array to sort
      int i, j, k;            // Loop index

      // Instantiate a Timer object
      Timer t = new Timer();

      // Start the timer running
      t.resetTimer();

      // Waste some time
      for ( i = 1; i <= 5000; i++ ) {
         for ( j = 1; j <= 5000; j++ ) {
            k = i + j;
         }
      }

      // Read and display elapsed time
      System.out.println( "Time = " + t.elapsedTime() + " s" );
   }
}
```

Figure 7.4. A program to test the `Timer` class

When this program is executed on my Pentium 133 MHz PC, the results are as follows:

```
D:\book\java\chap7>java TestTimer
Time = 0.401 s
```

The measured time will of course differ on computers of different speeds.

Note that this class saves the time of the last reset in an instance variable `savedTime`. Each time that an object is instantiated from a class, it receives its own copy of all instance variables defined in the class. This fact means that many `Timer` objects could be instantiated and used simultaneously in a program, and *they will not interfere with each other*, because each timer has its own private copy of the instance variable `savedTime`.

Notice that each class member in the program in Figure 7.2 is declared with either a **`public`** or **`private`** keyword. These keywords are known as **member access modifiers**. Any instance variable or method definition declared with the `public` member access modifier can be accessed any time that a program has access to an object of class `Timer`. Any instance variable or method declared with the `private` member access modifier is only accessible to methods of the object in which it is defined.

In this case, the instance variable `savedTime` is declared `private`, so it cannot be seen or modified by any method outside of the object in which it is defined. Since no method outside of `Timer` can see `savedTime`, it is not possible for some other method to accidentally modify the value stored there and so mess up the elapsed time measurement. The only way that a program can utilize the elapsed time measurement is through the `public` methods `resetTimer()` and `elapsedTime()`.

Every instance variable and method definition in a class should be preceded by an explicit member access modifier. It is good practice to group `public` and `private` member access modifiers together in a class for clarity and readability.

GOOD PROGRAMMING PRACTICES

Every instance variable and method definition in a class should be preceded by an explicit member access modifier.

The instance variables of a class are normally declared `private` and the methods of a class are normally declared `public`, so that the methods form an interface with the outside world, hiding the internal behavior of the class from any other parts of the program. This approach has many advantages, since it makes programs more modular. For example, suppose that we have written a program that makes extensive use of `Timer` objects. If necessary, we could completely re-design the internal behavior of the `Timer` class, and the program will continue to work properly as long as we have not changed the parameters or returned values from methods `resetTimer()` and `elapsedTime()`. This **public interface** isolates the internals of the class from rest of the program, making incremental modifications easier.

GOOD PROGRAMMING PRACTICES

The instance variables of a class should normally be declared `private`, and the class methods should be used to provide a standard interface to the class.

There are some exceptions to this general rule. Many classes contain `private` methods that perform specialized calculations in support of the `public` methods of the class. These are called **utility methods**, and since they are not intended to be called directly by users, they are declared with the `private` member access modifier. Also, some classes contain `public` instance variables that can be directly accessed by a user. This is relatively rare, but it is sometimes done for efficiency reasons.

7.3 CLASS SCOPE

The instance variables and methods of a class all have **class scope**, meaning that they can be accessed from any method within the class by name. For example, any method in class `Timer` can access the instance variable `savedTime` just by using its name. Similarly, any method in a class can call any other method in the same class by simply using its name[1].

Outside of a class's scope, those instance variables and methods that have been declared `public` can still be accessed, but *not* directly. Instead, they have to be accessed by reference to the object in which they are defined. Primitive data type instance variables would be accessed as `Reference.primitiveVariableName`, and object instance variables and methods would be accessed as `Reference.objectName` or `Reference.methodName`. For example, the program in Figure 7.4 created a `Timer` object using a reference `t`, and then invoked method `elapsedTime()` using that reference: `t.elapsedTime()`.

By contrast, variables defined within a method have **block scope**, with the block being the method body. These variables cannot be accessed from any other method, even if it is in the same class.

It is possible for a variable in a method to have the *same name* as an instance variable or method in the class in which the method is defined. If this happens, then the instance variable is *hidden* from the method by the local variable, and the method cannot access it directly even though the variable has class scope.

For example, Figure 7.5 shows a portion of a Java class defining a point in two-dimensional space. This portion of the class contains two instance variables `x` and `y`, two constructors, and a method to set the value of the point.

Notice that method `setPoint()` includes two local variables `x` and `y`, which have the same names as the class instance variables `x` and `y`. The instance variables have *class scope*, so they are accessible in any method within the class. However, the local variables in `setPoint()` that have the same names as the instance variables prevent the method from just using the names `x` and `y` to access the instance variables. Instead, *the instance variables are referred to as* `this.x` *and* `this.y`. The reference **`this`** always refers to the *current object*, which is the object within which the reference appears. Therefore, `this.x` refers to the instance variable `x` defined in the current object, and `this.y` refers to the instance variable `y` defined in the current object.

GOOD PROGRAMMING PRACTICE

If the name of a class member is hidden by a local variable name in a method within the same class, use the `this` reference to refer to the class member.

[1]This statement is not quite true for static methods, as we shall see later in the chapter.

```
public class Point {
   // Define instance data
   private double x;     // x position of point
   private double y;     // y position of point

   // Define constructors
   public void Point() {
      x = 0;
      y = 0;
   }

   public void Point(double x, double y) {
      this.x = x;
      this.y = y;
   }

   public void setPoint(double x, double y) {
      this.x = x;
      this.y = y;
   }

    ...
    ...
    ...
}
```

Figure 7.5. Partial definition of a class, illustrating access to class instance variables having the same name as method variables.

7.4 TYPES OF METHODS

Since instance variables are usually hidden within a class, the only way to work with them is through the interface formed by the class's methods. The methods are the public face of the class, providing a standard way to work with the information while hiding the unnecessary details of the implementation from the user.

A class's methods must perform certain common "housekeeping" functions, as well as the specific actions required by the class. These housekeeping functions fall into a few broad categories, and they are common to most classes regardless of their specific purpose. A class must usually provide a way to store data into its instance variables, read data from its instance variables, test the status of its instance variables, and display the contents of its instance variables in a human-readable form.

Since the instances variables in a class cannot be used directly, classes must define methods to store data into the instance variables and to read data from them. By Java convention, the names of methods that store data begin with "`set`" and are called **set methods**, while the names of methods that read data begin with "`get`" and are called **get methods**.

Set methods take information from the outside world and store the data into the class's instance variables. In the process, they *should also check the data for validity and consistency.* This checking prevents the instance variables of the class from being set into an illegal state.

For example, suppose that we have created a class `Date` containing instance variables `day` (with a range of 1–31), `month` (with a range of 1–12), and `year` (with a range of 1900–2100). If these instance variables were declared `public`, then any method in the program could modify them directly. For example, assume that a `Date` object was declared as

```
Date d1 = new Date();
```

With this declaration, any method in the program could directly set the day to an illegal value.

```
d1.day = 32;
```

Set methods and private instance variables prevent this sort of illegal behavior by testing the input parameters. If the parameters are valid, the method stores them in the appropriate instance variables. If the parameters are invalid, the method either modifies the inputs to be legal, or by raises an exception if it is not possible to modify the illegal values into acceptable ones.

Get methods are used to retrieve information from the instance variables and to format it properly for presentation to the outside world. For example, our `Date` class might include methods `getDay()`, `getMonth()`, and `getYear()` to recover the day, month, and year respectively.

Another type of method tests for the truth or falsity of some condition. These methods are called **predicate methods**. These methods typically begin with the word `is`, and they return a `boolean` (true/false) result. For example, a `Date` class might include a method `isLeapYear()`, which would return `true` if the specified year is a leap year, and `false` otherwise. In could also include methods like `isEqual()`, `isEarlier()`, and `isLater()` to compare two dates chronologically.

GOOD PROGRAMMING PRACTICE

1. Use set methods to check the validity and consistency of input data before it is stored in an object's instance variables.
2. Define predicate methods to test for the truth or falsity of conditions associated with any classes you create.

Another very important function of methods is to display the contents of the object in human-readable form. This is accomplished with a special method called `toString()`. The `toString()` method is defined for every Java class, since it is defined in the `Object` class, and every class is a subclass of `Object`. It is used to convert the data stored in an object into a form suitable for printing out. Every class should include a customized `toString()` method to properly format its data for display. The `toString()` method is automatically called whenever an object is concatenated with a string. For example, in the statement

```
System.out.println( "Date = " + d1);
```

the data in object `d1` is converted into a string and printed on the standard output stream. If `d1` is an object of the `Date` class, and if the `Date` class defines a `toString()` method, then that method will be used to convert the date into a string. The result might be something like

```
Date = 1/5/1999
```

GOOD PROGRAMMING PRACTICE

Override the `toString()` method of any classes the you define to create a reasonable display of the class's data.

EXAMPLE 7-1: CREATING A Date CLASS

We will illustrate the concepts described in this chapter by creating a Date class designed to hold and manipulate dates on the Gregorian calendar.

This class should be able to hold the day, month, and year of a date in instance variables that are protected from outside access. The class must include constructors to create dates, set and get methods to retrieve the stored information, predicate methods to recover information about date objects and to allow two Date objects to be compared, and a toString method to allow the information in a Date object to be displayed easily.

SOLUTION

The Date class will need three instance variables, day, month, and year. They will be declared private to protect them from direct manipulation by outside methods. The day variable should have a range of 1–31, corresponding to the days in a month. The month variable should have a range of 1–12, corresponding to the months in a year. The year variable will be greater than or equal to zero.

We will define two constructors for our class. One constructor will have no input parameters, and will initialize the date to January 1, 1900. The other constructor will have a day, month, and year as input arguments, and will initialize the date to the appropriate values.

We will also define a method setDate() to set a new date into a Date object, and three methods getDay(), getMonth(), and getYear() to return the day, month, and year from a given Date object.

The supported predicate methods will include isLeapYear() to test if a year is a leap year. This method will use the leap year test described in Example 4-2. In addition, we will create three methods isEqual(), isEarlier(), and isLater() to compare two Date objects. (For the purposes of comparison, an expression like d1.isEarlier(d2) should return true if d1 is an earlier date than d2.) Finally, method toString() will format the date as a string in the normal US style: dd/mm/yyyy.

The resulting class is shown in Figure 7.6.

```
/*
   Purpose:
     This class stores and manipulates dates on the
     Gregorian calendar.  It implements constructors,
     set methods, get methods, and predicate methods,
     and overrides the toString method.

     Method list:
        Date()                    Date constructor
        Date(day,month,year)      Date constructor
        setDate(day,month,year)   Set Date
        getDay()                  Get day
        getMonth()                Get month
        getYear()                 Get year
        isLeapYear()              Test for leap year
        isEqual()                 Test for equality
        isEarlier()               Is chronologically earler
        isLater()                 Is chronologically later
        toString()                Convert to string for display

   Record of revisions:
      Date       Programmer          Description of change
      ====       ==========          =====================
    4/16/98     S. J. Chapman        Original code
```

Figure 7.6. *(cont.)*

```
*/
public class Date {

   // Define instance variables
   private int year;        // Year (0 - xxxx)
   private int month;       // Month (1 - 12)
   private int day;         // Day (1 - 31)

   // Default constructor is January 1, 1900
   public Date() {
      year = 1900;
      month = 1;
      day = 1;
   }

   // Constructor for specified date
   public Date(int day, int month, int year) {
      setDate( day, month, year );
   }

   // Method to set a date
   public void setDate(int day, int month, int year) {
      this.year  = year;
      this.month = month;
      this.day   = day;
   }

   // Method to get day
   public int getDay() {
      return day;
   }

   // Method to get month
   public int getMonth() {
      return month;
   }

   // Method to get year
   public int getYear() {
      return year;
   }

   // Method to check for leap year
   public boolean leapYear() {
      boolean leapYear;
      if ( year % 400 == 0 )
         leapYear = true;
      else if ( year % 100 == 0 )
         leapYear = false;
      else if ( year % 4 == 0 )
         leapYear = true;
      else
         leapYear = false;
      return leapYear;
   }
```

Figure 7.6. *(cont.)*

```
// Method to check for equality
public boolean isEqual( Date d ) {
   boolean equal;
   if ( year == d.year && month == d.month && day == d.day )
      equal = true;
   else
      equal = false;
   return equal;
}

// Method to check if the date stored in this
// object is earlier than the Date d.
public boolean isEarlier( Date d ) {
   boolean earlier;

   // Compare years
   if ( year > d.year )
      earlier = false;
   else if ( year < d.year )
      earlier = true;
   else {

      // Years are equal.  Compare months
      if ( month > d.month )
         earlier = false;
      else if ( month < d.month )
         earlier = true;
      else {

         // Months are equal.  Compare days.
         if ( day >= d.day )
            earlier = false;
         else
            earlier = true;
      }
   }
   return earlier;
}

// Method to check if the date stored in this
// object is later than the Date d.
public boolean isLater( Date d ) {
   boolean later;

   // Compare years
   if ( year > d.year )
      later = true;
   else if ( year < d.year )
      later = false;
   else {

      // Years are equal.  Compare months
      if ( month > d.month )
         later = true;
      else if ( month < d.month )
         later = false;
      else {
```

Figure 7.6. *(cont.)*

```
            // Months are equal.  Compare days.
            if ( day > d.day )
               later = true;
            else
               later = false;
         }
      }
      return later;
   }

   // Method to convert a date to a string.
   public String toString() {

      return (month + "/" + day + "/" + year);
   }
}
```

Figure 7.6. The `Date` class

We must create a test driver class to test the `Date` class. Such a class is shown in Figure 7.7. Class `TestDate` instantiates four `Date` objects, and initializes them using both constructors. It then exercises all of the methods defined in the class (note that the `toString()` method is implicitly exercised by the `System.out.println()` statements).

```
/*
   Purpose:
     This class tests the Date class.

   Record of revisions:
      Date       Programmer          Description of change
      ====       ==========          =====================
     4/16/98    S. J. Chapman        Original code
*/
public class TestDate {

   // Define the main method to test class Date
   public static void main(String[] args) {

      // Declare variables, and define each variable
      Date d1 = new Date(4,1,1996);    // Date 1
      Date d2 = new Date(1,3,1998);    // Date 2
      Date d3 = new Date();            // Date 3
      Date d4 = new Date();            // Date 4

      // Set d3
      d3.setDate(3,1,1996);

      // Print out dates
      System.out.println ("Date 1 = " + d1);
      System.out.println ("Date 2 = " + d2);
      System.out.println ("Date 3 = " + d3);
      System.out.println ("Date 4 = " + d4);

      // Check isLeapYear
      if ( d1.isLeapYear() )
         System.out.println (d1.getYear() + " is a leap year.");
```

Figure 7.7. *(cont.)*

```
        else
           System.out.println (d1.getYear() + " is not a leap year.");
        if ( d2.isLeapYear() )
           System.out.println (d2.getYear() + " is a leap year.");
        else
           System.out.println (d2.getYear() + " is not a leap year.");

        // Check isEqual
        if ( d1.isEqual(d3) )
           System.out.println (d3 + " is equal to " + d1);
        else
           System.out.println (d3 + " is not equal to " + d1);

        // Check isEarlier
        if ( d1.isEarlier(d3) )
           System.out.println (d1 + " is earlier than " + d3);
        else
           System.out.println (d1 + " is not earlier than " + d3);

        // Check isLater
        if ( d1.isLater(d3) )
           System.out.println (d1 + " is later than " + d3);
        else
           System.out.println (d1 + " is not later than " + d3);

   }
}
```

Figure 7.7. Class `TestDate` to test the `Date` class

When this program is executed, the results are as follows:

```
D:\book\java\chap7>java TestDate
Date 1 = 1/4/1996
Date 2 = 3/1/1998
Date 3 = 1/3/1996
Date 4 = 1/1/1900
1996 is a leap year.
1998 is not a leap year.
1/3/1996 is not equal to 1/4/1996
1/4/1996 is not earlier than 1/3/1996
1/4/1996 is later than 1/3/1996
```

Note that the date strings are being written out in the order `month/day/year`. From the test results, this class appears to be functioning correctly.

This class works, but it could be improved. For example, there is no validity checking performed on the input values in the `setDate()` method, and the `toString()` method could be modified to produce dates with explicit month names such as "January 1, 1900".

PROFESSIONAL SUCCESS: RETURNING MULTIPLE VALUES FROM JAVA METHODS

In other programming languages such as Fortran, a single subroutine (which corresponds to a Java method) can calculate many different results and return them all to the calling program. Unfortunately, Java methods are designed to only return a *single value* to the calling program. This limitation can be quite severe, requiring multiple methods to perform the same function that a single subroutine can perform in another language.

For example, suppose we wanted to write a function to calculate the average, standard deviation, minimum, and maximum in a data set. In Fortran, we could create a single subroutine `calc_stats`, and that subroutine could calculate and return all four values. In contrast, Java would require four different methods to return the four values, resulting in duplicate calculations and taking more time.

It is possible to get around this limitation of Java by creating a class containing the four instance variables `average`, `stdDev`, `min`, and `max`, and designing the `calcStats` method to return an object of that class. The method is returning a *single* object, but it is actually calculating and returning all four values!

```
public class StatValues {
   double average;
   double stdDev;
   double min;
   double max;

// Constructor that does nothing
   public StatValues()
}
```

and the statements shown below define a method `calcStats` that returns a `StatValues` object.

```
public StatValues calcStats( double[] data ) {
   StatValues result;
   ...
   (Insert calculations here)
   ...
   return result;
}
```

7.5 STANDARD JAVA PACKAGES

Every program that we have written has used classes and methods imported from the Java API. These pre-defined classes make programming much easier by allowing us to take advantage of other people's work instead of having to "reinvent the wheel" each time that we set out to write a program.

The Java API consists of literally hundreds of pre-defined classes containing thousands of pre-defined methods. These classes and methods are organized into related groups called **packages**. A package consists of a set of classes and methods that share some related purpose. For example, we could use classes from the `java.io` package to read data into our programs. This package is a set of classes that allow programs to input or output data.

The standard Java API packages as of Java 1.1.x are summarized in Table 7-1. A large number of additional packages have been added in Java 2, mostly to do with graphics. We will discuss some of the additional packages in Chapters 8 and 9.

TABLE 7-1 The Java API Packages

JAVA API PACKAGE	EXPLANATION
`java.applet`	The Java Applet Package. This package contains the `Applet` class and several interfaces that allow programmers to create applets and control their interactions with a browser.
`java.awt`	The Abstract Windowing Toolkit (AWT) Package. This package contains the classes and interfaces required to create Graphical User Interfaces. The term "abstract" is applied to this packages because in can create GUI windows on any type of computer regardless of the underlying operating system type.
`java.awt.datatransfer`	The Java Data Transfer Package. This package contains classes and interfaces that allow a program to transfer data between a Java program and a computer's clipboard (a temporary storage area used for cut and paste operations).
`java.awt.event`	The Java AWT Event Package. This package contains classes and interfaces that support event handling for GUI components.
`java.awt.image`	The Java AWT Event Package. This package contains classes and interfaces that enable storing and manipulating images in a program.
`java.awt.peer`	The Java AWT Peer Package. This package contains interfaces that allow Java's GUI components to interact with their platform-specific versions (for example, a button is actually implemented differently on a Macintosh than it is on a Windows or X-Windows machine). This package should never be used directly by Java programmers.
`java.beans`	The Java Beans Package. This package contains classes and interfaces that enable programmers to create reusable software components.
`java.io`	The Java Input/Output Package. This package contains classes that enable programs to input and output data.
`java.lang`	The Java Language Package. This package contains the basic classes and interfaces required to make Java programs work. It is automatically imported into all Java programs.
`java.lang.reflect`	The Java Core Reflection Package. This package contains classes and interfaces that allow a program to discover the accessible methods and variables of a class dynamically during the execution of a program.
`java.net`	The Java Networking Package. This package contains classes that enable a program to communicate over a network (the Internet or an intranet).

`java.rmi`
`java.rmi.dgc`
`java.rmi.registry`
`java.rmi.server`
	The Java Remote Method Invocation Packages.
	These packages contains classes and interfaces that enable a programmer to create distributed Java programs. A program can use RMI to call methods in other programs, whether they are located on the same computer or on another computer somewhere else on the network.

TABLE 7-1 (continued)

JAVA API PACKAGE	EXPLANATION
`java.security` `java.security.acl` `java.security.interfaces`	The Java Security Packages These packages contains classes and interfaces that enable a program to encrypt data, and to control the access privileges provided to a Java program for security purposes.
`java.sql`	The Java Database Connectivity Package. This package contains classes and interfaces that enable a Java program to communicate with a database.
`java.text`	The Java Text Package. This package contains classes and interfaces that enable a Java program to manipulate numbers, dates, characters, and strings. This package provides many of Java's internationalization capabilities. For example, date and time strings can be automatically displayed in the proper format for the country in which a Java program is running.
`java.util`	The Java Utilities Package. This package contains classes and interfaces that perform important utility functions in a program. Examples include: date and time manipulations, random number generation, storing an processing large amounts of data, and certain string manipulations.
`java.util.zip`	The Java Utilities Zip Package. This package contains classes and interfaces that allow a Java program to create and read compressed archives called Java Archive (JAR) files. These archives can hold pre-compiled `.class` files as well as audio and image information.

A Java API class must be **imported** before it can be used in a program. There are two ways to import a Java class. The most common way is to include an **`import`** statement in the program before the class definition in which the API class will be used. For example, to use the `Arrays` class from the `java.util` package, we would include the line

```
import java.util.Arrays; // import java.util.Array class
```

at the start of the class in which it is used. It is also possible to import all of the classes and interfaces in an entire package with a statement of the form

```
import java.util.*; // import entire java.util package
```

The package `java.lang` contains classes that are fundamental to the operation of all Java programs, and it is automatically imported into every program. No `import` statement is required for this package.

A detailed description of all of the classes in the Java API can be found in the Java Development Kit (JDK) documentation. The JDK may be downloaded for free from http://java.sun.com, and the documentation may be viewed with any Web browser.

7.6 CREATING YOUR OWN PACKAGES

One of the great strengths of Java is the ability to re-use classes written for one project on other projects. Packages are a very useful way to bundle groups of related classes and to make them easy to re-use. If a set of useful classes is created and placed in a package, then those classes can be re-used on other projects by simply importing them into your new programs in the same manner as you import the packages built into the Java API. In fact, we have already been doing this with the `chapman.io`, `chapman.graphics`, and `chapman.math` packages, which were written for this book.

GOOD PROGRAMMING PRACTICE

Create packages containing of groups of related classes to make it easy to re-use those classes in other programs.

It is easy for a programmer to create his or her own packages. Any class that a programmer writes can be included in a package by adding a **package statement** to the file defining the class. For example, the class `Class1` shown in Figure 7-8 can be placed in package `chapman.testpackage` by including a `package` statement in the file before the beginning of the class definition. This statement indicates that the class defined in this file is a part of the specified package. (Note that the `package` and `import` statements are the only two statements in Java that can occur outside of a class definition.)

```
// Class to test creating and using a package
package chapman.testpackage;    // Place in testpackage
public class Class1 {

   // Method mySum
   public int mySum(int a, int b) {
      return a + b;
   }
}
```

Figure 7.8. A sample class containing a `package` statement

When a class containing a `package` statement is compiled, the resulting `.class` file is automatically placed in the directory indicated by the `package` statement. The series of names separated by periods in the statement is actually the directory hierarchy leading to the directory in which the file will be placed. Thus, the compiled output of this file will be placed in directory `<classroot>\chapman\testpackage` on a PC system, or directory `<classroot>/chapman/testpackage` on a Unix system. The `<classroot>` is the starting directory for the package directory structure on a particular computer, as described in Section 7.6.1.

When a class containing a `package` statement is compiled with the compiler in the Java Development Kit, the root directory for the package structure `<classroot>` must be specified using the `-d` compiler option. For example, the statement

```
javac -d c:\packages Class1.java
```

specifies that the root directory of the package structure is `c:\packages`. The root directory must already exist when the file is compiled. When this command is executed, the file `Class1.java` will be compiled, and the resulting `.class` file will be placed in directory `c:\packages\chapman\testpackage`. If necessary, the subdirectories will be created to hold the package. Other classes can be added to this package in the same way.

7.6.1 Setting the Class Path

Before a user-defined package can be used, the root directory of the package directory structure `<classroot>` must be inserted into your computer's **class path**. The class path is defined by an environment variable called `CLASSPATH`. When the Java compiler or Java interpreter needs to locate a class in a package, it looks at each directory defined in `CLASSPATH` to see if directory tree containing the package can be found there.

On a PC running Windows 95/98 the class path can be defined by adding the following line to the `autoexec.bat` file:

```
set CLASSPATH=c:\packages
```

If the `CLASSPATH` variable already exists in the file, add a semicolon followed by `c:\packages` to the end of the existing path. You will need to restart your computer before this change takes effect.

On a Windows NT computer, the class path is set through the System option of the Control Panel. On a Unix machine, the class path is set in different ways depending on the particular shell that you are using. If necessary, see you instructor for help in setting the class path on these computers.

7.6.2 Using User-Defined Packages

Once a package is created and the class path has been set, the package can be imported into any class desiring to use the package. For example, the class in Figure 7.9 imports package `testpackage`, and then uses `Class1` from it.

```
// Class to test using a package
import chapman.testpackage.*;
public class TestClass1 {

   // Define the main method to test Class1
   public static void main(String[] args) {

      // Declare variables
      int i = 8, j = 6;

      // Instantiate a Class1 object
      Class1 c = new Class1();

      // Use the object
      System.out.println("i + j = " + c.mySum(i,j));
   }
}
```

Figure 7.9. Importing and using a class from a user-defined package

This class can be compiled and executed just like any other class. The computer automatically searches the specified class path to locate `testpackage` and import it.

```
C:\book\java\chap7>javac TestClass1.java
C:\book\java\chap7>java TestClass1
i + j = 14
```

7.7 MEMBER ACCESS MODIFIERS

There are four types of member access modifiers in Java: `public`, `private`, `protected`, and package. The first three are defined by explicit keywords in an instance variable or method definition, and the last is the default access that results if no access modifier is explicitly selected.

We have already seen the two member access modifiers `public` and `private`. A class member that is declared `public` may be accessed by any method anywhere within a program. A class member that is declared `private` may only be accessed by methods within the same class as the class member.

If no access modifiers are included in a definition, then the class member has **package access**. The member may be accessed by methods in all classes within the same package (that is, within the same directory) as the class in which the member is defined, but not by methods in other classes. This type of access is often convenient, because the classes in a package are usually related and must work closely together, while the details of their interactions will be hidden from methods outside the package.

This type of access is illustrated in Figure 7.10. Note that we are able to modify the instance variables `a.x` and `a.s` directly, because classes `TestPackageAccess` and `AccessTest` both reside in the same directory.

```
// Class to test package access
public class TestPackageAccess {

   // Define the main method to test Class1
   public static void main(String[] args) {

      // Instantiate a AccessTest object
      AccessTest a = new AccessTest();

      // Write out the value of the instance variable
      System.out.println(a.toString());

      // Modify the instance variables of class
      // AccessTest directly
      a.x = 3;
      a.s = "After: ";

      // Write out the value of the instance variable
      System.out.println(a.toString());
   }
}

// Define class AccessTest. Note that this class has
// package access.
class AccessTest {

   // Instance variables
   int x = 1;
   String s = "Before: ";

   // Method toString()
   public String toString() {
   return (s + x);
   }
}
```

Figure 7.10. A program illustrating package access

When this program is compiled and executed, the results are as follows:

```
C:\book\java\chap7>javac TestPackageAccess.java
C:\book\java\chap7>java TestPackageAccess
Before: 1
After: 3
```

Package access is inherently dangerous, since the methods of one class may be modifying the members of another class directly. If the class whose members are being modified directly is changed, then that modification may "break" the other classes that are accessing the class members directly. Thus, package access weakens the information hiding and the inherent modifiability of Java classes.

On the other hand, package access is efficient, because the classes' instance variables are being used and set without going through the overhead of `get` and `set` methods. This reduction in overhead allows the whole package to execute faster, while still hiding implementation details from the outside world. If a package only contains a few closely-related classes that can be modified and maintained together, then package access can provide an acceptable compromise between speed and safety.

GOOD PROGRAMMING PRACTICE

Package access is inherently dangerous, since it weakens information hiding in Java. Do not use it unless it is absolutely necessary for performance reasons. If it is necessary, restrict the package to a few closely-related classes.

The final access modifier is **`protected`**. Members declared with a `protected` modifier can be accessed by methods in all classes within the same package as the class in which the member is defined, and also by all *subclasses* of the class.

PRACTICE!

This quiz provides a quick check to see if you have understood the concepts introduced in Sections 7.1 through 7.7. If you have trouble with the quiz, reread the section, ask your instructor, or discuss the material with a fellow student. The answers to this quiz are found in the back of the book.

1. Name the major components of a class, and describe their purposes.
2. What types of member access modifiers may be defined in Java, and what access does each type give? What member access modifier should normally be used for instance variables? for methods?
3. What is the difference between class scope and block scope?
4. What happens in a program if a method contains a local variable with the same name as an instance variable in the method's class?
5. What statement(s) do you have to include in a program before you can use classes in Java API packages other than `java.lang`?
6. How do you create a user-defined package? How do you use the package?
7. What is the function of the `CLASSPATH` environment variable?
8. Explain the difference between `public`, `private`, `protected`, and package access.

7.8 FINALIZERS AND GARBAGE COLLECTION

Just before an object is destroyed, it makes a call to a special method called a **finalizer**, which performs any necessary clean-up (releasing resources, etc.) before the object is destroyed. This special method is always named `finalize`. There can be at most one finalizer in a class, and many classes do not need a finalizer at all.

When a class is instantiated with the `new` operator, the class constructor creates a new object. The constructor allocates memory for defined instance variables and objects, and may acquire other system resources such as open files, network sockets, etc. These resources are used by the object for as long as it is needed in the program.

When an object is no longer needed, it should be destroyed and its resources should be returned to the system for re-use. This function is performed automatically in Java by the **garbage collector**. The garbage collector is a low-priority thread[2] within the Java interpreter that normally runs in the background whenever the interpreter is executing. It constantly scans the list of objects created by the program. Any object that no longer has a reference pointing to it is a candidate for garbage collection, *because once all references to an object are gone, the object can no longer be used in any way by the program*. When the garbage collector spots such an object, it makes a call to the object's `finalize` method, and then destroys the object, releasing its memory to the system.

A program can force the garbage collector to run at high priority by making an explicit call to the `System.gc()` method. We might want to do this to ensure that garbage collection occurs at a specific time in a program.

The call to the object's `finalize` method is an opportunity for the object to close any files that it might have open and to perform any other required terminal housekeeping. Once the `finalize` method completes, the object is destroyed by the garbage collector.

The `finalize` method is normally declared `protected`, so that it is protected against accidental calls from outside methods. We will illustrate the use of `finalize` methods in Example 7-2.

7.9 STATIC CLASS MEMBERS

It is possible for both variables and methods in a class to be declared `static`. Such variables and methods have special properties, which are explained in this section.

7.9.1 Static Variables

Each object of a class has its own copy of all of the instance variables defined in the class. If a class defines an instance variable `x`, and two objects `a` and `b` are created from the class, then `a.x` and `b.x` will be two separate variables.

However, it is sometimes useful to have a particular variable in a class shared by all of the objects created from the class. If a variable is declared **`static`**, then one copy of that variable will be created the first time that a class is loaded, and that copy will remain in existence until the program stops running. Any objects instantiated from that class will share the single copy of that variable. Static variables are also known as **class variables**.

This concept is illustrated in Figure 7.11, which shows a class `A` defining two fields `x` and `y` and a `static` or class variable `z`. Two objects `a1` and `a2` are instantiated from this class. These two objects contain separate instance variables `x` and `y`, but they share a single copy of `static` variable `z`.

[2]A *thread* or *thread of execution* is a part of a program that can execute in parallel with other parts of the same program. Java supports multithreading as a part of its basic structure, but the discussion of that topic is beyond the scope of this book.

Figure 7.11. Two objects `a1` and `a2` instantiated from a single class `A`. These two objects have their own instance variables `x` and `y`, but share a single copy of the static variable `z`.

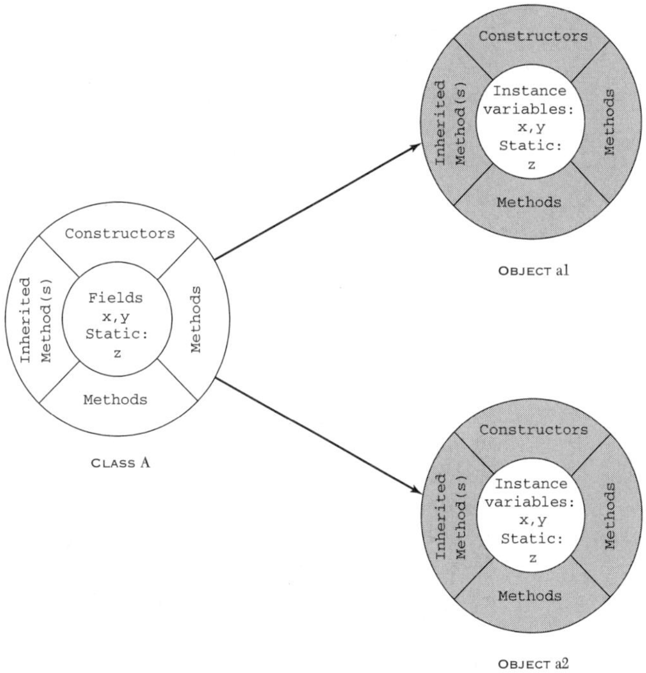

`Static` variables are useful for keeping track of global information such as the number of objects instantiated from a class, or the number of those objects still surviving at any given time. They are also useful for defining single copies of final variables that will be shared among all objects of the class. For example, the speed of light in a vacuum is $c = 2.99792458 \times 10^8$ meters per second. This value could be declared in a class as

```
static final double C = 2.99792458e8;
```

and a single copy of this constant will be created and shared among all of the objects instantiated from the class.

GOOD PROGRAMMING PRACTICE

Use `static` class variables to share a single copy of a variable among all of the objects instantiated from the class.

The *duration* of a `static` variable is different from the duration of an instance variable. Instance variables are created each time that an object is instantiated, and destroyed when the object is destroyed by the garbage collector. In contrast, `static` variables are created as soon as the class is loaded into memory, and persist until program execution ends.

Although `static` variables are global in the sense that they are shared among all of the objects instantiated from a class, they still have *class scope*. They are only visible outside of the class by reference to the class in which they are defined.

If a `static` variable is declared `public`, it may be accessed through any object of the class, or directly through the class name using the dot operator. For example, suppose that a class `Class1` defines a `public static` variable `z`, and an object `c` is instantiated from that class. Then the variable `z` can be accessed by other objects in the program either by `Class1.z` or by `c.z`.

Note that *the* `static` *variables in a class can be accessed without first creating an object from that class*. A good example of a `static` variable is `out` in the `System` class. We have used `System.out` in all of our programs since Chapter 1 without having to first instantiate a `System` object.

7.9.2 Static Methods

`Static` methods are methods that are declared `static` within a class. These methods can directly access the `static` variables in the class, but *cannot* directly access instance variables. Static methods are also known as **class methods**.

`Static` methods are commonly used to perform calculations that are independent of any instance data that might be defined in a class. For example, the method `sqrt()` in the `Math` class calculates the square root of any value passed to it, and returns the result. This method is independent of any data stored in the `Math` class, so it is declared `static`.

`Static` methods may be accessed by reference to a class's name without first creating an object from the class. Thus, we are able to use `Math.sqrt()` in any program without first instantiating an object from the `Math` class.

Every method in Java must be defined within a class, but some methods like `sqrt()`, `sin()`, `cos()`, etc. are not closely tied to any data within the class. These methods are usually declared `static` so that they can be accessed without having to instantiate an object first.

A very famous `static` method is `main`, the starting point of any Java application. The `main` method must be `static` so that it can be executed without first instantiating any object.

GOOD PROGRAMMING PRACTICE

Utility methods that perform functions independent of the instance data within a class may be declared `static` to make them easier to access and use.

EXAMPLE 7-2: USING STATIC VARIABLES AND FINALIZERS

To illustrate the use of `static` variable and finalizers, we will create a class containing two `static` variables. One of the variables will keep track of the number of times that an object is instantiated from the class, and the other one will keep track of the number of those objects still in existence. We will make these variables `private` so that no outside methods can tamper with them, and will provide `static` get methods to recover their current values when necessary.

In addition, the class will contain a single instance variable, which will contain an integer value, and a method to recover that value.

The resulting class is shown in Figure 7.12.

```
/*
   Purpose:
     This class illustrates the use of static variables to keep track of
      the number of objects created from the class.

   Record of revisions:
      Date       Programmer        Description of change
      ====       ==========        =====================
     4/20/98    S. J. Chapman      Original code
*/
public class Widget {

   // Define class variables
   private int value;           // Instance variable containing a value
   private static int created;  // Total number of objects created
   private static int alive;    // Total number of objects still alive

   // Constructor
   public Widget(int value) {
      this.value = value;
      created++;
      alive++;
   }

   // Method to get number of Widgets created
   public static int getCreated() {
      return created;
   }

   // Method to get number of Widgets still alive
   public static int getAlive() {
      return alive;
   }

   // Method to get value of object
   public int getValue() {
      return value;
   }

   // Finalizer
   protected void finalize() {
      alive--;
      System.out.println("Finalizer running ...");
   }
}
```

Figure 7.12. The `Widget` class

We must create a test driver class to test the `Widget` class. Such a class is shown in Figure 7.13.

This class creates three objects of type `Widget`, initializing them to 10, 20, and 30 respectively. It then calls the `static` methods `Widget.getCreated()` and `Widget.getAlive()` to show how many objects of this type have been created, and how many objects are still alive.

Next the method calls the method `getValue()` for each object, showing that the instance variables in each object are unique to that object.

Next, the program nullifies the references to objects `w1` and `w2`. Once these references are gone, the objects are inaccessible, and they become candidates for garbage collection. In theory, the two objects could disappear at any time after this point in the program. However, the garbage collector is a low priority thread, so the objects are not likely to disappear until we explicitly force the garbage collector to run.

When the garbage collector runs, it calls the finalizers for the objects being destroyed before actually destroying them. After starting the garbage collector running,

```
/*
   Purpose:
     This class tests the Widget class.

   Record of revisions:
     Date        Programmer          Description of change
     ====        ==========          =====================
    4/20/98     S. J. Chapman        Original code
*/
public class TestWidget {

      // Define the main method to test class Widget
      public static void main(String[] args) {

       // Create three new Widgets
      Widget w1 = new Widget(10);     // Widget 1
      Widget w2 = new Widget(20);     // Widget 2
      Widget w3 = new Widget(30);     // Widget 3

      // Check to see how many Widget are created and alive
      System.out.println (Widget.getCreated() + " Widgets created");
      System.out.println (Widget.getAlive() + " Widgets alive");

      // Print values of widgets
      System.out.println ("Widget 1 = " + w1.getValue());
      System.out.println ("Widget 2 = " + w2.getValue());
      System.out.println ("Widget 3 = " + w3.getValue());

      // Nullify references to two Widgets
      w1 = null;
      w2 = null;

      // Check to see how many Widget are created and alive
      System.out.println (Widget.getCreated() + " Widgets created");
      System.out.println (Widget.getAlive() + " Widgets alive");

      // Run the garbage collector
      System.gc();

      // Check to see how many Widget are created and alive
      System.out.println (Widget.getCreated() + " Widgets created");
      System.out.println (Widget.getAlive() + " Widgets alive");
   }
}
```

Figure 7.13. Test driver for the `Widget class`

the `main` program continues executing and calls the `static` methods `Widget.getCreated()` and `Widget.getAlive()` again to show the status of the `Widget` objects.

Note that the garbage collector is running at the *same time* as the `main` program. If the garbage collector has had time to destroy the two objects with nullified references before the `main` program calls `Widget.getAlive()`, the call will show only one object left. However, the timing of the garbage collector will vary from run to run.

When this program was executed twice on my computer, the results were:

```
C:\book\java\chap7>java TestWidget
3 Widgets created
3 Widgets alive
Widget 1 = 10
Widget 2 = 20
Widget 3 = 30
3 Widgets created
3 Widgets alive
Finalizer running...
Finalizer running...
3 Widgets created
```

```
1 Widgets alive
D:\book\java\chap7>java TestWidget
3 Widgets created
3 Widgets alive
Widget 1 = 10
Widget 2 = 20
Widget 3 = 30
3 Widgets created
3 Widgets alive
3 Widgets created
3 Widgets alive
Finalizer running...
Finalizer running...
```

Note that each object's finalizer ran before the object was destroyed. On the first execution, the garbage collector destroyed the two objects *before* the second call to `Widget.getAlive()`. On the second execution, the garbage collector destroyed the two objects *after* the second call to `Widget.getAlive()`. Since the garbage collector and the main program are "racing", the number of objects left can vary from run to run.

EXAMPLE 7-3: EXTENDED MATH METHODS

Class `Math` in package `java.lang` contains a number of common mathematical functions such as `abs()`, `sin()`, `cos()`, `sqrt()`, and so forth. While these functions are useful, the list is limited compared to the elementary functions available in other languages such as Fortran. Write an extended mathematics class `ExMath` containing static methods implementing the following additional functions:

Hyperbolic sine: $$\sinh(x) = \frac{e^x - e^{-x}}{2},$$

Hyperbolic cosine: $$\cosh(x) = \frac{e^x + e^{-x}}{2},$$

Hyperbolic tangent: $$\tanh(x) = \frac{e^x - e^{-x}}{e^x + e^{-x}},$$

Log to the base 10: $$\log_{10}(x) = \frac{\log_e x}{\log_e 10}.$$

Insert the class into package `chapman.math`.

SOLUTION

This class will contain only `static` methods and the `static` final variable $\log_e 10$, so that the methods in it can be called without first instantiating an object.

1. **State the problem**. The problem is clearly stated above. We will write four different methods, implementing the four functions just described.
2. **Define the inputs and outputs**. The input to each method will be a single value of type `double`, and the result will be of type `double`. Numeric promotion will allow these methods to be used with data of other types.
3. **Decompose the program into classes and their associated methods**. There will be a single class, with four methods to calculate the hyperbolic sine, hyperbolic cosine, hyperbolic tangent, and logarithm to the base 10. We will call the class `ExMath`, and make it a subclass of the root class `Object`.

4. **Design the algorithm that you intend to implement for each method.** The pseudocode for the `sinh()` method is:

```
return ( (Math.exp(x) - Math.exp(-x)) / 2 );
```

The pseudocode for the `cosh()` method is:

```
return ( (Math.exp(x) + Math.exp(-x)) / 2 );
```

The pseudocode for the `tanh()` method is:

```
double exp = Math.exp(x);
double exm = Math.exp(-x);
return ( (exp - exm) / (exp + exm) );
```

The pseudocode for the `log10(x)` method is:

```
return ( Math.log(x) / LOGE_10 );
```

where `LOGE_10` is the constant $\log_e 10$.

5. **Turn the algorithm into Java statements.** The resulting Java methods are shown in Figure 7.14.

```
/*
   Purpose:
      This class defines an extended library of mathematical func-
      tions beyond those built into the java.lang.Math class.

   Record of revisions:
      Date        Programmer       Description of change
      ====        ==========       =====================
     4/22/98    S. J. Chapman    Original code
*/

// Specify package for class
package chapman.math;

public class ExMath {

   // Define class variables
   final static private double LOGE_10 = 2.302585092994046;

   // Hyperbolic sine method
   public static double sinh ( double x ) {
      return ( (Math.exp(x) - Math.exp(-x)) / 2 );
   }

   // Hyperbolic cosine method
   public static double cosh ( double x ) {
      return ( (Math.exp(x) + Math.exp(-x)) / 2 );
   }

   // Hyperbolic tangent method
   public static double tanh ( double x ) {
      double exp = Math.exp(x);
      double exm = Math.exp(-x);
      return ( (exp - exm) / (exp + exm) );
   }

   // Logarithm to the base 10
   public static double log10 ( double x ) {
      return ( Math.log(x) / LOGE_10 );
   }
}
```

Figure 7.14. Class `ExMath`

6. **Test the resulting Java program**. To test this program, we can manually calculate the values of the hyperbolic sine, hyperbolic cosine, hyperbolic tangent, and logarithm to the base 10 for $x = 1$ and $x = 10$, and compare the results with the output of a test driver program. The results of calculations on a scientific hand calculator are as follows:

X	SINH(*X*)	COSH(*X*)	TANH(*X*)	LOG10(*X*)
1	1.175201193	1.543080634	0.7615941559	0.0
10	11013.23287	11013.23292	0.9999999959	1.0

An appropriate test driver program is shown in Figure 7.15.

```
/*
   Purpose:
     This class tests the ExMath class.

   Record of revisions:
      Date        Programmer        Description of change
      ====        ==========        =====================
     4/22/98    S. J. Chapman       Original code
*/

import chapman.math.*;    // Get ExMath class

public class TestExMath {

   // Define the main method to test class Date
   public static void main(String[] args) {

      // Call the various methods
      System.out.println("sinh( 1)  = " + ExMath.sinh( 1));
      System.out.println("sinh(10)  = " + ExMath.sinh(10));
      System.out.println("cosh( 1)  = " + ExMath.cosh( 1));
      System.out.println("cosh(10)  = " + ExMath.cosh(10));
      System.out.println("tanh( 1)  = " + ExMath.tanh( 1));
      System.out.println("tanh(10)  = " + ExMath.tanh(10));
      System.out.println("log10( 1) = " + ExMath.log10( 1));
      System.out.println("log10(10) = " + ExMath.log10(10));
   }
}
```

Figure 7.15. Test driver for class `ExMath`

When this program is executed, the results are as follows:

```
C:\book\java\chap7>java TestExMath
sinh( 1)  = 1.1752011936438014
sinh(10)  = 11013.232874703397
cosh( 1)  = 1.5430806348152437
cosh(10)  = 11013.232920103328
tanh( 1)  = 0.7615941559557649
tanh(10)  = 0.9999999950776926
log10( 1) = 0.0
log10(10) = 1.0
```

The results of the program agree with our hand calculations to the number of significant digits that we performed the calculation.

PRACTICE!

This quiz provides a quick check to see if you have understood the concepts introduced in Sections 7.8 and 7.9. If you have trouble with the quiz, reread the section, ask your instructor, or discuss the material with a fellow student. The answers to this quiz are found in the back of the book.

1. What is the garbage collector? How does it operate? When are objects eligible for garbage collection?
2. What are `static` variables? What are they typically used for?
3. What are `static` methods typically used for?

SUMMARY

- The members of a class are instance variables and methods. Members of a class are accessed using the member access operator—the dot operator.
- Class definitions begin with the keyword `class`. The body of a class definition is included within braces (`{}`).
- An instance variable or method that is declared `public` is visible to any method with access to an object of the class.
- An instance variable or method that is declared `private` is only visible to other members of the class.
- An instance variable or method that is declared `protected` is visible to any method in the same package with access to an object of the class, and also to any method of a subclass of the class.
- An instance variable or method that is has no member access modifier is visible to any method in the same package with access to an object of the class.
- A constructor is a special method used to initialize a new object. Constructors may be overloaded to provide multiple ways to initialize a new object.
- A finalizer is a special method used to release resources just before an object is destroyed.
- Within a class's scope, class members may be referenced by their names alone. Outside a class's scope, accessible class members are referenced through a reference to an object plus the dot operator.
- The instance variables in a class are normally declared `private`, and `public` set and get methods are used to control access to them.
- Predicate methods are methods used to test the truth or falsity of some condition relating to an object.
- Java packages are convenient ways to create libraries of reusable software. Classes are placed in packages using the `package` statement, and are imported from packages using the `import` statement.
- The `CLASSPATH` environment variable must be set properly before user-defined packages can be imported into programs.
- The `this` reference may be used to reference both methods and instance variables from within an object.
- A `static` variable is a variable that is common to all objects created from a given class. `Static` variables are created when the class is first loaded, and

remain in existence until the program terminates. `Static` class variables have class scope.

- A `static` method cannot access non-`static` class members. `Static` methods and variables exist independently of any objects instantiated from a class. `Static` methods are commonly used for utility operations that are basically independent of objects, such as `Math.sqrt()`.

APPLICATIONS: SUMMARY OF GOOD PROGRAMMING PRACTICES

The following guidelines introduced in this chapter will help you to develop good programs:

1. Every instance variable and method definition in a class should be preceded by an explicit member access modifier.
2. The instance variables of a class should normally be declared private, and the class methods should be used to provide a standard interface to the class.
3. If the name of a class member is hidden by a local variable name in a method within the same class, use the this reference to refer to the class member.
4. Use set methods to check the validity and consistency of input data before it is stored in an object's instance variables.
5. Define predicate methods to test for the truth or falsity of conditions associated with any classes you create.
6. Override the toString() method of any classes the you define to create a reasonable display of the class's data.
7. Create packages containing of groups of related classes to make it easy to re-use those classes in other programs.
8. Package access is inherently dangerous, since it weakens information hiding in Java. Do not use it unless it is absolutely necessary for performance reasons. If it is necessary, restrict the package to a few closely-related classes
9. Use static class variables to share a single copy of a variable among all of the objects instantiated from the class.
10. Utility methods that perform functions independent of the instance data within a class may be declared static to make them easier to access and use.

KEY TERMS

block scope
`CLASSPATH`
class hierarchy
class members
class methods
class scope
class variables
constructor
dot operator
field
`finalizer`
garbage collector
immediate superclass
inheritance
information hiding
instance variables
member access operator
method
package access
`package` statement
predicate methods
`private`
programmer-defined types
`public`
set methods
`System.currentTimeMillis()`
`static` method
`static` variable
subclass
superclass
`this`
utility method

Problems

1. List and describe the major components of a class.
2. What is the difference between instance variables and methods and `static` variables and methods? When should instance variables and methods be used? When should `static` variables and methods be used?
3. What types of member access modifiers exist in Java? What restriction does each modifier place on access to a class member?
4. **Complex Data Type** Create a class called `Complex` to perform arithmetic with complex numbers. The class should have two `private` instance variables for the real and imaginary parts of the number. In addition, it should have class constructors, a set method to store a complex value, two get methods to recover the real and imaginary parts of the complex number, and methods for addition, subtraction, multiplication, division, and the absolute value function. In addition, the class should override the `toString()` method to print a complex number as a string of the form $a + b\ i$, where a is the real part of the number and b is the imaginary part of the number.

   If complex numbers c_1 and c_2 are defined as $c_1 = a_1 = b_1 i$ and $c_2 = a_2 + b_2 i$, then the addition, subtraction, multiplication, and division of c_1 and c_2 are defined as:

$$c_1 + c_2 = (a_1 + a_2) + (b_1 + b_2)i, \tag{7-1}$$

$$c_1 - c_2 = (a_1 - a_2) + (b_1 - b_2)i, \tag{7-2}$$

$$c_1 \times c_2 = (a_1 a_2 - b_1 b_2) + (a_1 b_2 + b_1 a_2)i, \tag{7-3}$$

$$\frac{c_1}{c_2} = \frac{a_1 a_2 + b_1 b_2}{a_2^2 + b_2^2} + \frac{b_1 a_2 - a_1 b_2}{a_2^2 + b_2^2} i. \tag{7-4}$$

   The absolute value of c_1 is defined as

$$|c_1| = \sqrt{a_1^2 + b_1^2}. \tag{7-5}$$

   Create a test driver program to test your class and confirm that all methods are working properly.

5. Determine whether the following class is correct or not. If it is in error, specify what is wrong with it. If it is correct, describe what the program does.

```
public class Norm {

   // Define instance data
   private double x;      // x position of point
   private double y;      // y position of point

   // Define constructor
   public void Norm(double x, double y) {
      this.x = x;
      this.y = y;
   }
   public void calcNorm() {
      return Math.sqrt(x*x + y*y);
   }

   public static void main(String s[]) {
      x = 3;
      y = 4;
      System.out.println("The norm is " + Norm.calcNorm () );
   }
}
```

6. **Extended Math Class** Expand the extended math class created in this chapter to include the following additional functions:

$$\text{Inverse hyperbolic sine: asinh}(x) = \log_e[x + \sqrt{x^2 + 1}]$$
$$\text{for all } x, \tag{7-6}$$

$$\text{Inverse hyperbolic cosine: acosh}(x) = \log_e[x + \sqrt{x^2 - 1}]$$
$$x \ge 1, \tag{7-7}$$

$$\text{Inverse hyperbolic tangent: atanh}(x) = \frac{1}{2}\log_e\left(\frac{1+x}{1-x}\right)$$
$$-1 < x < 1. \tag{7-8}$$

Create a test driver program to verify that these functions work properly. The program should verify that the inverse functions work by showing that the inverse of a function undoes the action of the function itself. For example, it could show that asinh(sinh x) is just x.

7. Enhance the `Date` class created in this chapter by adding:
   1. A method to calculate the day-of-year for the specified date.
   2. A method to calculate the number of days since January 1, 1900 for the specified date.
   3. A method to calculate the number of days between the date in the current `Date` object and the date in another `Date` object.

   Also, convert the `toString` method to generate the date string in the form Month dd, yyyy. Generate a test driver program to test all of the methods in the class.

8. **Comparing Sort Algorithms** Write a program to compare the sorting speed of the selection sort method developed in Example 6-1 with the quicksort sorting method included in the `java.util.Arrays` class. Use method `Math.random()` to generate two arrays containing 1000 and 10,000 random values between 0.0 and 1.0. Then, use both sorting methods to sort copies of these arrays. How does the sorting time compare for these two methods? Use the `Timer` class to compare the times of the two sorting algorithms.

9. **Three-Dimensional Vectors** The study of the dynamics of objects in motion in three dimensions is an important area of engineering. In the study of dynamics, the position and velocity of objects, forces, torques, and so forth are usually represented by three-component vectors $\mathbf{v} = x\,\hat{\mathbf{i}} + y\,\hat{\mathbf{j}} + z\,\hat{\mathbf{k}}$, where the three components (x, y, z) represent the projection of the vector $\mathbf{v}$ along the x-, y-, and z- axes respectively, and $\hat{\mathbf{i}}$, $\hat{\mathbf{j}}$, and $\hat{\mathbf{k}}$ are the unit vectors along the x-, y-, and z- axes respectively (see Figure 7.16). The solutions of many mechanical problems involve manipulating these vectors in specific ways. The most common operations performed on these vectors are:
   1. **Addition**. Two vectors are added together by separately adding their x-, y-, and z-components. If

   $$\mathbf{v}_1 = x_1\,\hat{\mathbf{i}} + y_1\,\hat{\mathbf{j}} + z_1\,\hat{\mathbf{k}} \tag{7-9}$$

   and

   $$\mathbf{v}_2 = x_2\,\hat{\mathbf{i}} + y_2\,\hat{\mathbf{j}} + z_2\,\hat{\mathbf{k}} \tag{7-10}$$

   then

   $$\mathbf{v}_1 + \mathbf{v}_2 = (x_1 + x_2)\hat{\mathbf{i}} + (y_1 + y_2)\hat{\mathbf{j}} + (z_1 + z_2)\hat{\mathbf{k}}. \tag{7-11}$$

   2. **Subtraction**. Two vectors are subtracted by separately subtracting their x-, y-, and z-components. If

   $$\mathbf{v}_1 = x_1\,\hat{\mathbf{i}} + y_1\,\hat{\mathbf{j}} + z_1\,\hat{\mathbf{k}} \tag{7-12}$$

Figure 7.16. A three-dimensional vector

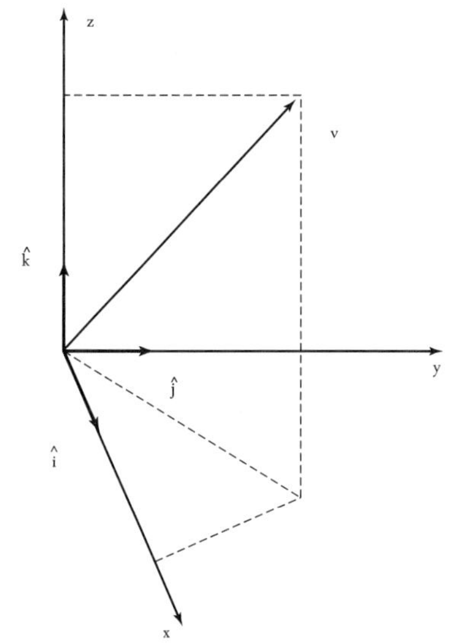

and

$$\mathbf{v}_2 = x_2\,\hat{\mathbf{i}} + y_2\,\hat{\mathbf{j}} + z_2\,\hat{\mathbf{k}} \tag{7-13}$$

then

$$\mathbf{v}_1 - \mathbf{v}_2 = (x_1 - x_2)\hat{\mathbf{i}} + (y_1 - y_2)\hat{\mathbf{j}} + (z_1 - z_2)\hat{\mathbf{k}}. \tag{7-14}$$

3. **Multiplication by a Scalar**. A vector is multiplied by a scalar by separately multiplying each component by the scalar. If

$$\mathbf{v} = x\,\hat{\mathbf{i}} + y\,\hat{\mathbf{j}} + z\,\hat{\mathbf{k}} \tag{7-15}$$

then

$$a\mathbf{v} = ax\,\hat{\mathbf{i}} + ay\,\hat{\mathbf{j}} + az\,\hat{\mathbf{k}}. \tag{7-16}$$

4. **Division by a Scalar**. A vector is divided by a scalar by separately dividing each component by the scalar. If

$$\mathbf{v} = x\,\hat{\mathbf{i}} + y\,\hat{\mathbf{j}} + z\,\hat{\mathbf{k}}, \tag{7-17}$$

then

$$\frac{\mathbf{v}}{a} = \frac{x}{a}\,\hat{\mathbf{i}} + \frac{y}{a}\,\hat{\mathbf{j}} + \frac{z}{a}\,\hat{\mathbf{k}}. \tag{7-18}$$

5. **The Dot Product**. The dot product of two vectors is one form of multiplication operation performed on vectors. It produces a scalar which is the sum of the products of the vector's components. If

$$\mathbf{v}_1 = x_1\,\hat{\mathbf{i}} + y_1\,\hat{\mathbf{j}} + z_1\hat{\mathbf{k}} \tag{7-19}$$

and

$$\mathbf{v}_2 = x_2\,\hat{\mathbf{i}} + y_2\,\hat{\mathbf{j}} + z_2\hat{\mathbf{k}}, \tag{7-20}$$

then the dot product of the vectors is

$$\mathbf{v}_1 \cdot \mathbf{v}_2 = x_1x_2 + y_1y_2 + z_1z_2 \tag{7-21}$$

6. **The Cross Product**. The cross product is another multiplication operation that appears frequently between vectors. The cross product of two vectors is another vector whose direction is perpendicular to the plane formed by the two input vectors. If

$$\mathbf{v}_1 = x_1\,\hat{\mathbf{i}} + y_1\,\hat{\mathbf{j}} + z_1\,\hat{\mathbf{k}} \tag{7-22}$$

and

$$\mathbf{v}_2 = x_2\,\hat{\mathbf{i}} + y_2\,\hat{\mathbf{j}} + z_2\,\hat{\mathbf{k}}, \tag{7-23}$$

then the cross product of the two vectors is defined as

$$\mathbf{v}_2 \times \mathbf{v}_2 = (y_1z_2 - y_2z_1)\hat{\mathbf{i}} + (z_1x_2 - z_2x_1)\hat{\mathbf{j}} + (x_1y_2 - x_2y_1)\hat{\mathbf{k}}. \tag{7-24}$$

Create a class called `Vector3D`, having three components `x`, `y`, and `z`. Define methods to create vectors from three-element arrays, to convert vectors to arrays, and to perform the six vector operations defined above. Define a `toString` method that creates an output string of the form $x\,\mathbf{i} + y\,\mathbf{j} + z\,\mathbf{k}$. Then, create a program to test all of the functions of your new class.

10. **Derivative of a Sampled Function** The *derivative* of a continuous function $f(x)$ is defined by the equation

$$\frac{d}{dx}f(x) = \lim_{\Delta x \to 0} \frac{f(x+\Delta x) - f(x)}{\Delta x} \tag{7-25}$$

In a sampled function, this definition becomes

$$f'(x_i) = \frac{f(x_{i+1}) - f(x_i)}{\Delta x}, \tag{7-26}$$

where $\Delta x = x_{i+1} - x_i$. Assume that an array `samples` contains a series of samples of a function taken at a spacing of `dx` per sample. Create a class `Derivative`, and write a method that will calculate the derivative of this array of samples from Equation 7-7.

To check your method, you should generate a data set whose derivative is known, and compare the result of the method with the known correct answer. A good choice for a test function is sin x. From elementary calculus, we know that d/d_x (sin x) = cos x. Generate an input array containing 100 values of the function sin x starting at $x = 0$, and using a step size Δx of 0.05. Take the derivative of the vector with your method, and plot the function and its derivative on the same set of axes. Compare the derivative calculated by your method to the known correct answer. How close did your method come to calculating the correct value for the derivative?

11. **Derivative in the Presence of Noise** We will now explore the effects of input noise on the quality of a numerical derivative. First, generate an input array containing 100 values of the function sin x starting at $x = 0$, and using a step size Δx of 0.05, just as you did in the previous problem. Next, use method `Math.random()` to generate a small amount of uniform random noise with a maximum amplitude of ±0.02, and add that random noise to the samples in your input vector. Note that the peak amplitude of the noise is only 2% of the peak amplitude of your signal, since the maximum value of sin x is 1.0 (see Figure 7.17). Now take the derivative of the function using the `derivative` method that you developed in the last problem. Plot the derivative with and without noise on the same set of axes. How close to the theoretical value of the derivative did you come?

12. **Histograms** A *histogram* is a plot that shows how many times a particular measurement falls within a certain range of values. For example, consider the students in this class. Suppose that there are 30 students in the class, and that their scores on the last exam had a spread described by the following table:

Figure 7.17. (*a*) A plot of sin *x* as a function of *x* with no noise added to the data. (*b*) A plot of sin *x* as a function of *x* with a 2% peak amplitude uniform random noise added to the data.

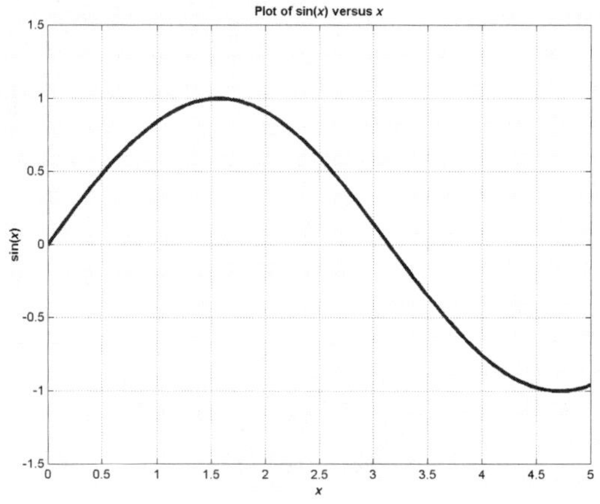

(*a*)

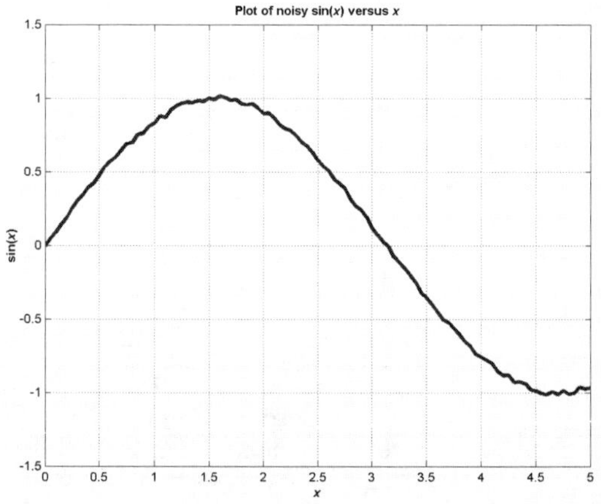

(*b*)

RANGE	NO. OF STUDENTS
100 – 95	3
94 – 90	6
89 – 85	9
84 – 80	7
79 – 75	4
74 – 70	2
69 – 65	1

Figure 7.18 is a histogram plot of the number of students scoring in each range of numbers.

To create this histogram, we started with a set of data consisting of 30 student grades. We divided the range of possible grades on the test (0 to 100) into 20 bins, and

Figure 7.18. Histogram of Student Scores on Last Test

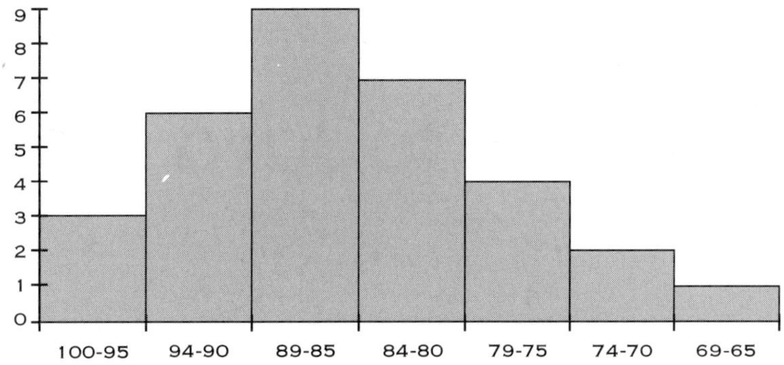

then counted how many student scores fell within each bin. Then we plotted the number of grades in each bin. (Since no one scored below 65 on the exam, we didn't bother to plot all of the empty bins between 0 and 64 in Figure 7.8.)

Create a `Histogram` class that contains an instance array of bins to accumulate data in. The number of bins should be specified in the class constructor. The class should include a method `hist` that accepts an array of double values, determines the bin each value falls in, and increments the count in the appropriate bin. It should also include two methods `getBins()` and `getCount()` to return arrays containing the centers of each bin and the number of counts in each bin respectively.

Finally, create a test driver program that reads input data from a user-specified disk file, calculates a histogram from the data using the `Histogram` class, and uses the plotting classes in package `chapman.graphics` to create a plot of the histogram.

8

Introduction to Java Graphics

We will now begin our discussion of one of Java's most interesting and important features: its built-in device- and platform-independent graphics. Java's graphics system has evolved rapidly from Java Development Kit 1.0 through JDK 1.1 to JDK 1.2 (now renamed Java 2), expanding dramatically in terms of power, flexibility, and capability. This growth has been accomplished while maintaining backward compatibility with earlier versions of the graphics system, which has unfortunately resulted in a complex mishmash of old and new graphics classes and interfaces.

In this book, we will restrict ourselves to a small fraction of Java's graphics classes and methods, concentrating on only the most recent and capable techniques, which use the Java 2D geometry package and the Swing Graphical User Interface (GUI). The Java 2D geometry package is an improved set of classes for creating high-quality 2D graphics. It is the principal topic of this chapter. The Swing GUI is a new, more efficient and more flexible graphical user interface system that is the preferred way to create GUIs in Java 2. It will be discussed in in the next chapter.

Note that if you are modifying older pre-existing programs, you will need to consult other texts or the on-line JDK documentation for the details of how the earlier GUI and graphics systems worked.

SECTIONS

OBJECTIVES

After reading this chapter, you should be able to:

- Understand how to create and display Java graphics.
- Draw lines, rectangles, ellipses, and arcs.
- Control the line style and color used to draw objects.
- Fill objects with selected colors.
- Display text in various fonts.

TABLE 8-1 Selected Java AWT Packages

JAVA API PACKAGE	EXPLANATION
`java.awt`	*The Abstract Windowing Toolkit (AWT) Package* This package contains the classes and interfaces required to create Graphical User Interfaces. The term "abstract" is applied to this packages because in can create GUI windows on any type of computer regardless of the underlying operating system type.
`java.awt.event`	*The Java AWT Event Package* This package contains classes and interfaces that support event handling for GUI components.
`java.awt.font`	*The Java AWT Font Package* This package contains classes and interfaces relating to fonts.
`java.awt.geom`	*The Java AWT Geometry Package* Provides the Java 2D classes for defining and performing operations on objects related to two-dimensional geometry.
`javax.awt.swing`	*The Swing Package* This package contains many of the classes and interfaces required to support the newer Swing Graphical User Interface.

Java's graphics system can be found in the Abstract Windowing Toolkit (AWT) Package, the Swing Package, and in several subordinate packages, the most important of which are summarized in Table 8-1.

In this chapter, we will learn about the basic graphics concepts of a **container** and a **component,** and then concentrate on learning how to draw graphical elements on the computer screen. In the following chapter, we will expand on this beginning by learning how to create a Graphical User Interface (GUI), complete with buttons, text boxes, etc., that can respond to input from the keyboard or the mouse.

8.1 CONTAINERS AND COMPONENTS

Two of the most important graphics objects are components and containers. A **component** is a visual object containing text or graphics, which can respond to keyboard or mouse inputs. All Swing components are subclasses of class `javax.swing.JComponent`. Examples of components include buttons, labels, text boxes, check boxes, and lists. A completely blank component is known as a **canvas** (like an artist's canvas); it can be used as a drawing area for text or graphics.[1] All components share a common set of methods, the most important of which is **`paintComponent`**. The `paintComponent` method causes a component to be drawn or redrawn whenever it is called. This method is called automatically whenever a component is made visible, or in response to such actions as dragging or resizing with a mouse.

A **container** is a graphical object that can hold components or other containers. The most important type of container is a **frame**, which is an area of the computer screen surrounded by borders and a title bar. Frames are implemented by class `javax.swing.JFrame`.

8.1.1 Creating and Displaying Graphics

Every component has a special **`paintComponent`** method associated with it. When this method is called, the component issues graphics commands to draw or re-draw

[1]There is a standard `Canvas` class in the older Java AWT GUI, but there is no equivalent `JCanvas` class in the Swing GUI. However, the `chapman.graphics` package includes a `JCanvas` class, and we will use that class to create all of the graphical examples in this chapter.

itself. By default, the `paintComponent` method of a `JCanvas` sets the background color and exits without doing anything useful. The way to create useful graphics is to create a *subclass* of `JCanvas`, and to override the `paintComponent` method in the subclass to display the data you are interested in.

A `paintComponent` method always has the calling sequence

```
paintComponent ( Graphics g );
```

where `g` is a reference to the `java.awt.Graphics` object used to draw lines, figures, text, etc. To use the modern Java graphics features, this `Graphics` object must be immediately cast to a `java.awt.Graphics2D` object, and then all of the tools in the `java.awt.geom` package can be applied to draw graphics on the screen.[2]

A sample class that extends `JCanvas` and draws a single line on a white background is shown in Figure 8.1, together with the resulting output. Note that this class immediately casts the `Graphics` reference to a `Graphics2D` reference, and uses that reference to draw a white background and a single black line on the screen. Also note that the class imports package `java.awt.geom`. All of Java's 2D drawing tools are in this package, so you must always import it into your graphics programs.

```
import java.awt.*;
import java.awt.event.*;
import java.awt.geom.*;
import javax.swing.*;
import chapman.graphics.JCanvas;
public class DrawLine extends JCanvas {

   // This method draws a line on a blank canvas.
   public void paintComponent ( Graphics g ) {

      // Cast the graphics object to Graph2D
      Graphics2D g2 = (Graphics2D) g;

      // Set background color
      Dimension sz = getSize();
      g2.setColor( Color.white );
      g2.fill(new Rectangle2D.Double(0,0,sz.width,sz.height));

      // Draw line
      g.setColor( Color.black );
      Line2D line = new Line2D.Double(10., 10., 360., 360.);
      g2.draw(line);
   }

   // This method illustrates how to create graphics.
   // It creates a new JFrame, attaches a JCanvas to it,
   // and makes the frame and canvas visible.
   public static void main(String s[]) {

      // Create a Window Listener to handle "close" events
      MyWindowListener l = new MyWindowListener();

      // Create a DrawLine object (a canvas)
      DrawLine c = new DrawLine();

```

Figure 8.1. *(cont.)*

[2]This rather silly business of forcing every `paintComponent` method to accept a `Graphics` object and immediately casting it to a `Graphics2D` object is for backward compatibility with earlier versions of Java. The first JDK had rather primitive graphics and only supported `Graphics` objects—if the `paintComponent` methods were changed to pass the new `Graphics2D` objects as parameters, all of those older programs would no longer work.

```
36          // Create a frame and place the object in the center
37          // of the frame.
38          JFrame f = new JFrame("Test Line ...");
39          f.addWindowListener(l);
40          f.getContentPane().add(c, BorderLayout.CENTER);
41          f.setSize(400,400);
42          f.setVisible( true );
43      }
44  }
```

```

class MyWindowListener extends WindowAdapter {

   // This method implements a simple listener that detects
   // the "window closing event" and stops the program.
   public void windowClosing(WindowEvent e) {
      System.exit(0);
   };
}
```

Figure 8.1. A program that draws a single line on a canvas. This program creates a frame, places the canvas within the frame, and displays the result. This is the basic core structure required to display Java graphics.

Line 6 of this program defines a new canvas called `DrawLine`. The graphics commands in the `paintComponent` method paint the canvas white (lines 15–17), and then draw a single black line on it (lines 20–22).

The `main` method of this program creates and displays the graphics. Line 34 of this program creates a new `DrawLine` object. Line 38 creates a new `JFrame`, and sets its title bar to display the words "`Test Line...`". Line 40 adds the `DrawLine` object to the center of the frame, and the statement "`f.setVisible(true)`" on line 42 makes the frame and its contents visible.

Line 31 creates a `MyWindowListener` object, whose job is to listen for mouse clicks on the "close window" box within the frame, and to stop the program if such a mouse click occurs. The definition of the `MyWindowListener` class appears at the bottom of the figure. The listener is very important, because without it we could never close a frame once it is created. Line 39 adds this listener to the `JFrame`, so that it will monitor mouse clicks that happen within the frame.

When this program is executed, it produces the result shown in Figure 8.2.

8.1.2 The Graphics Coordinate System

Java employs a coordinate system whose origin is in the upper left hand corner of the screen, with positive x values to the right and positive y values down (see Figure 8.3). By default, the units of the coordinate system are **pixels**, with 72 pixels to an inch. However, we shall see later that a programmer can create his or her own mapping for the screen.

In class `DrawLine`, we drew a line from (10,10) to (360,360). The resulting line extended from the upper left-hand corner of the `Canvas` (10,10) to the lower right-hand corner of the `Canvas` (360,360).

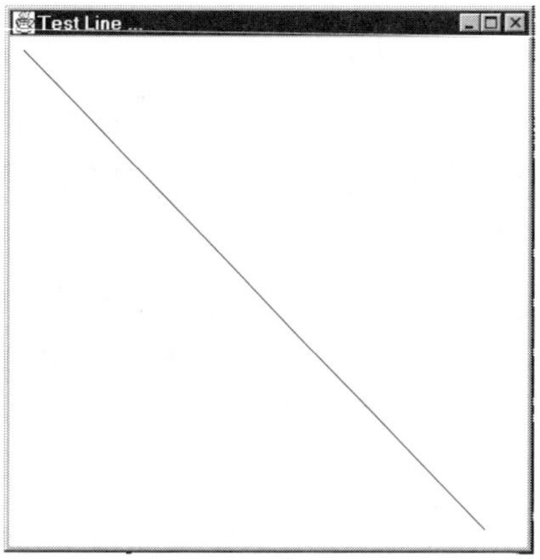

Figure 8.2. Class `DrawLine` extends `JCanvas`, and draws a single line on a white background.

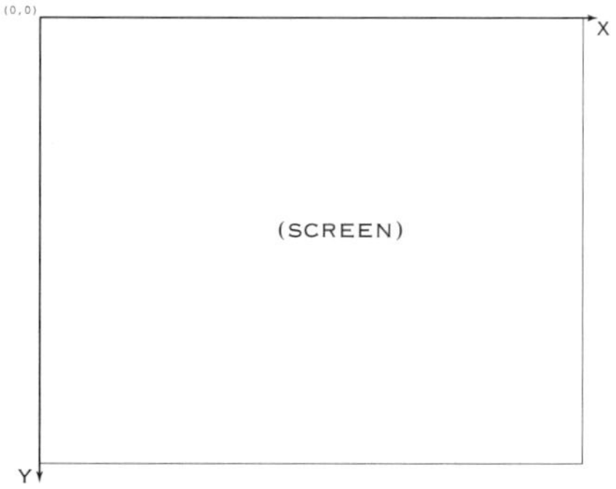

Figure 8.3. The graphics coordinate system begins in the upper left-hand corner of the display device, with the x-axis extending horizontally and the y-axis extending downward.

GOOD PROGRAMMING PRACTICE

To display Java graphics:

1. Create the component to display. Override the `paintComponent` method to create the desired graphics.
2. Create a `JFrame` to hold the component, and place the component into the frame.
3. Create a "listener" object to detect and respond to mouse clicks, and assign the listener to the `JFrame`.

8.2 DRAWING LINES

In this section we will learn how to draw lines on a graphics device. In the process of doing so, we will learn about controlling line color, line width, and line style, all of which will apply to other shapes as well. Finally, we will learn how to eliminate jagged edges from lines using Java's antialiasing technology.

TABLE 8-2 Predefined Java Colors

JAVA API PACKAGE	EXPLANATION
Color.black	Color.magneta
Color.blue	Color.orange
Color.cyan	Color.pink
Color.darkGray	Color.red
Color.green	Color.white
Color.lightGray	Color.yellow

8.2.1 Drawing Simple Lines

The class used to draw a line in Java is **java.awt.geom.Line2D.Double**. The most common constructor for a Line2D object has the form

```
Line2D.Double( double x1, double y1, double x2, double y2 )
```

where the line is defined from point (x_1, y_1) to point (x_2, y_2) on the display. Once a line object is created, the actual line can be drawn by calling the Graphics2D draw method with a reference to the line object. For example, the following statements create a line going from (10,10) to (360,360), and draw the line on the current graphics object, as we saw above.

```
Line2D line = new Line2D.Double (10., 10., 360., 360.);
g2.draw(line);
```

8.2.2 Controlling Line Color, Width, and Style

The color, width, and style of any line (or any other Java2D object) may be easily controlled. The color of a line is set by a call to the Graphics2D method **setColor**. The form of this method call is

```
g2.setColor(color)
```

where color is any object of class java.awt.Color. This class includes many predefined color constants (see Table 8-2), and you can also create your own custom colors. (To create your own custom colors, use the methods in class java.awt.Color. They are described in the JDK on-line documentation.)

The width, style, and ends of a line are controlled by a special class called **java.awt.BasicStroke**. This class defines four basic attributes of lines

- Line width in pixels
- The shape of line end caps
- The shape of decorations where two line segments meet
- The style of the line (solid, dashed, dotted, etc.)

The two most common constructors for a BasicStroke object have the following form:

```
BasicStroke(float width);
BasicStroke(float width, int cap, int join, float miterlimit,
            float[] dash, float dash_phase);
```

The meanings of these parameters are listed in Table 8-3.

TABLE 8-3 **BasicStroke** Parameters

PARAMETER	DESCRIPTION
width	A float value representing the width of the line in pixels.
cap	An int value representing the type of caps to draw on the ends of the lines. Possible choices are CAP_BUTT, CAP_SQUARE (default), and CAP_ROUND.
join	An int value representing the connection to be made between line segments. Possible choices are JOIN_BEVEL, JOIN_MITER (default), and JOIN_ROUND.
dash	A float array representing the dashing pattern in pixels. Even-numbered ([0], [2], . . .) elements in the array represent the lengths of visible segments, in pixels, and odd-numbered ([1], [3], . . .) elements in the array represent the lengths of transparent segments, in pixels.
dash phase	A float value containing the offset in pixels at which to start the dash pattern

The program in Figure 8.4 illustrates the use of these features to control the way a line is displayed. This program creates and displays two lines. The first line is red, solid, and 2 pixels wide, while the second line is blue, dashed, and 4 pixels wide.

```
import java.awt.*;
import java.awt.event.*;
import java.awt.geom.*;
import javax.swing.*;
import chapman.graphics.JCanvas;
public class DrawLine2 extends JCanvas {

   // This method draws two lines with color and styles.
   public void paintComponent ( Graphics g ) {

      BasicStroke bs;                       // Ref to BasicStroke
      Line2D line;                          // Ref to line
      float[] solid = {12.0f,0.0f};         // Solid line style
      float[] dashed = {12.0f,12.0f};       // Dashed line style

      // Cast the graphics object to Graph2D
      Graphics2D g2 = (Graphics2D) g;

      // Set background color
      Dimension size = getSize();
      g2.setColor( Color.white );
      g2.fill(new Rectangle2D.Double (0,0,size.width,size.height));

      // Set the Color and BasicStroke
      g2.setColor(Color.red);
      bs = new BasicStroke( 2.0f, BasicStroke.CAP_SQUARE,
                            BasicStroke.JOIN_MITER, 1.0f,
                            solid, 0.0f );
      g2.setStroke(bs);

      // Draw line
      line = new Line2D.Double (10., 10., 360., 360.);
      g2.draw(line);
```

Figure 8.4. *(cont.)*

```
      // Set the Color and BasicStroke
      g2.setColor(Color.blue);
      bs = new BasicStroke( 4.0f, BasicStroke.CAP_SQUARE,
                            BasicStroke.JOIN_MITER, 1.0f,
                            dashed, 0.0f );
      g2.setStroke(bs);

      // Draw line
      line = new Line2D.Double (10., 300., 360., 10.);
      g2.draw(line);
   }

   public static void main(String s[]) {

      // Create a Window Listener to handle "close" events
      MyWindowListener l = new MyWindowListener();

      // Create a DrawLine2 object
      DrawLine2 c = new DrawLine2();

      // Create a frame and place the object in the center
      // of the frame.
      JFrame f = new JFrame("DrawLine2 ...");
      f.addWindowListener(l);
      f.getContentPane().add(c, BorderLayout.CENTER);
      f.setSize(400,400)
      f.setVisible( true );
   }
}
```

Figure 8.4. (a) Class `DrawLine2`

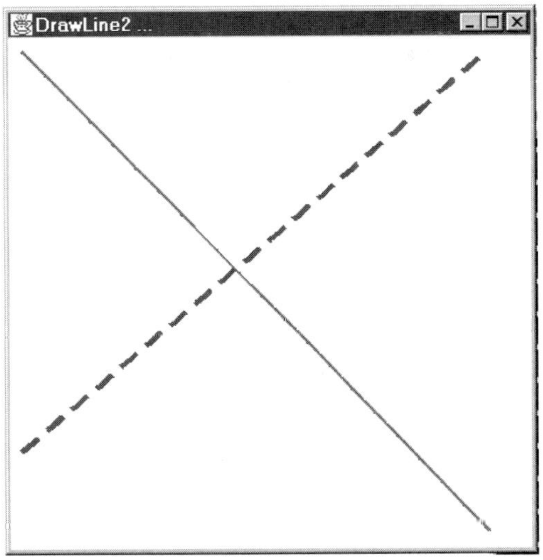

Figure 8.4. (b) The output from class `DrawLine2`

8.2.3 Eliminating Jagged Edges from Lines

If you look closely at the lines drawn in Figure 8.4, you may notice that the edges of the lines have a slightly jagged appearance. This jaggedness happens because the canvas on which the lines are drawn has only a finite number of pixels, and each pixel is either fully on or off. When a line jumps over by one pixel, the jump can leave a rough edge.

Java graphics includes a special technology know as **antialiasing** to eliminate these rough edges. When it is turned on, it allows pixels at the edges of the line to be partially on or off, causing the edge of the line to appear smooth to a human observer. This technology is controlled by a `Graphics2D` method called `setRenderingHints`. The command to turn on antialiasing is

```
// Set rendering hints to improve display quality
g2.setRenderingHint(RenderingHints.KEY_ANTIALIASING,
                    RenderingHints.VALUE_ANTIALIAS_ON);
```

When this command is included in the `paintComponent` method of class `DrawLine2`, the results are as shown Figure 8.5. Note how much smoother and cleaner the lines appear to be in that figure compared to Figure 8.4.

GOOD PROGRAMMING PRACTICES

To draw lines in Java:

1. Select a line color using the Graphics2D `setColor` method.
2. Select a line width and line style using a `BasicStroke` object, and associate that basic stroke with the line using Sthe Graphics2D `setBasicStroke` method.
3. Set the endpoints of the line using a `Line2D.Double` method, and draw the line with a call to the Graphics2D `draw` method.

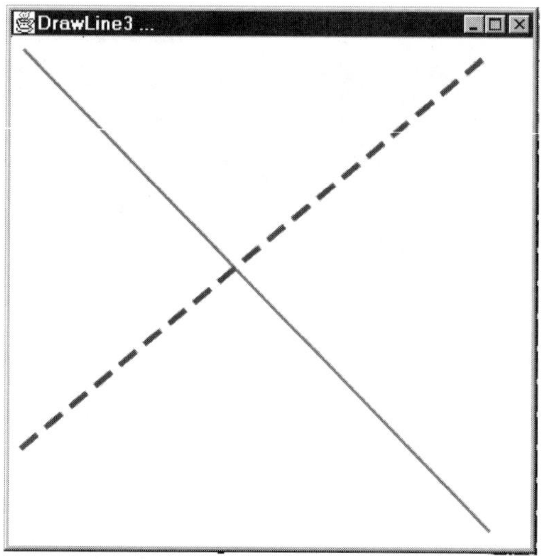

Figure 8.5. The output of program `DrawLine2` with antialiasing turned on. Note how much smoother the edges of the lines are here compared to Figure 8-4.

GOOD PROGRAMMING PRACTICE

Eliminate jagged edges from your lines by turning on antialiasing with the Graphics2D method `setRenderingHints`.

EXAMPLE 8-1: PLOTTING THE FUNCTION SIN θ

The `Line2D.Double` class can be used to plot curves of arbitrary shape by breaking each curve into small, straight line segments, and plotting each segment separately. To illustrate this operation, we will create a plot of the function sin θ over the range $0 \leq \theta \leq 2\pi$. The plotted curve should be a solid blue line four pixels wide. Use anti-aliasing to smooth the edges of the curve.

SOLUTION

To create an overall sinusoidal shape, we will divide this curve into 40 separate line segments and plot each segment separately. The code required to generate the 41 points bounding the 40 line segments is

```
double theta[] = new double[41];
double sin[] = new double[41];
delta = 2 * Math.PI / 40;
for ( i = 0; i < theta.length; i++ ) {
   theta[i] = delta * i;
   sin[i] = Math.sin(theta[i]);
}
```

Once we have the ends of the line segments, it is necessary to plot them in the space provided. If we assume that the space available for the plot is about 380 × 380 pixels, then the range of possible values of θ must be mapped into 380 horizontal pixels, and the range of possible values of sin θ must be mapped into 380 vertical pixels. The range of θ is $0 \leq \theta \leq 2\pi$, so $\theta = 0$ should correspond to pixel 0, and $\theta = 2\pi$ should correspond to pixel 380. A suitable mapping function would be the following:

$$xpos = \left(\frac{380}{2\pi}\right)\theta = \left(\frac{190}{\pi}\right)\theta \tag{8-1}$$

Similarly, the range of possible value of sin θ is $-1 \leq \sin\theta \leq 1$, so the 380 vertical pixels must be mapped to that range. The y-axis mapping is trickier, though, because y values *start at zero at the top of the display and increase downward.* To make the plot come out right, we must make $y = -1$ correspond to pixel 380, and $y = +1$ correspond to pixel 0. A suitable mapping function would be

$$ypos = 180 - 180\sin\theta. \tag{8-2}$$

Note that this function produces 0 when $\sin\theta = 1$, and 360 when $\sin\theta = -1$. The code required to apply these mappings to the function can be displayed is:

```
for ( i = 0; i < theta.length; i++ ) {
   theta[i] = (190/Math.PI) * theta[i];
   sin[i] = 180 - 180 * sin[i];
}
```

Finally, the code required to plot the 40 line segments is:

```
for ( i = 0; i < theta.length-1; i++ ) {
   line = new Line2D.Double (theta[i], sin[i],
                             theta[i+1], sin[i+1]);
   g2.draw(line);
}
```

```
   // This method plots one cycle of a sine wave.
   public void paintComponent ( Graphics g ) {

      BasicStroke bs;                        // Ref to BasicStroke
      double delta;                          // Step between points
      int i;                                 // Loop index
      Line2D line;                           // Ref to line
      double sin[] = new double[41];         // sin(theta)
      float[] solid = {12.0f,0.0f};          // Solid line style
      double theta[] = new double[41];       // Angles in radians

   // Cast the graphics object to Graph2D
   Graphics2D g2 = (Graphics2D) g;

   // Set rendering hints to improve display quality
   g2.setRenderingHint(RenderingHints.KEY_ANTIALIASING,
                       RenderingHints.VALUE_ANTIALIAS_ON);

   // Set background color
   Dimension size = getSize();
   g2.setColor( Color.white );
   g2.fill(new Rectangle2D.Double(0,0,size.width,size.height));

   // Set the Color and BasicStroke
   g2.setColor(Color.blue);
   bs = new BasicStroke( 4.0f, BasicStroke.CAP_SQUARE,
                         BasicStroke.JOIN_MITER, 1.0f,
                         solid, 0.0f );
   g2.setStroke(bs);

   // Calculate points on curve
   delta = 2 * Math.PI / 40;
   for ( i = 0; i < theta.length; i++ ) {
      theta[i] = delta * i;
      sin[i] = Math.sin(theta[i]);
   }

   // Translate curve position to pixels
   for ( i = 0; i < theta.length; i++ ) {
      theta[i] = (190/Math.PI) * theta[i];
      sin[i] = 180 - 180 * sin[i];
   }

   // Plot curve
   for ( i = 0; i < theta.length-1; i++ ) {
      line = new Line2D.Double (theta[i], sin[i],
                                theta[i+1], sin[i+1]);
      g2.draw(line);
   }
}
```

Figure 8.6. The `paintComponent` method from a program to plot the function $\sin\theta$ for $0 \le \theta \le 2\pi$

The `paintComponent` method required to generate this curve is shown in Figure 8.6. The rest of the program is not shown, because it is essentially the same as the previous two programs.

When this program is executed, the results are as shown in Figure 8.7.

Figure 8.7. The output of program `PlotSine`

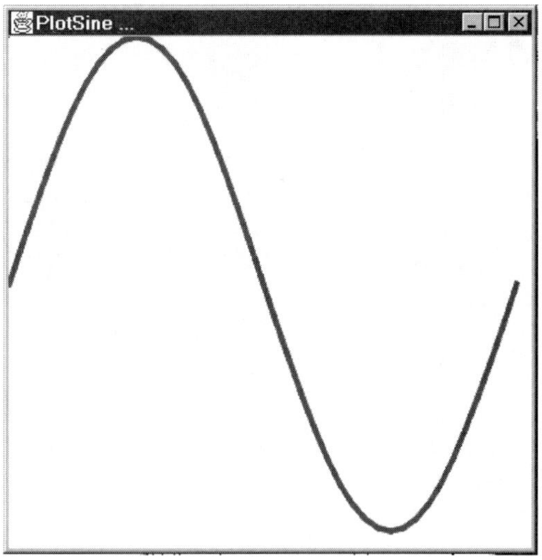

EXAMPLE 8-2: AUTOMATIC SCALING OF PLOTS

The program in Example 8-1 contains a serious flaw. Note that we designed the plot to occupy a space of 380 × 380 pixels. What would happen to this plot if we changed the size of the frame it was plotted in? For example, suppose that we used the mouse to make the frame larger or smaller. What would we see? If the frame is made smaller, then only a portion of the curve will be displayed. If the frame is made larger, then the curve will only occupy a portion of the available space. These problems are illustrated in Figure 8.8.

What we need is a way to determine the size of the canvas that we a plotting on, so that the plot can automatically re-scale whenever the size changes. Fortunately, every Java `Component` includes a method **`getSize()`** to recover the size of the `Component`. Since `JCanvas` is a subclass of `Component`, it automatically inherits the `getSize()` method.

The method `getSize()` is used as follows:

```
Dimension size = getSize();
```

This method returns a `Dimension` object, which has two `public` instance variables `height` and `width`. Thus the height of the component in pixels will be `size.height` and the width of the component in pixels will be `size.width`.

We can use this information to create a `paintComponent` method that automatically re-sizes whenever its container re-sizes by changing the mappings to be

$$xpos = \left(\frac{\texttt{size.width}}{2\pi}\right)\theta \tag{8-3}$$

and

$$ypos = \frac{\texttt{size.height}}{2} - \frac{\texttt{size.height}}{2}\sin\theta \tag{8-4}$$

The `paintComponent` method with automatic resizing is shown in Figure 8.9.

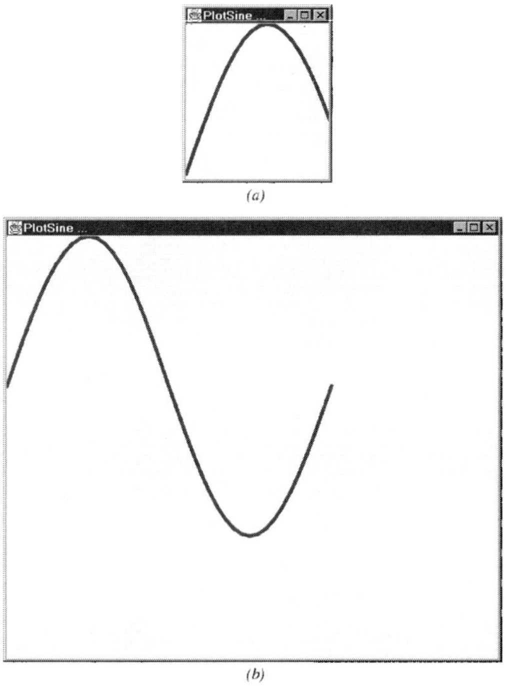

Figure 8.8. When the frame containing the sinusoidal plot is resized, the plot does not change size to match: (*a*) small frame, (*b*) large frame.

```
// This method plots one cycle of a sine wave.
public void paintComponent ( Graphics g ) {

   BasicStroke bs;                     // Ref to BasicStroke
   double delta;                       // Step between points
   int i;                              // Loop index
   Line2D line;                        // Ref to line
   double sin[] = new double[41];      // sin(theta)
   float[] solid = {12.0f,0.0f};       // Solid line style
   double theta[] = new double[41];    // Angles in radians

   // Cast the graphics object to Graph2D
   Graphics2D g2 = (Graphics2D) g;

   // Set rendering hints to improve display quality
   g2.setRenderingHint(RenderingHints.KEY_ANTIALIASING,
                       RenderingHints.VALUE_ANTIALIAS_ON);

   // Get plot size
   Dimension size = getSize();

   // Set background color
   g2.setColor( Color.white );
   g2.fill(new Rectangle2D.Double
   (0,0,size.width,size.height));

   // Set the Color and BasicStroke
   g2.setColor(Color.blue);
   bs = new BasicStroke( 4.0f, BasicStroke.CAP_SQUARE,
                         BasicStroke.JOIN_MITER, 1.0f,
                         solid, 0.0f );
```

Figure 8.9. *(cont.)*

```
      g2.setStroke(bs);

      // Calculate points on curve
      delta = 2 * Math.PI / 40;
      for ( i = 0; i < theta.length; i++ ) {
         theta[i] = delta * i;
         sin[i] = Math.sin(theta[i]);
      }

      // Translate curve position to pixels
      for ( i = 0; i < theta.length; i++ ) {
         theta[i] = (size.width/(2*Math.PI)) * theta[i];
         sin[i] = size.height/2 - size.height/2 * sin[i];
      }

      // Plot curve
      for ( i = 0; i < theta.length-1; i++ ) {
         line = new Line2D.Double (theta[i], sin[i],
                                   theta[i+1], sin[i+1]);
         g2.draw(line);
      }
   }
```

Figure 8.9. The `paintComponent` method to plot sin θ with automatic re-sizing

Execute this program and re-size its frame with a mouse. Notice how the plot changes size to take advantage of the available space.

This plot program is still not perfect. Note that the curve is clipped a bit at the top and the bottom of the plot. You should rewrite it again to make the curve scale properly, while still leaving a small amount of space around each of the edges.

8.3 DRAWING OTHER SHAPES

The `java.awt.geom` package includes classes to draw several other shapes, including rectangles, rounded rectangles, ellipses, and arcs. All of these shapes function in a manner basically similar to the `Line2D` class that we saw in the previous section. They all use the same techniques to set color, line width, and line style, so we already know most of what we need to know to use them.

8.3.1 Rectangles

The class used to draw a rectangle is **`java.awt.geom.Rectangle2D.Double`**. The most common constructors for a `Rectangle2D.Double` object have the form

```
Rectangle2D.Double( double x, double y, double w, double h )
```

where the upper left-hand corner of the rectangle is a point (x,y), and the rectangle is w pixels wide and h pixels high. For example, the following statements create a rectangle starting at position (30,40) that is 200 pixels wide and 150 pixels high, and draw the rectangle on the current graphics device:

```
Rectangle2D.Double rect;
rect = new Rectangle2D.Double (30., 40., 200., 150.);
g2.draw(rect);
```

Unlike a line, a rectangle is a closed shape that has an interior and a border. The method `g2.draw(rect)` draws the *border* of the rectangle, but leaves the interior

Figure 8.10. The rectangle created by the statements in Section 8.3.1

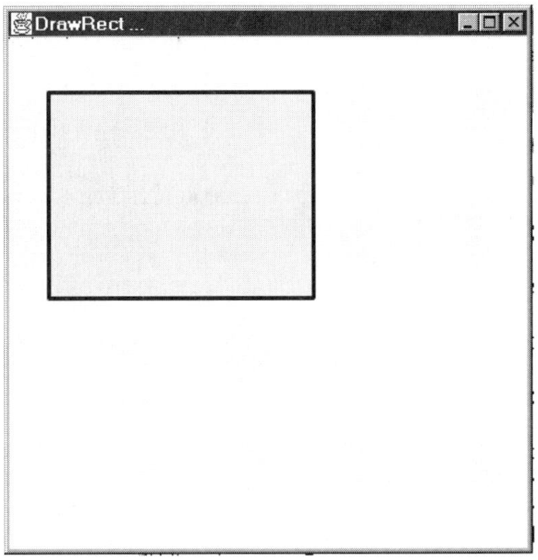

empty. It is also possible to fill the interior of a rectangle with the `Graphics2D` method `fill`. For example, the following statements create a 200 × 150 rectangle object, fill its interior with yellow, and draw a black border around it.

```
bs = new BasicStroke( 3.0f, BasicStroke.CAP_SQUARE,
                      BasicStroke.JOIN_MITER, 1.0f,
                      solid, 0.0f );
g2.setStroke(bs);
Rectangle2D.Double rect =
                new Rectangle2D.Double (30., 40., 200., 150.);
g2.setColor(Color.yellow);
g2.fill(rect);
g2.setColor(Color.black);
g2.draw(rect);
```

The resulting shape is shown in Figure 8.10. Note that the `fill` method has been used with a `Rectangle2D.Double` object in all of our examples to paint the background color of each canvas.

GOOD PROGRAMMING PRACTICE

Use the `Rectangle2D.Double` class to create rectangles.

8.3.2 Ellipses

The class used to draw circles and ellipses is **`java.awt.geom.Ellipse2D.Double`**. The constructor for an `Ellipse2D.Double` object has the form

```
Ellipse2D.Double( double x, double y, double w, double h);
```

Figure 8.11. The ellipse created by the statements in Section 8.3.2

where the upper left-hand corner of the rectangular box in which the ellipse is drawn is point (x,y), and the ellipse is w pixels wide and h pixels high. Note that if w and h are equal, this class draws a circle. For example, the following statements create an ellipse starting at position (30,40) that is 200 pixels wide and 150 pixels high.

```
Ellipse2D.Double ell
ell = new Ellipse2D.Double (30.,40.,200.,150.);
g2.draw(ell);
```

Similarly, the following statements create a 200 × 150 ellipse object, and fill its interior with black.

```
Ellipse2D.Double ell = new Ellipse2D.Double
                (30., 40., 200., 150.);
g2.setColor(Color.black);
g2.fill(ell);
```

The resulting shape is shown in Figure 8.11.

GOOD PROGRAMMING PRACTICE

Use the `Ellipse2D.Double` class to create circles and ellipses.

8.3.3 Arcs

An *arc* is a portion of an ellipse. An arc is drawn from a *starting angle* and covers an *extent*, both of which are given in degrees. The starting angle is the angle at which the arc begins, and the extent is the is the number of degrees covered by the arc. For this

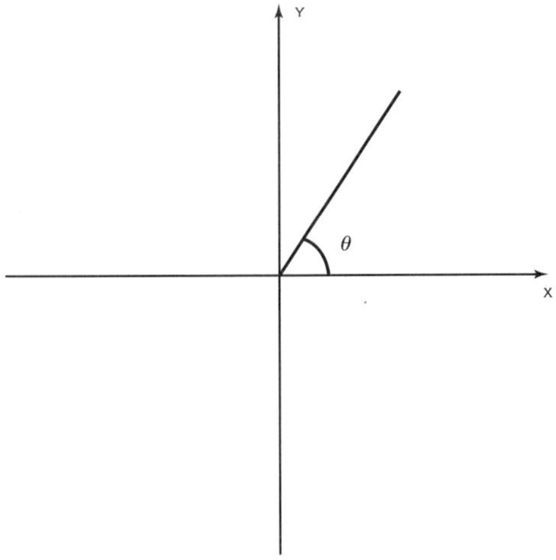

Figure 8.12. Arc starting angles are measured from the positive *x*-axis, and are considered to be positive counterclockwise. Arc extents are positive if they are counterclockwise, and negative if they are clockwise.

purpose, angles are defined as they are on a Cartesian coordinate plane, positive counterclockwise from the positive *x* axis (see Figure 8.12). Arcs with a positive extent sweep clockwise from the starting angle, while arcs with a negative extent sweep counterclockwise from the starting angle.

The class used to draw arcs is **`Arc2D.Double`**. The constructor for `Arc2D.Double` objects has the form

```
Arc2D.Double( double x, double y, double w, double h,
              double start, double extent, int type );
```

where the upper left-hand corner of the rectangular box in which the arc is drawn is point (x,y), and the arc is w pixels wide and h pixels high. The starting angle of the arc is `start` degrees, measured counterclockwise from the positive *x* axis, and the extent of the arc is `extent` degrees. Finally, `type` is the closure type for the arc. There are three possible closure types: `Arc2D.OPEN`, `Arc2D.CHORD`, and `Arc2D.PIE`. The `Arc2D.OPEN` type leaves the end of the arc open, while the `Arc2D.CHORD` type connects the ends with a straight line, and the `Arc2D.PIE` type connects the end with a pie slice.

For example, the following statements create an ellipse starting at position (30,40) that is 200 pixels wide and 150 pixels high. The starting angle is 0°, and the extent is 90°. The closure type for this arc is `Arc2D.OPEN`.

```
Arc2D.Double arc = new Arc2D.Double (30.,40.,200.,150.,
                                     0., 90., Arc2D.OPEN);
g2.draw(arc);
```

It is also possible to fill an arc with the `fill` method.

Figure 8.13 shows the `paintComponent` method of a program that illustrates more of the arc options. This method creates four arcs with various combinations of starting angles, extents, fills, and closures.

```
// This method draws several arcs with different options.
public void paintComponent ( Graphics g ) {

   BasicStroke bs;                     // Ref to BasicStroke
   Arc2D.Double arc;                   // Ref to arc
   float[] solid = {12.0f,0.0f};       // Solid line style

   // Cast the graphics object to Graph2D
   Graphics2D g2 = (Graphics2D) g;

   // Set rendering hints to improve display quality
   g2.setRenderingHint(RenderingHints.KEY_ANTIALIASING,
                       RenderingHints.VALUE_ANTIALIAS_ON);

   // Set background color
   Dimension size = getSize();
   g2.setColor( Color.white );
      g2.fill(new Rectangle2D.Double
      (0,0,size.width,size.height));

   // Set the basic stroke
   bs = new BasicStroke( 3.0f, BasicStroke.CAP_SQUARE,
                         BasicStroke.JOIN_MITER, 1.0f,
                         solid, 0.0f );
   g2.setStroke(bs);

   // Define arc1
   arc = new Arc2D.Double (20., 40., 100., 150.,
                0., 60., Arc2D.PIE);

   g2.setColor(Color.yellow);
   g2.fill(arc);
   g2.setColor(Color.black);
   g2.draw(arc);

   // Define arc2
   arc = new Arc2D.Double (10., 200., 100., 100.,
                        90., 180., Arc2D.CHORD);
   g2.setColor(Color.black);
   g2.draw(arc);

   // Define arc3
   arc = new Arc2D.Double (220., 10., 80., 200.,
                0., 120., Arc2D.OPEN);
   g2.setColor(Color.lightGray);
   g2.fill(arc);
   g2.setColor(Color.black);
   g2.draw(arc);

   // Define arc4
   arc = new Arc2D.Double (220., 220., 100., 100.,
                        -30., -300., Arc2D.PIE);
   g2.setColor(Color.orange);
   g2.fill(arc);
}
```

Figure 8.13. A `paintComponent` method to draw four arcs

Figure 8.14. Miscellaneous arcs created with `Arc2D`.

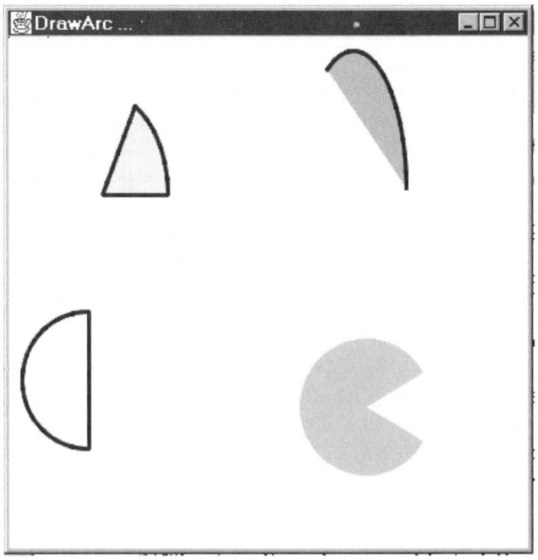

The resulting arcs are shown in Figure 8.14.

GOOD PROGRAMMING PRACTICE

Use the `Arc2D.Double` class to create elliptical and circular arcs.

PROGRAMMING PITFALLS

The angles in class `Arc2D.Double` are given in *degrees*, while the angles in almost every other Java class are given in *radians*. Be careful not to confuse the angle units when using this class.

8.4 DISPLAYING TEXT

Text may be displayed on a graphics device using the `Graphics2D` method **`drawString`**. The most common forms of this methods are

```
drawString(String s, int x, int y);
drawString(String s, float x, float y);
```

where `s` is the string to display, and *(x,y)* is the *lower-left-hand corner* of region where the `String` will be displayed. When this method is executed, the characters in `s` will be displayed on the screen in the current color, and using the current `Font`.

The `paintComponent` method from an example program that displays a `String` is shown in Figure 8.15, together with the result produced on the display.

```
// This method displays a string on the graphics device.
public void paintComponent ( Graphics g ) {

   // Cast the graphics object to Graph2D
   Graphics2D g2 = (Graphics2D) g;

   // Set rendering hints to improve display quality
   g2.setRenderingHint(RenderingHints.KEY_ANTIALIASING,
                       RenderingHints.VALUE_ANTIALIAS_ON);

   // Set background color
   Dimension size = getSize();
   g2.setColor( Color.white );
   g2.fill(new Rectangle2D.Double
   (0,0,size.width,size.height));

   // Display string
   g2.setColor( Color.black );
   g2.drawString("This is a test!",20,40);
}
```

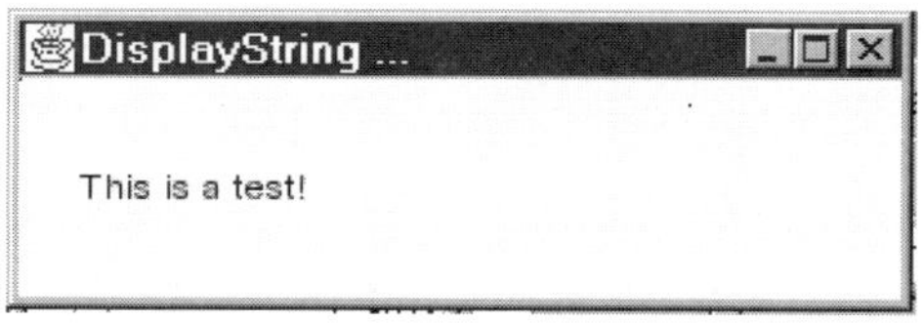

Figure 8.15. Method demonstrating how to write a `String` to a graphics device

Note that if you specify a *y* value of 0, no text will be visible, since 0 corresponds to the *top* of the display area, and it also marks the *bottom* of the region where the `String` will be displayed! This is a common mistake made by novice Java programmers.

PROGRAMMING PITFALLS

Do not to display text at an (*x*,*y*) position with a y value of 0. It will be displayed above the top of the display window, and so will be invisible!

8.4.1 Selecting and Controlling Fonts

The font used to display text on a graphics device can be controlled by defining a **`java.awt.Font`** object, and then specifying that object to be the current font. A `Font` object is declared with a constructor of the form

```
Font( String s, int style, int size );
```

where `s` is the name of the font to use, `style` is the style of the font (plain, italic, bold, or bold italic), and `size` is the point size of the font. There may be many fonts available on a system, but Java guarantees that the fonts shown in Table 8-4 will *always* be available on every Java implementation.

The font style may be specified by one or more of the constants `Font.PLAIN`, `Font.BOLD`, and `Font.ITALIC`. Note that it is possible to use the `BOLD` and `ITALIC` styles at the same time by simply adding the two constants together. The font

TABLE 8-4 Standard Font Names

FONT NAME	DESCRIPTION
`Serif`	This is the standard serif font for a particular system. Examples are Times and Times New Roman.
`SansSerif`	This is the standard sansserif font for a particular system. Examples are Helvetica and Arial.
`Monospaced`	This is the standard monospaced font for a particular system. Examples are Courier and Courier New.
`Dialog`	This is the standard font for dialog boxes on a particular system
`DialogInput`	This is the standard font for dialog inputs on a particular system.

size may be any integer point size. For example, the statement shown below creates a 14-point bold italic monospaced font.

```
Font font1 = new Font( "Monospaced", Font.BOLD+Font.ITALIC, 14);
```

Once a font has been created, it can be set to be the current font with the `Graphics2D` method **`setFont`**. The form of this method is

```
g2.setFont( font1 );
```

After this method is executed, any subsequent text will be displayed in the specified font.

The program in Figure 8.16 illustrates various combinations of font names, styles, and sizes.

```
// This method tests various fonts
public void paintComponent ( Graphics g ) {

   // Cast the graphics object to Graph2D
   Graphics2D g2 = (Graphics2D) g;

   // Set rendering hints to improve display quality
   g2.setRenderingHint(RenderingHints.KEY_ANTIALIASING,
                       RenderingHints.VALUE_ANTIALIAS_ON);

   // Set background color
   Dimension size = getSize();
   g2.setColor( Color.white );
   g2.fill(new Rectangle2D.Double
   (0,0,size.width,size.height));

   // Define several fonts ...
   Font f1 = new Font("Serif",Font.PLAIN,12);
   Font f2 = new Font("SansSerif",Font.ITALIC,16);
   Font f3 = new Font("Monospaced",Font.BOLD,14);
   Font f4 = new Font("Serif",Font.BOLD+Font.ITALIC,20);

   // Display fonts
   g2.setColor( Color.black );
   g2.setFont(f1);
   g2.drawString("12-point plain Serif",20,40);
   g2.setFont(f2);
   g2.drawString("16-point italic SansSerif",20,80);
   g2.setFont(f3);
   g2.drawString("14-point bold Monospaced",20,120);
   g2.setFont(f4);
   g2.drawString("20-point bold italic Serif",20,160);
}
```

Figure 8.16. *(cont.)*

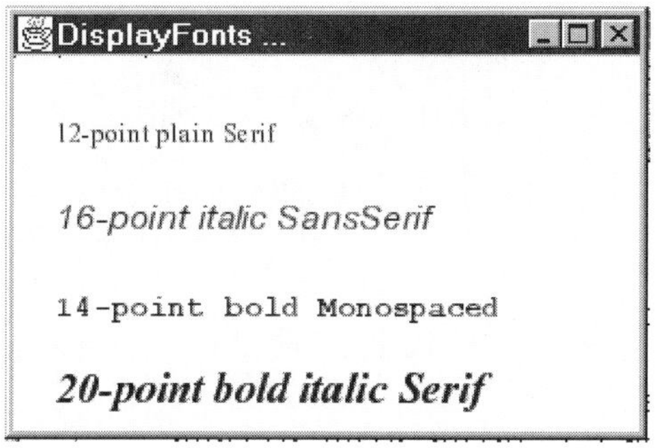

Figure 8.16. Method displaying various fonts

A summary of useful methods in the `java.awt.Font` class can be found in Table 8-5.

TABLE 8-5 Selected methods in the `java.awt.Font` Class

METHOD NAME	DESCRIPTION
`public int getStyle()`	Returns an integer value containing the current font style.
`public int getSize()`	Returns an integer value containing the current font size.
`public String getName()`	Returns the current font name as a `String`.
`public String getFamily()`	Returns the current font family as a `String`.
`public boolean isBold()`	Returns `true` if the font is bold.
`public boolean isItalic()`	Returns `true` if the font is italic.
`public boolean isPlain()`	Returns `true` if the font is plain.

8.4.2 Getting Information About Fonts

It is sometimes necessary to get precise information about a font that it being used in an application. For example, we may need to place two or more lines of text under each other at a comfortable spacing. How can we tell precisely how tall and long a particular line is, so that we can place other information above, below, or to the side of it? Java contains a special class called **`java.awt.FontMetrics`** to provide this information about any specified font.

There are several types of metrics associated with a font. These metrics include the font's *height*, *ascent* (the amount a normal character rises above the baseline), *descent* (the amount a character dips below the baseline), and *leading* (the amount above the ascent line occupied by especially tall characters). These quantities are illustrated in Figure 8.17.

A new `FontMetrics` object can be declared with a constructor of the form:

```
FontMetrics fm = new FontMetrics( Font f );
```

In addition, a `FontMetrics` object for the current font can be created using the `getFontMetrics()` method of the `Graphics2D` class:

```
FontMetrics fm = g2.getFontMetrics();
```

Once the object has been created by one of these techniques, information about the font can be retrieved with any of the methods in Table 8-6.
The `paintComponent` method in Figure 8.18 illustrates the use of these methods to recover information about the current font.

In addition, a `FontMetrics` object for the current font can be created using the `getFontMetrics()` method of the `Graphics2D` class:

```
FontMetrics fm = g2.getFontMetrics();
```

Once the object has been created by one of these techniques, information about the font can be retrieved with any of the methods in Table 8-6.

TABLE 8-6 Methods in the `FontMetrics` Class

METHOD NAME	DESCRIPTION
`public int getAscent()`	Returns the ascent of a font in pixels.
`public int getDescent()`	Returns the descent of a font in pixels.
`public int getHeight()`	Returns the height of a font in pixels.
`public int getLeading()`	Returns the leading of a font in pixels.

The `paintComponent` method in Figure 8.18 illustrates the use of these methods to recover information about the current font.

```
// This method illustrates the use of FontMetrics
public void paintComponent ( Graphics g ) {

   // Cast the graphics object to Graph2D
   Graphics2D g2 = (Graphics2D) g;

   // Define a font . . .
   Font f1 = new Font("Serif",Font.PLAIN,14);

   // Set font
   g2.setFont(f1);

   // Get information about the font
   FontMetrics fm = g2.getFontMetrics();

   // Get information about the current font
   System.out.println("Font metrics:");
   System.out.println("Font height  = " + fm.getHeight());
   System.out.println("Font ascent  = " + fm.getAscent());
   System.out.println("Font descent = " + fm.getDescent());
   System.out.println("Font leading = " + fm.getLeading());
}
```

Figure 8.18. Method displaying font metrics

When this program is executed, the results are as follows:

```
D:\book\java\chap8>java ShowFontMetrics
Font metrics:
Font height = 20
Font ascent = 15
Font descent = 4
Font leading = 1
```

Thus the spacing between successive lines of this font must be greater than 20 pixels.

EXAMPLE 8-3: DISPLAYING MULTIPLE LINES OF TEXT

Write a program that will display three lines of text, leaving the proper vertical spacing between lines. Calculate the proper spacing using the `FontMetrics` methods.

SOLUTION

The program in Figure 8.19 displays the required data. Note that it uses the height returned from the `getHeight()` method to set the spacing between successive lines.

```
import java.awt.*;
import java.awt.event.*;
import java.awt.geom.*;
import javax.swing.*;
import chapman.graphics.JCanvas;
class DisplayStrings extends JCanvas {

   // This method illustrates the use of FontMetrics to automatically
   // set the proper spacing between lines.
   public void paintComponent ( Graphics g ) {

      // Cast the graphics object to Graph2D
      Graphics2D g2 = (Graphics2D) g;

      // Set rendering hints to improve display quality
      g2.setRenderingHint(RenderingHints.KEY_ANTIALIASING,
                          RenderingHints.VALUE_ANTIALIAS_ON);

      // Set background color
      Dimension size = getSize();
      g2.setColor( Color.white );
      g2.fill(new Rectangle2D.Double(0,0,size.width,size.height));

      // Define a font ...
      Font f1 = new Font("Serif",Font.BOLD,16);

      // Set font
      g2.setFont(f1);

      // Get font height
      int height = g2.getFontMetrics().getHeight();

      // Display the text
      g2.setColor( Color.black );
      g2.drawString("This is line 1.",20, height+20);
      g2.drawString("This is line 2.",20,2*height+20);
      g2.drawString("This is line 3.",20,3*height+20);
   }

      // Create a Window Listener to handle "close" events
      MyWindowListener l = new MyWindowListener();

      // Create a DisplayStrings object
      DisplayStrings c = new DisplayStrings();

      // Create a frame and place the object in the center of the frame.
      JFrame f = new JFrame("DisplayStrings ...");
      f.addWindowListener(l);
      f.getContentPane().add(c, BorderLayout.CENTER);
      f.setSize(300,200);
      f.setVisible( true );
   }
}
```

Figure 8.19. *(cont.)*

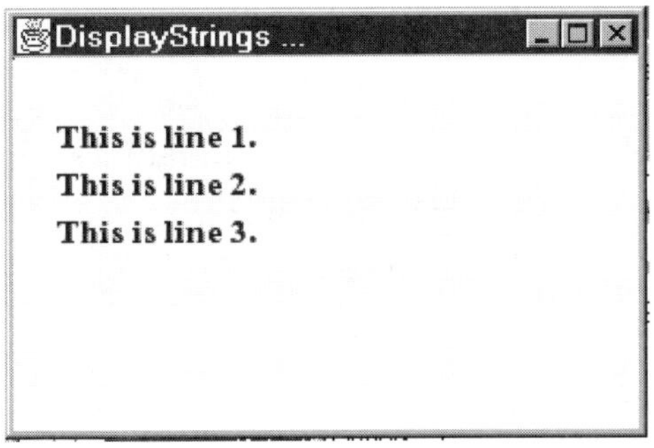

Figure 8.19. A program to display three lines of text with proper spacing, together with the program's output

The display produced when this program is executed is also shown in Figure 8.19.

PRACTICE!

This quiz provides a quick check to see if you have understood the concepts introduced in Sections 8.1 through 8.4. If you have trouble with the quiz, reread the section, ask your instructor, or discuss the material with a fellow student. The answers to this quiz are found in the back of the book.

1. What is a container? Which type of container are we using in this chapter?
2. What is a component? Which type of component are we using in this chapter?
3. What steps are required to display graphics in Java?
4. What coordinate systems is used to display graphics in Java? Where is the point (0,0) in this system?
5. What class does the method `getSize()` belong to? How is it used?
6. What class controls the style of lines and borders in Java?
7. How is text displayed on a Java graphics device? What classes are used to set and get information about the font being used?

SUMMARY

- The `java.awt` and `java.awt.geom` packages contain the classes necessary to work with graphics. The `javax.swing` package contains the classes necessary to create Swing GUIs.
- A component is a visual object containing text or graphics, which can respond to keyboard or mouse inputs.
- A canvas is a special component that is a blank area for drawing text or graphics.
- A container is a graphical object that can hold components or other containers.
- A frame is a container with borders and a title bar. In the Swing GUI, frames are implemented by class `JFrame`.
- Every component includes a `paintComponent` method, which is called whenever the object is to be displayed on a graphics device. Every `paintComponent` method has a single parameter, which is an object of the `java.awt.Graphics` class.

- Every `paintComponent` method should immediately cast the `Graphics` object to a `Graphics2D` object, so that the "2D" classes can be used with it.
- The `Graphics2D` methods `draw` and `fill` are used to draw and fill the interior of an object on a graphics device.
- Class `Line2D.Double` is used to draw lines on a graphics device.
- Class `Rectangle2D.Double` is used to draw rectangles on a Java graphics device.
- Class `Ellipse2D.Double` is used to draw circles and ellipses on a Java graphics device.
- Class `Arc2D` is used to draw arcs on a Java graphics device.
- The width, style, and ends of a "2D" line or border are controlled by an object of the `BasicStroke` class.
- The color of an object displayed on a graphics device may be controlled with the `Graphics2D` method `setColor`. This method has a single parameter, which is an object of the `java.awt.Color` class.
- Antialiasing may be used to eliminate jagged edges from objects displayed on a Java graphics device. Antialiasing is controlled with the `Graphics2D` method `setRenderingHints`.
- Text may be displayed on a graphics device with the `Graphics2D` method `drawString`. The font used to display the text is set with the `Graphics2D` method `setFont`.
- Class `Font` is used to specify the characteristics of a particular font.
- Class `FontMetrics` is used get information about the characteristics of a particular font.
- A summary of the `Graphics2D` methods discussed in the chapter is presented in Table 8-7.

TABLE 8-7 Summary of `Graphics2D` Methods Discussed in Chapter 8

PARAMETER	DESCRIPTION
`draw(Shape s);`	Draws the outline of an object, such as `Rectangle2D`, `Arc2D`, etc.
`drawString(String s,float x, float y)` `drawString(String s, int x, int y)`	Draws the specified text at location *(x,y)*, using the current color and font.
`fill(Shape s);`	Fills the interior of a `Shape` object, such as `Rectangle2D`, `Arc2D`, etc.
`getFontMetrics();`	Returns a `FontMetrics` object containing the current font metrics.
`setColor(Color c)`	Sets the color with which to draw objects on this graphics device.
`setFont(Font f)`	Sets the specified font as the current font to use when rendering text on this graphics device.
`setRenderingHint(String hintKey,Object hintValue)`	Sets the preferences for the rendering algorithms. Used to specify the use of antialiasing algorithms
`setStroke(BasicStroke b)`	Sets the basic stroke to use when drawing the outlines of objects

APPLICATIONS: SUMMARY OF GOOD PROGRAMMING PRACTICES

The following guidelines introduced in this chapter will help you to develop good programs:

1. To display Java graphics, (1) create a component to display, overriding the `paintComponent` method of the component to create your graphics; (2) Create a `JFrame` to hold the component, and place the component into the frame; and (3) create a "listener" object to detect and respond to mouse clicks, and assign the listener to the `JFrame`.
2. To draw lines in Java: (1) select a line color using the `Graphics2D setColor` method; (2) Select a line width and line style using a `BasicStroke` object; (3) Set the end-points of the line using a `Line2D.Double` method, and (4) draw the line with a call to the `Graphics2D draw` method.
3. Eliminate jagged edges from your lines by turning on antialiasing with the `Graphics2D` method `setRenderingHints`.
4. Use the `Rectangle2D.Double` class to create rectangles.
5. Use the `Ellipse2D.Double` class to create circles and ellipses.
6. Use the `Arc2D.Double` class to create elliptical and circular arcs.

KEY TERMS

component
container
chapman.graphics.JCanvas class
draw method
drawString method
fill method
java.awt.BasicStroke class
java.awt.Container class
java.awt.Font class
java.awt.FontMetrics class
java.awt.geom.Arc2D.Double class
java.awt.geom.Ellipse2D.Double class
java.awt.geom.Ellipse2D.Double class
java.awt.geom.Line2D.Double class
java.awt.geom.Rectangle2D.Double class
java.awt.Graphics2D class
javax.swing.JComponent class
javax.swing.JFrame class
paintComponent method
setColor method
setRenderingHints method
setStroke method

Problems

1. Explain the steps required to generate graphics in Java.
2. Write a program that draws a series of 5 concentric circles. The radii of the circles should differ by 20 pixels, with the innermost circle having a 20 pixel radius and the outer circle having a 100 pixel radius. Fill each circle with a different color.
3. Modify the program of Exercise 8.2 so that each circle is surrounded by a 2-pixel-wide black border. Be sure to use antialiasing to make the borders smooth.
4. Modify the program created in Exercise 8.3 so that it automatically re-scales whenever the frame containing it is re-sized.
5. Create a program that plots the function e^x over the range $-1 \leq x \leq 1$. Use a solid 4-pixel-wide blue line to plot the curve, and use solid 1-pixel-wide black curves for the x- and y-axes.
6. Write a program that plots a two-pixel-wide red dashed spiral. A spiral can be specified by two values r and θ, where r is the distance from the origin to a point and θ is the angle

counterclockwise from the positive x-axis to the point. In these terms, the spiral will be specified by the following equation:

$$r = \frac{\theta}{2\pi}, \text{ for } 0 \leq \theta \leq 6\pi.$$

Draw in the x- and y-axis as one-pixel-wide black lines.

7. Create a plot of cos x versus x for $-2\pi \leq x \leq 2\pi$. Use a solid six-pixel-wide line for the plot. Add a thin black box around the plot, and create a title and x- and y-axis labels.
8. Modify the plot in Exercise 8.7 to add thin dotted grid lines in both the horizontal and vertical directions.
9. Modify the plot in Exercise 8.7 to re-size properly whenever the frame containing the plot is re-sized.
10. Create a program that displays samples of all the standard Java fonts on your computer.
11. What is the height of a line of 24-point `SansSerif` text? What class and method did you use to learn this information?
12. Create a program that plots four ellipses of random size and shape. Each ellipse should be a distinct color.
13. Write a program that creates a bar plot, using `Rectangle2D.Double` objects to create each bar.

9

Graphical User Interfaces and Applets

A Graphical User Interface (GUI) is a pictorial interface to a program. A good GUI can make programs easier to use by providing them with a consistent appearance, and with intuitive controls like push buttons, sliders, pull-down lists, menus, etc. The GUI should behave in an understandable and predictable manner, so that a user knows what to expect when he or she performs an action. For example, when a mouse click occurs on a push button, the GUI should initiate the action described on the label of the button.

The Java API contains two different graphical user interfaces. The "Old GUI" is generally known as the Abstract Windowing Toolkit (AWT) GUI; it was introduced with JDK 1.0. The "New GUI" is known as the Swing GUI; it became a part of the standard JDK with the release of Java 2. The Swing GUI consists of additional classes built on top of the older AWT classes. It is faster and more flexible than the AWT GUI, and is recommended for all new program development. This book teaches the Swing GUI only—if you must work with older programs containing the AWT GUI, please refer to the JDK on-line documentation.

SECTIONS

OBJECTIVES

After reading this chapter, you should be able to:

- Understand the operation of Graphical User Interfaces.
- Build basic Graphical User Interfaces.
- Understand the role events and event handlers play in GUI operation.
- Understand `ActionEvents` and the `ActionListener` interface.
- Create and manipulate labels, buttons, check boxes, radio buttons, text fields, password fields, combo boxes, and panels.
- Understand and be able to use layout managers.
- Create applets, including applets that can also run as applications.

This chapter contains an introduction to the basic elements of the Swing GUIs. It does *not* contain a complete description of components or GUI features, but it does provide us with the basics required to create functional GUIs for our programs.

The chapter also introduces applets, and shows how a single program can be both an applet and an application.

9.1 HOW A GUI WORKS

A graphical user interface provides the user with a familiar environment in which to work. It contains push buttons, drop down lists, menus, text fields, and so forth, all of which are already familiar to the user, so that he or she can concentrate on the purpose of the application instead of the mechanics involved in doing things. However, GUIs are harder for the programmer, because a GUI-based program must be prepared for mouse clicks (or possibly keyboard input) for any GUI element at any time. Such inputs are known as **events**, and a program that responds to events is said to be *event driven*.

The four principal elements required to create a Java Graphical User Interface are as follows:

1. **Components**. Each item on a Java Graphical User Interface (push buttons, labels, text fields, etc.) is a **component**. Regardless of their function, components all share a common set of methods to set size, color, etc. Each component also has methods specific to its function. These components are found in package **`javax.swing`**.
2. **Container**. The components of a GUI must be arranged within a **container**. In this chapter, we will work with two types of containers: **`JPanel`** and **`JFrame`**. A `JPanel` is a very simple container to which components can be attached. A `JFrame` is a more complex container with borders and a title bar. These containers are found in package `javax.swing`.
3. **Layout Manager**. When the components of a GUI are added to a container, a **layout manager** controls the location at which they will be placed within the container. Java provides six standard layout managers, each of which lays out components in a different fashion. A layout manager is automatically associated with each container when it is created, but the programmer can freely change the layout manager, if desired. The layout managers are found in packages `java.awt` and `javax.swing`.
4. **Event Handlers**. Finally, there must be some way to perform an action if a user clicks a mouse on a button or types information on a keyboard. A mouse click or a key press creates an event, which is an object like everything else in Java. Events are handled by creating **listener classes**, which listen for a specific type of event, and execute a specific method (called an **event handler**) if the event occurs. Listener classes implement **listener interfaces**, which specify the names of the event handler methods required to handle specific types of events. The listener interfaces that we will discuss are found in package `java.awt.event`.

A subset of the basic GUI elements are summarized in Table 9-1. We will be studying examples of these elements, and then build working GUIs from them.

TABLE 9-1 Some Basic GUI Elements

ELEMENT	DESCRIPTION
	Components
`JButton`	A graphical object that implements a push button. It triggers an event when clicked with a mouse while the cursor is over it.
`JCheckBox`	A graphical object that is either selected or not selected. This element creates checkboxes.
`JComboBox`	A drop-down list of items, one of which may be selected. Single-clicking an item selects it, while double-clicking an item generates an action event.
`JLabel`	An area to display a label (text and/or images that a user cannot change).
`JPasswordField`	Displays a text field that can be used to enter passwords. Asterisks are printed out in the field as characters are entered.
`JRadioButton`	A graphical object that implements a radio button: one of a set of buttons, only one of which can be selected at a time.
`JTextField`	An area (surrounded by a box) where a program can display text data and a user can optionally enter text data.
	Containers
`JFrame`	A container with borders or title bar.
`JPanel`	A simple container with no borders or title bar that uses the `FlowLayout` manager.
	Layout Managers
`BorderLayout`	A layout manager that lays out elements in a central region and four surrounding borders. This is the default layout manager for a `JFrame`.
`BoxLayout`	A layout manager that allows multiple components to be laid out either vertically or horizontally, without wrapping.
`FlowLayout`	A layout manager that lays out elements left-to-right and top-to-bottom within a container. This is the default layout manager for a `JPanel`.
`GridLayout`	A layout manager that lays out elements in a rigid grid.

9.2 CREATING AND DISPLAYING A GRAPHICAL USER INTERFACE

The basic steps required to create a Java GUI are:

1. Create a container class to hold the GUI components. We will use subclasses of `JPanel` as the basic container.
2. Select a layout manager for the container, if the default layout manager is not acceptable.
3. Create components and add them to the container.
4. Create "listener" objects to detect and respond to the events expected by each GUI component, and register the listeners with appropriate components.
5. Create a `JFrame` object, and place the completed container in the center of **content pane** associated with the frame. (The content pane is the location where GUI objects are attached to a `JFrame` or `JApplet` container. The method `getContentPane()` returns a reference to a container's content pane.)

Figure 9.1 shows a program that creates a simple GUI with a single button and a single label field. The label field contains the number of times that the button has been pressed since the program started. Note that class `FirstGUI` is based on class `JPanel`, and this class serves as the container for our GUI components (step 1 from the previous list of steps).

```
// A first GUI.  This class creates a label and
// a button.  The count in the label is incremented
// each time that the button is pressed.
import java.awt.*;
import java.awt.event.*;
import javax.swing.*;
public class FirstGUI extends JPanel {

   // Instance variables
   private int count = 0;       // Number of pushes
   private JButton pushButton; // Push button
   private JLabel label;       // Label

   // Initialization method
   public void init() {

      // Set the layout manager
      setLayout( new BorderLayout() );

      // Create a label to hold push count
      label = new JLabel("Push Count: 0");
      add( label, BorderLayout.NORTH );
      label.setHorizontalAlignment( label.CENTER );

      // Create a button
      pushButton = new JButton("Test Button");
      pushButton.addActionListener(new ButtonHandler(this));
      add( pushButton, BorderLayout.SOUTH );
   }

   // Method to update push count
   public void updateLabel() {
      label.setText( "Push Count: " + (++count) );
   }

   // Main method to create frame
   public static void main(String s[]) {

      // Create a frame to hold the application
      JFrame fr = new JFrame("FirstGUI ...");
      fr.setSize(200,100);

      // Create a Window Listener to handle "close" events
      WindowHandler l = new WindowHandler();
      fr.addWindowListener(l);

      // Create and initialize a FirstGUI object
      FirstGUI fg = new FirstGUI();
      fg.init();

```

Figure 9.1. *(cont.)*

```
51        // Add the object to the center of the frame
52        fr.getContentPane().add(fg, BorderLayout.CENTER);
53
54        // Display the frame
55        fr.setVisible( true );
56     }
57  }
58  class ButtonHandler implements ActionListener {
59     private FirstGUI fg;
60
61     // Constructor
62     public ButtonHandler ( FirstGUI fg1 ) {
63        fg = fg1;
64     }
65
66     // Execute when an event occurs
67     public void actionPerformed( ActionEvent e ) {
68        fg.updateLabel();
69     }
70  }
```

Figure 9.1. A program that creates a container, sets a layout manager for the container, adds components and listeners for the components, and places the whole container within a `JFrame`. This is the basic core structure required to create Java GUIs.

This program contains two classes: class `FirstGUI` to create and display the GUI, and class `ButtonHandler` to respond to mouse clicks on the button. Class `FirstGUI` contains three methods: `init()`, `updateLabel()`, and `main`. Method `init()` initializes the GUI.[1] It specifies which layout manager to use with the container (line 18), creates the `JButton` and `JLabel` components (lines 21 and 26), and adds them to the container (lines 22 and 28). In addition, it creates a "listener" object of class `ButtonHandler` to listen for and handle events generated by mouse clicks, and assigns that object to monitor mouse clicks on the button (line 27).

Method `updateLabel()` (lines 32–34) is the method that should be called every time that a button click occurs. It updates the label with the number of button clicks that have occurred.

The `main` method creates a new `JFrame` (line 40), creates and initializes a `FirstGUI` object (lines 48–49), places the `FirstGUI` object in the center of the frame (line 52), and makes the frame visible (line 55).

Class `ButtonHandler` is a "listener" class designed to listen for and handle mouse clicks on the button. It implements the `ActionListener` interface, which guarantees that the class will have a method called `actionPerformed`. If an object of this class is associated (*registered*) with a GUI button, then the `actionPerformed` method in the object will be called every time that a user clicks on the button with the mouse. An object of this class is registered to handle events from the button in class `FirstGUI` (line 27), so whenever this button is clicked, the `ButtonHandler` method `actionPerformed` is called. Since that method calls the `FirstGUI` method `updateLabel()`, each mouse click causes the count displayed by the label to increase by one.

[1] It is customary (but not required) to use the name `init()` for the method that sets up a GUI in a Java application. It is required to use the name `init()` for the method that sets up a GUI in a Java applet.

Figure 9.2. The display produced by program `FirstGUI` after three button clicks

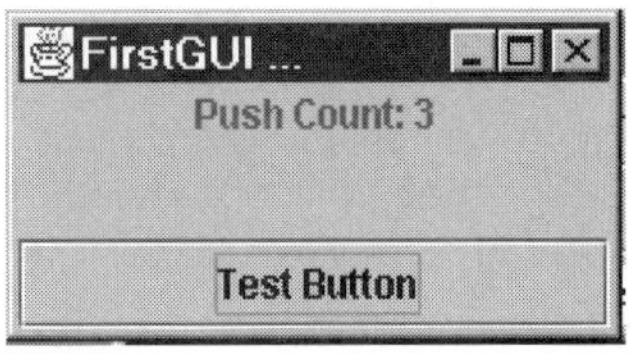

When this program is executed, the `main` method creates a new `JFrame`, creates and initializes a `FirstGUI` object, places the object in the center of the content pane associated with the frame, and makes the frame visible. The `FirstGUI` object creates a label and a button, as well as an `ButtonHandler` to listen for button clicks. At this point, program execution stops, and it will *only* resume if an external event such as a mouse click occurs. If a mouse click occurs on the button, Java automatically calls method `actionPerformed`, because the `ButtonHandler` object was set to listen for mouse clicks on the button. This method calls method `updateLabel()`, which increases the button click count and updates the display (see Figure 9.2).

9.3 EVENTS AND EVENT HANDLING

An **event** is an object that is created by some external action, such as a mouse click, key press, etc. When an event such as a mouse click occurs, *Java automatically sends that event to the GUI object that was clicked on.* For example, if a mouse click occurs over a button, Java sends the event to the GUI object that created that button.

When the mouse click event is received, the button checks to see if an object has *registered* with it to receive mouse events, and it forwards the event to the `actionPerformed` method of that object. The `actionPerformed` method is known as an **event handler** because it performs whatever steps are required to process the event. In many cases, this event handler makes a call to a **callback method** in the object that created the GUI, since such methods can update instance variables within the object directly.

This process is illustrated in Figure 9.3 for the `FirstGUI` program. When a mouse click occurs on the button, an event is created and sent to the `JButton` object. Since the `ButtonHandler` object is registered to handle mouse clicks on the button, the `JButton` object calls the `actionPerformed` method of the `ButtonHandler` object. This method makes a callback to `FirstGUI` method `updateLabel`, which actually performs the required work (updating the instance variable `count` and the label text).

This basic procedure works for all types of Java events.

9.4 SELECTED GRAPHICAL USER INTERFACE COMPONENTS

This section summarizes the basic characteristics of some common graphical user interface components. It describes how to create and use each component, as well as the types of events each component can generate. The components discussed in this section are as follows:

Figure 9.3. Event handling in program `FirstGUI`. When a user clicks on the button with the mouse, an event is sent to the `JButton` object. The `JButton` object calls the `actionPerformed` method of the `ButtonHandler` object, because that object is registered to handle the button's mouse click events. The `actionPerformed` method calls the `update` method of the `FirstGUI` object, which in turn calls the `setText` method of the `JLabel` object change the push count.

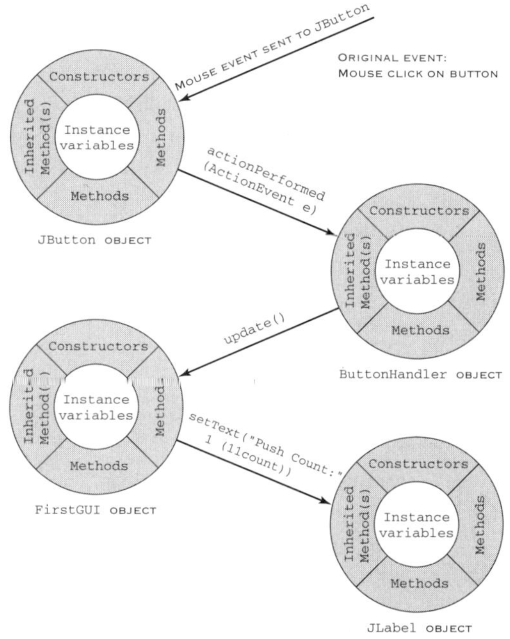

- Labels
- Push Buttons
- Text Fields and Password Fields
- Combo Boxes (Drop-down Lists)
- Check boxes
- Radio Buttons

9.4.1 LABELS

A **label** is an object that displays a single line of *read-only text and/or an image.* A `JLabel` object can display either text, or an image, or both. You can specify how the text and image are aligned in the display area by setting the vertical and horizontal alignment, and you can specify whether the text is placed left, right, above, or below the image. By default, text-only labels are left-aligned, and image-only labels are horizontally centered. If both text and image are present, text is to the right of the image by default.

A label is created with the `JLabel` class using one of the following constructors

```
public JLabel(String s);
public JLabel(String s, int horizontalAlignment);
public JLabel(Icon image);
public JLabel(Icon image, int horizontalAlignment);
public JLabel(String s, Icon image, int horizontalAlignment);
```

TABLE 9-2 `JLabel` Methods

METHOD	DESCRIPTION
`public Icon getIcon()`	Returns the image from a `JLabel`.
`public String getText()`	Returns the text from a `JLabel`.
`public void setIcon(Icon image)`	Sets the `JLabel` image.
`public void setText(String s)`	Sets the `JLabel` text.
`public void setHorizontalAlignment(int alignment)`	Sets the horizontal alignment of the `JLabel` text and image. Legal values are `LEFT`, `CENTER`, and `RIGHT`.
`public void setHorizontalTextPosition(int textPosition)`	Sets the position of the text relative to the image. Legal values are `LEFT`, `CENTER`, and `RIGHT`.
`public void setVerticalAlignment(int alignment)`	Sets the vertical alignment of the `JLabel` text and images. Legal values are `TOP`, `CENTER`, and `BOTTOM`.
`public void setVerticalText Position(int textPosition)`	Sets the position of the text relative to the image. Legal values are `TOP`, `CENTER`, and `BOTTOM`.

In these constructors, `s` is a string containing the text to display on the label, `image` is the image to display on the label, and `horizontalAlignment` is the alignment of the text and image within the label. Possible alignment values are `JLabel.LEFT`, `JLabel.RIGHT`, and `JLabel.CENTER`.

Some of the methods in class `JLabel` are described in Table 9-2. Note that `JLabels` do not generate any events, so there are no listener interfaces or classes associated with them.

The following program (Figure 9.4) shows how to create labels with and without images. The first label consists of an image followed by text, left justified in its field. The second label consists of text followed by an image, right justified in its field. The third label consists of text only, centered in its field. Note that the images may be stored in GIF or JPEG files, which can be read by creating an `ImageIcon` object.

```
// Test labels.  This program creates a GUI containing
// three labels, with and without images.
import java.awt.*;
import java.awt.event.*;
import javax.swing.*;
public class TestLabel extends JPanel {

   // Instance variables
   private JLabel l1, l2, l3;  // Labels

   // Initialization method
   public void init() {

     // Set the layout manager
      setLayout( new BorderLayout() );

      // Get the images to display
      ImageIcon right = new ImageIcon("BlueRightArrow.gif");
      ImageIcon left = new ImageIcon("BlueLeftArrow.gif");

      // Create a label with icon and text
      l1 = new JLabel("Label 1", right, JLabel.LEFT);
      add( l1, BorderLayout.NORTH );
```

Figure 9.4. *(cont.)*

```
      // Create a label with text and icon
      l2 = new JLabel("Label 2", left, JLabel.RIGHT);
      add( l2, BorderLayout.CENTER );
      l2.setHorizontalTextPosition( JLabel.LEFT );

      // Create a label with text only
      l3 = new JLabel("Label 3 (Text only)", JLabel.CENTER);
      add( l3, BorderLayout.SOUTH );
   }

   // Main method to create frame
   public static void main(String s[]) {

      // Create a frame to hold the application
      JFrame fr = new JFrame("TestLabel ...");
      fr.setSize(200,100);

      // Create a Window Listener to handle "close" events
      MyWindowListener l = new MyWindowListener();
      fr.addWindowListener(l);

      // Create and initialize a TestLabel object
      TestLabel tl = new TestLabel();
      tl.init();

      // Add the object to the center of the frame
      fr.getContentPane().add(tl, BorderLayout.CENTER);

      // Display the frame
      fr.setVisible( true );
   }
}
```

Figure 9.4. A program illustrating the use of labels, with and without images

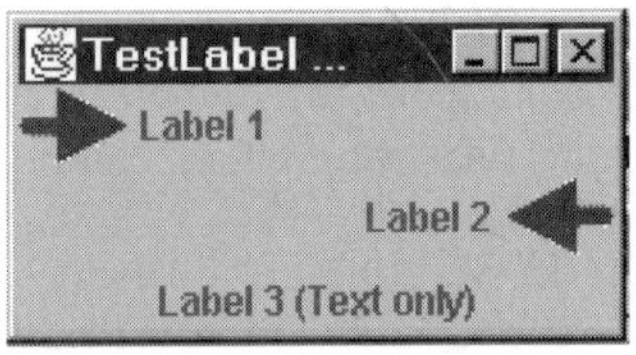

Figure 9.5. The display produced by program `TestLabel`

When this program is executed, the results are as shown in Figure 9.5.

9.4.2 Push Buttons and Associated Events

A **push button** is a component that a user can click on to trigger a specific action. Push buttons are created by the **`JButton`** class. Each `JButton` has a label and/or icon printed on its face to identify the purpose of the button. A GUI can have many buttons, but each button label should be unique, so that a user can tell them apart.

Java buttons are very flexible. Each button may be labeled with text, an image, or both, and the location of the text and image on the button can be controlled. Buttons may be enabled and disabled during program execution; when a button is disabled, it is grayed out and cannot be pressed. The icon displayed by a button can be automatically

TABLE 9-3 `JButton` Methods

METHOD	DESCRIPTION
`public void addActionListener(ActionListener l)`	Adds the specified action listener to receive action events from this button.
`public Icon getIcon()`	Returns the image from a `JButton`.
`public String getLabel()`	Returns the label of this button.
`public void setDisabled Icon(Icon icon)`	Sets the icon to display when the button is disabled.
`public void setEnabled(boolean b)`	Enables or disables this button.
`public void setHorizontalAlignment(int alignment)`	Sets the horizontal alignment of the text and images. Legal values are `LEFT`, `CENTER`, and `RIGHT`.
`public void setHorizontalTextPosition(int textPosition)`	Sets the position of the text relative to the images. Legal values are `LEFT`, `CENTER`, and `RIGHT`.
`public void setLabel(String s)`	Sets the label of this button to the specified `String`.
`public void setIcon(Icon icon)`	Sets the default icon for this button.
`public void setMnemonic(char mnemonic)`	Set the keyboard character combination used to activate the button from the keyboard.
`public void setPressed Icon(Icon icon)`	Sets the icon to display when the button is pressed.
`public void setToolTipText(String text)`	Sets the tool tip text to display when the cursor rests over the button.
`public void setVertical Alignment(int alignment)`	Sets the vertical alignment of the text and images. Legal values are `TOP`, `CENTER`, and `BOTTOM`.
`public void setVertical TextPosition(int text Position)`	Sets the position of the text relative to the images. Legal values are `TOP`, `CENTER`, and `BOTTOM`.

changed whenever it is pressed or disabled. In addition, **keyboard shortcuts** (mnemonic keys) can be defined to allow a user to activate the button via the keyboard. Finally, Java buttons support **tool tips**, which are messages that are displayed in a popup windows whenever the cursor rests over the top of the button. Tool tips are used to explain the function of the button to the program's user.

A push button is created with one of the following constructors:

```
public JButton(String s);
public JButton(Icon image);
public JButton(String s, Icon image);
```

These constructors create new buttons with a text label, an image, or both. Some of the methods in class `Button` are described in Table 9-3.

Events Associated with Buttons

When the mouse button is *both pressed and released* while the cursor is over a button, the button generates an **`ActionEvent`** and sends that event to any objects that have been registered with it as listeners. An `ActionEvent` is a special kind of event that means that the usual action associated with the component has occurred. For example, buttons are meant to be clicked on, so an `ActionEvent` from a button means that the button has been clicked on. (`ActionEvents` from other components will have different meanings.)

TABLE 9-4 `ActionEvent` Methods

METHOD	DESCRIPTION
`public String getActionCommand()`	Returns the command string associated with this action. By default, this method returns the *text label printed on the button.*
`public int getModifiers()`	Returns the modifier keys held down during this action event. Possible modifiers are `ALT_MASK`, `CTRL_MASK`, `META_MASK`, and `SHIFT_MASK`.
`public String paramString()`	Returns a parameter string identifying this action event. This method is useful for event logging and for debugging.

TABLE 9-5 `ActionListener` Interface Method

METHOD	DESCRIPTION
`void actionPerformed(ActionEvent e)`	Method invoked when an `ActionEvent` occurs.

The `ActionEvent` class includes several methods that allow a program to recover information about the triggering event. The most important of these methods are listed in Table 9-4.

The `ActionListener` *Interface*

`ActionEvents` are processed by classes that implement the **`ActionListener` interface** (Table 9-5), which defines the single event handling method **`actionPerformed`**. When an `ActionListener` object is registered with a button using the `JButton` method `addActionListener`, method `actionPerformed` will be automatically called whenever a click occurs on that button.

The following program (Figure 9.6) illustrates the use of push buttons. It defines three buttons, with text and images on each button. The left button enables the center button when it is clicked, and the right button disables the center button when it is clicked. If the center button is enabled, each click on the button will increment the displayed count by one. Note that disabled buttons are "grayed out", and mouse clicks on them are ignored.

Each button has a keyboard shortcut (mnemonic) assigned to it, so the key combination `ALT+e` will press the left button, `ALT+c` will press the center button, and `ALT+d` will press the right button.

This application also illustrates the display of multiple images depending on the state of a button. When the middle button is disabled, it displays a red ball. When it is enabled, it displays a green ball. Finally, when the button is clicked, the ball turns yellow.

This program also shows how a class can handle its own events. Note that the `TestPushButtons` class implements the `ActionListener` interface, and this class is registered with each button using the method `addActionListener(this)`. When an enabled button is clicked, it sends an `ActionEvent` to the `actionPerformed` method of the `TestPushButtons` class. This method checks to see which button generated the event, using the string returned by the `ActionEvent` method `getActionCommand()` to distinguish among them. Note that for buttons the default string returned by `getActionCommand()` is the text on the label of the button. As long as this text is different for every button, we can use that label to tell which button was pressed.

When the left button is pressed, the `actionPerformed` method is called automatically and the result of `e.getActionCommand()` will be `"Enable"`. When the right button is pressed, the `actionPerformed` method is called automatically and

```
// This program tests push buttons.
import java.awt.*;
import java.awt.event.*;
import javax.swing.*;
public class TestPushButtons extends JPanel
                          implements ActionListener {

   // Instance variables
   private int c = 0;              // Count
   private JButton b1, b2, b3;     // Buttons

   // Initialization method
   public void init() {

      // Set the layout manager
      setLayout( new FlowLayout() );

      // Get the images to display
      ImageIcon right = new ImageIcon("RightArrow.gif");
      ImageIcon left = new ImageIcon("LeftArrow.gif");
      ImageIcon green = new ImageIcon("green-ball.gif");
      ImageIcon red = new ImageIcon("red-ball.gif");
      ImageIcon yellow = new ImageIcon("yellow-ball.gif");

      // Create buttons
      b1 = new JButton("Enable",right);
      b1.addActionListener( this );
      b1.setMnemonic('e');
      b1.setToolTipText("Enable middle button");
      add(b1);

      String s = "Count = " + c;
      b2 = new JButton(s,green);
      b2.addActionListener( this );
      b2.setMnemonic('c');
      b2.setEnabled(false);
      b2.setToolTipText("Press to increment count");
      b2.setPressedIcon(yellow);
      b2.setDisabledIcon(red);
      add(b2);

      b3 = new JButton("Disable",left);
      b3.addActionListener( this );
      b3.setMnemonic('d');
      b3.setEnabled(false);
      b3.setToolTipText("Disable middle button");
      add(b3);
   }

   // Event handler to handle button pushes
   public void actionPerformed(ActionEvent e) {
      String button = e.getActionCommand();

     if (button.equals("Enable")) {
         b1.setEnabled(false);
         b2.setEnabled(true);
         b3.setEnabled(true);
      }
```

Figure 9.6. *(cont.)*

```
         else if (button.substring(0,5).equals("Count")) {
            b2.setText("Count = " + (++c));
         }
         else if (button.equals("Disable")) {
            b1.setEnabled(true);
            b2.setEnabled(false);
            b3.setEnabled(false);
         }
      }

      // Main method to create frame
      public static void main(String s[]) {

         // Create a frame to hold the application
         JFrame fr = new JFrame("TestPushButtons ...");
         fr.setSize(400,80);

         // Create a Window Listener to handle "close" events
         MyWindowListener l = new MyWindowListener();
         fr.addWindowListener(l);

         // Create and initialize a TestPushButtons object
         TestPushButtons ob = new TestPushButtons();
         ob.init();

         // Add the object to the center of the frame
         fr.getContentPane().add(ob, BorderLayout.CENTER);

         // Display the frame
         fr.setVisible( true );
      }
   }
```

Figure 9.6. A program to test the operation of push buttons

the result of `e.getActionCommand()` will be `"Disable"`. When the center button is pressed, the `actionPerformed` method is called automatically and the first five letters returned by `e.getActionCommand()` will be `"Count"`. An `if` structure can test the value of the string `button` and perform the proper action, depending on which button was pressed.

Figure 9.7*a* shows the initial state of the program, with the center button disabled. Figure 9.7*b* shows the program after the center button has been enabled and pressed twice.

9.4.3 Text Fields and Password Fields

A text field is a single-line area in which text can be entered by a user from the keyboard. When a user types information into a `JTextField` and presses the Enter key, an `ActionEvent` is generated. If an `ActionListener` has been registered with the text field, then the event will be handled by the `actionPerformed` method, and the data typed by the user is available for use in the program. A `JTextField` may also be used to display read-only text that a user cannot modify. A `JPasswordField` field is identical to a text field, except that asterisks are displayed instead of the characters that are typed on the keyboard.

Figure 9.7. *(a)* The display produced by program `TestPushButtons` when it is first started. *(b)* The display produced by the program after one button click on the left button and two clicks on the center button.

(a)

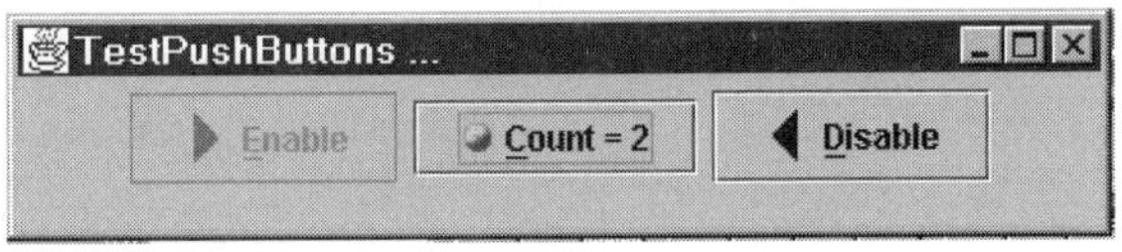

(b)

GOOD PROGRAMMING PRACTICES

1. Buttons generate `ActionEvents` when pressed. To handle button events, use a class that implements the `ActionListener` interface, and contains an `actionPerformed` method. Register an object from this class with each button, and code the `actionPerformed` method to do whatever is required when the button is pressed.
2. One `ActionListener` object can monitor many buttons, using the result of the `getActionCommand` method to determine the button that created a particular event.

A text field or password is created with one of the following constructors:

```
public JTextField(int cols);
public JTextField(String s);
public JTextField(String s, int cols);
public JPasswordField(int columns);
public JPasswordField(String s);
public JPasswordField(String s, int cols);
```

The first form of constructor creates a new blank `JTextField` or `JPasswordField` large enough to contain `cols` characters. The second form of the constructor creates a new `JTextField` or `JPasswordField` initialized with the specified string, and the third form of the constructor creates a new `JTextField` or `JPasswordField` large enough to contain `columns` characters and initialized with the specified string. Some of the methods in these classes are described in Table 9-6.

An `ActionEvent` object is created whenever a user presses Enter on the keyboard after typing data into a text field. The `getActionCommand()` method of the `ActionEvent` object will return a string containing the data that was typed in. As before, this event may be handled by the `actionEvent` method of an `ActionListener`.

The program shown in Figure 9.8 illustrates the use of editable and read-only text fields, plus a password field. The first text field in this program is an ordinary text field in which the data that a user types is visible. The second field is a `JPasswordField`, in which the data that a user types is replaced by asterisks. The third text field is a "read-only" text field. It is used to display the data returned by the event handler after a user presses the Enter key on one of the other text fields.

TABLE 9-6 `JTextField` and `JPasswordField` Methods

METHOD	DESCRIPTION
`public String getText()`	Gets the text currently displayed in this component.
`public String getSelectedText()`	Gets the selected text from the text currently displayed in this component.
`public void setEditable(boolean b)`	Sets the editability status of the field. If `true`, a user can change the data in the field. If `false`, the user cannot change the data in the field.
`public void setText(String t)`	Displays the text in string `t`.
`public void setToolTipText(String text)`	Sets the tool tip text to display when the cursor rests over the text field.

```
// This program tests text fields.
import java.awt.*;
import java.awt.event.*;
import javax.swing.*;
public class TestTextField extends JPanel {

   // Instance variables
   private JLabel l1, l2, l3;           // Labels
   private JTextField t1, t3;           // Text Fields
   private JPasswordField t2;           // Password Field
   private TextFieldHandler handler;    // ActionEvent handler

   // Initialization method
   public void init() {

      // Set background color
      setBackground( Color.lightGray );

      // Set the layout manager
      setLayout( new FlowLayout() );

      // Create ActionEvent handler
      handler = new TextFieldHandler( this );

      // Create first Text Field
      l1 = new JLabel("Visible text here:",JLabel.RIGHT);
      add( l1 );
      t1 = new JTextField("Enter Text Here",25);
      t1.addActionListener( handler );
      add( t1 );

      // Create Password Field
      l2 = new JLabel("Hidden text here:",JLabel.RIGHT);
      add( l2 );
      t2 = new JPasswordField("Enter Text Here",25);
      t2.addActionListener( handler );
      add( t2 );

      // Create third Text Field
      l3 = new JLabel("Results:",JLabel.RIGHT);
      add( l3 );
      t3 = new JTextField(25);
      t3.setEditable( false );
      add( t3 );
   }
```

Figure 9.8. *(cont.)*

```
      // Method to update t3
      public void updateT3( String s ) {
         t3.setText( s );
      }

      // Main method to create frame
      public static void main(String s[]) {

         // Create a frame to hold the application
         JFrame fr = new JFrame("TestTextField ...");
         fr.setSize(400,130);

         (rest of main is the same as previous examples ...)
      }
   }

   class TextFieldHandler implements ActionListener {
      private TestTextField ttf;

      // Constructor
      public TextFieldHandler ( TestTextField t ) {
         ttf = t;
      }

      // Execute when an event occurs
      public void actionPerformed( ActionEvent e ) {
         ttf.updateT3(e.getActionCommand());
      }
   }
```

Figure 9.8. A program to test `JTextField` and `JPasswordField` objects

When information is typed in `JTextField` 1 and the Enter key is pressed, an ActionEvent is generated and handled by the actionPerformed method. This method makes a callback to method `updateT3` to display the typed information in the read-only field.

When this program is executed, the results after information is typed into `JTextField` 1 are shown in Figure 9.9*a*, and the results after information is typed into `JPasswordField` 2 are shown in Figure 9.9*b*.

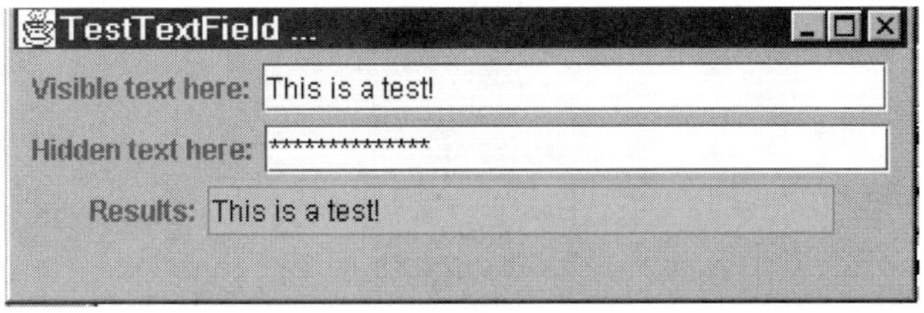

(a)

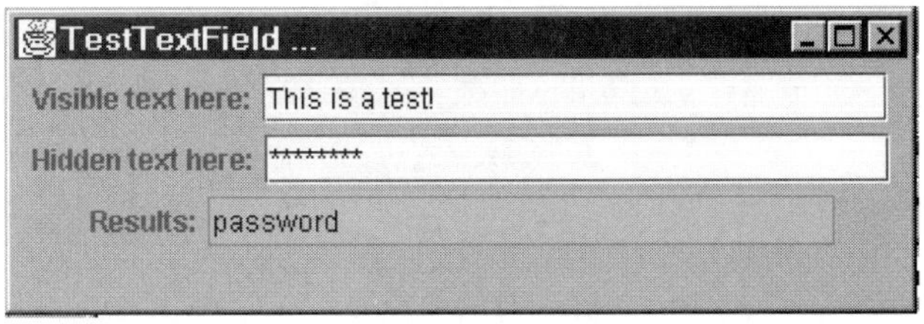

(b)

Figure 9.9. Results when information is typed in the text and password fields.

GOOD PROGRAMMING PRACTICES

1. Use `JTextFields` to accept single lines of input data from a user, or to display single lines of read-only data to the user.
2. Use `JPasswordFields` to accept input data from a user that you do not wish to have echoed to the screen.

EXAMPLE 9-1: TEMPERATURE CONVERSION

Write a program that converts temperature from degrees Fahrenheit to degrees Celsius and vice versa using a GUI to accept data and display results.

SOLUTION

To create this program, we will need a label and text field for the temperature in degrees Fahrenheit and another label and text field for the temperature in degrees Celsius. We will also need a method to convert degrees Fahrenheit to degrees Celsius, and a method to convert degrees Celsius to degrees Fahrenheit. Finally, we will need two event handlers to accept text entry in the two text fields.

The `init()` method for this program must create two labels and two text fields to hold the temperature in degrees Celsius and degrees Fahrenheit. In addition, it must create `ActionListener` objects for both text fields. The code for these steps is as

```
// Create ActionEvent handlers
cHnd = new DegCHandler( this );
fHnd = new DegFHandler( this );

// Create degrees Celsius field
l1 = new JLabel("deg C:", JLabel.RIGHT);
add( l1 );
t1 = new JTextField("0.0",15);
t1.addActionListener( cHnd );
add( t1 );

// Create degrees Celsius field
l2 = new JLabel("deg F:", JLabel.RIGHT);
add( l2 );
t2 = new JTextField("32.0",15);
t2.addActionListener( fHnd );
add( t2 );
```

Method `toC` will convert temperature from degrees Fahrenheit to degrees Celsius. It must implement the equation

$$\deg C = \frac{5}{9}(\deg F - 32). \tag{9-1}$$

It must also update the text fields with this information. The pseudocode for these steps is as follows:

```
degC ← (5. / 9.) * (degF - 32);
t1.setText( Fmt.sprintf("%5.1f",degC) );
t2.setText( Fmt.sprintf("%5.1f",degF) );
```

Note that we are using the `Fmt.sprintf` method in the `chapman.io` package to format the temperatures for display. Method `toF` will convert temperature from degrees Celsius to degrees Fahrenheit. It must implement the equation

$$\deg F = \frac{9}{5} \deg C + 32. \quad (9\text{-}2)$$

It must also update the text fields with this information. The pseudocode for these steps is as follows:

```
degF ← (9. / 5.) * degC + 32;
t1.setText( Fmt.sprintf("%5.1f",degC) );
t2.setText( Fmt.sprintf("%5.1f",degF) );
```

The `ActionListeners` must listen for inputs in a text field, convert the input `String` into a `double` value, and call the appropriate `toC` or `toF` method. For example, the `actionPerformed` method that monitors the degrees Celsius text field would be coded as follows:

```
public void actionPerformed( ActionEvent e ) {
   String input = e.getActionCommand();
   double degC = new Double(input).doubleValue();
   tc.toF( degC );
}
```

The final program is shown in Figure 9.10.

```
/*
   Purpose:
     This GUI-based program converts temperature in
     degrees Fahrenheit to degrees Celsius, and vice versa.

   Record of revisions:
       Date        Programmer           Description of change
       ====        ==========           =====================
     10/14/98    S. J. Chapman          Original code
*/
import java.awt.*;
import java.awt.event.*;
import javax.swing.*;
import chapman.io.*;
public class TempConversion extends JPanel {

   // Instance variables
   private JLabel l1, l2;      // Labels
   private JTextField t1, t2; // Text Fields
   private DegCHandler cHnd;  // ActionEvent handler
   private DegFHandler fHnd;  // ActionEvent handler

   // Initialization method
   public void init() {

      // Set the layout manager
      setLayout( new FlowLayout() );

      // Create ActionEvent handlers
      cHnd = new DegCHandler( this );
      fHnd = new DegFHandler( this );

      // Create degrees Celsius field
      l1 = new JLabel("deg C:", JLabel.RIGHT);
      add( l1 );
      t1 = new JTextField("0.0",15);
      t1.addActionListener( cHnd );
      add( t1 );
```

Figure 9.10. *(cont.)*

```
      // Create degrees Celsius field
      l2 = new JLabel("deg F:", JLabel.RIGHT);
      add( l2 );
      t2 = new JTextField("32.0",15);
      t2.addActionListener( fHnd );
      add( t2 );
   }

   // Method to convert deg F to deg C
   // and display result
   public void toC( double degF ) {
      double degC = (5. / 9.) * (degF - 32);
      t1.setText( Fmt.sprintf("%5.1f",degC) );
      t2.setText( Fmt.sprintf("%5.1f",degF) );
   }

   // Method to convert deg C to deg F
   // and display result
   public void toF( double degC ) {
      double degF = (9. / 5.) * degC + 32;
      t1.setText( Fmt.sprintf("%5.1f",degC) );
      t2.setText( Fmt.sprintf("%5.1f",degF) );
   }

   // Main method to create frame
   public static void main(String s[]) {

      // Create a frame to hold the application
      JFrame fr = new JFrame("TempConversion ...");
      fr.setSize(250,100);

      // Create a Window Listener to handle "close" events
      MyWindowListener l = new MyWindowListener();
      fr.addWindowListener(l);

      // Create and initialize a TempConversion object
      TempConversion tf = new TempConversion();
      tf.init();

      // Add the object to the center of the frame
      fr.getContentPane().add(tf, BorderLayout.CENTER);

      // Display the frame
      fr.setVisible( true );
   }
}

class DegCHandler implements ActionListener {
   private TempConversion tc;

   // Constructor
   public DegCHandler( TempConversion t ) { tc = t; }
   // Execute when an event occurs
   public void actionPerformed( ActionEvent e ) {
      String input = e.getActionCommand();
      double degC = new Double(input).doubleValue();
      tc.toF( degC );
   }
}

class DegFHandler implements ActionListener {
   private TempConversion tc;
```

Figure 9.10. *(cont.)*

```
      // Constructor
      public DegFHandler( TempConversion t ) { tc = t; }

      // Execute when an event occurs
      public void actionPerformed( ActionEvent e ) {
         String input = e.getActionCommand();
         double degF = new Double(input).doubleValue();
         tc.toC( degF );
      }
   }
```

Figure 9.10. A GUI-based temperature conversion program

When this program is executed, the results are as shown in Figure 9.11. Try this program for yourself with several different temperature values.

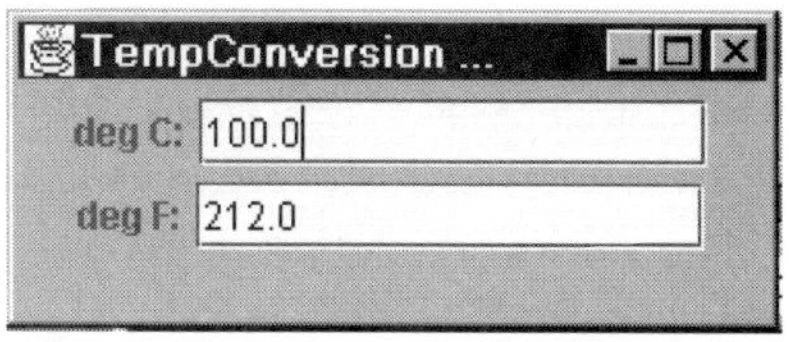

(a)

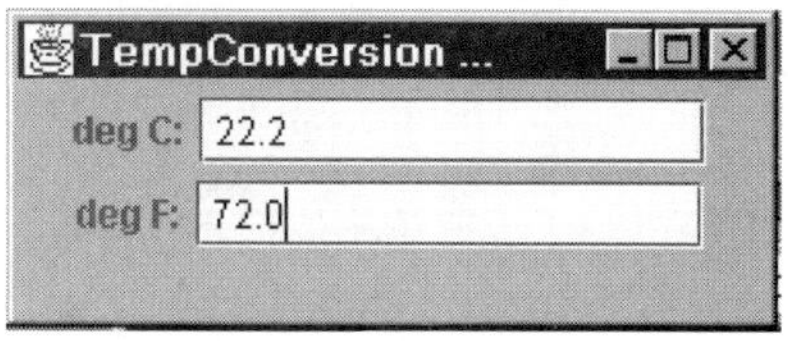

(b)

Figure 9.11. *(a)* Result when the user enters 100°C. *(b)* Result when the user enters 72°F.

9.4.4 Combo Boxes

A **combo box** is a field in which a user can either type an entry or select an entry from a *drop-down list* of choices. If desired, the combo box can be restricted so that only choices in the drop-down list may be selected. The selected choice is displayed in the combo box field after a selection has been made.

Combo boxes are implemented by the `JComboBox` class. A `JComboBox` can be created with the following constructor:

```
public JComboBox( Object[] );
```

This constructor builds a new combo box with the array of objects (such as `Strings`) used to initialize the choices in the box. Some of the methods in class `JComboBox` are described in Table 9-7.

Class `JComboBox` implements the `ActionListener` interface, which means that it generates `ActionEvents`. When an item is accepted in a `JComboBox` field, the `JComboBox` generates an `ActionEvents` and sends that event to any objects that have been registered with it as listeners.

TABLE 9-7 Selected `JComboBox` Methods

METHOD	DESCRIPTION
`public void addActionListener(ActionListener l)`	Adds the specified listener to receive `ActionEvents` from this `JComboBox` list. These events happen when selection is complete.
`public void addItem(Object o)`	Adds an item to the `JComboBox` list.
`public Object getItemAt(int index)`	Returns the `JComboBox` item at location `index`.
`public int getItemCount()`	Returns the number of items in this `JComboBox` list.
`public int getSelectedIndex()`	Returns the index of the selected `JComboBox` item.
`public Object getSelectedItem()`	Returns the selected `JComboBox` item.
`public boolean isEditable()`	Returns the editable state of the `JComboBox`.
`public void insertItemAt(Object o, int i)`	Inserts an item at position `i` in the `JComboBox` list.
`public void removeItem(Object o)`	Removes the specified `JComboBox` item.
`public void removeItemAt(int index)`	Removes the `JComboBox` item at location `index`.
`public void setEditable(boolean b)`	If `true`, user can type in the combo box (default is `false`).
`public void setToolTipText(String text)`	Sets the tool tip text to display when the cursor rests over the text field.

The program in Figure 9.12 shows how to create a `JComboBox`, and to implement choices based on selections in that field. This class creates a combo box and a read-only text field. The user selects a font name from the choice list (Serif, SansSerif, Monospaced, or Dialog), and the sample text is displayed with appropriate formatting in the text field. (Note that this example uses an *uneditable* combo box, because only the valid font names supplied should be selectable.)

```
// This program tests combo boxes.
import java.awt.*;
import java.awt.event.*;
import javax.swing.*;
public class TestComboBox extends JPanel {

   // Instance variables
   private JComboBox c1;            // Combo box
   private JTextField t1;           // TextField
   private ComboHandler handler;    // ActionEvent handler

   // Initialization method
   public void init() {

      // Set background color
      setBackground( Color.lightGray );

      // Set the layout manager
      setLayout( new FlowLayout() );

      // Create ActionEvent handler
      handler = new ComboHandler( this );

      // Create the JComboBox
      String[] s = {"Serif","SansSerif","Monospaced",
                    "Dialog"};
```

Figure 9.12. *(cont.)*

```
         c1 = new JComboBox(s);
         c1.addActionListener( handler );
         add( c1 );

         // Create the text field with default font
         Font font = new Font(c1.getItemAt(0).toString(),
                              Font.PLAIN, 14);
         t1 = new JTextField("Test string",30);
         t1.setEditable( false );
         t1.setFont( font );
         add( t1 );
      }


      // Method to update font
      public void updateFont() {
         int valBold, valItalic;

         // Get current font info
         int fontStyle = t1.getFont().getStyle();
         int fontSize  = t1.getFont().getSize();

         // Get new font name
         String fontName = (String) c1.getSelectedItem();

         // Set new font
         t1.setFont( new Font(fontName, fontStyle, fontSize) );

         // Repaint the JTextField
         t1.repaint();
      }

      // Main method to create frame
      public static void main(String s[]) {

         // Create a frame to hold the application
         JFrame fr = new JFrame("TestComboBox ...");
         fr.setSize(400,100);

         (rest of main is the same as previous examples ...)
      }
   }

   class ComboHandler implements ActionListener {
      private TestComboBox tcb;

      // Constructor
      public ComboHandler( TestComboBox t ) { tcb = t; }

      // Execute when an event occurs
      public void actionPerformed( ActionEvent e ) {

         // State has changed, so call updateFont
         tcb.updateFont();
      }
   }
```

Figure 9.12. A program showing how different fonts can be selected with a `JComboBox`

When the combo box is clicked, a drop-down list appears with the four possible font names. When the user selects one of these font names by clicking on the name, an `ActionEvent` is created and the `actionPerformed` method is called. This method in turn calls `updateFont()` to display the new font. Method `updateFont()` gets the new font name from the choice list using the `getSelectedItem()` method,

Figure 9.13. The results of program `TestComboBox` as different fonts are selected from the choice list

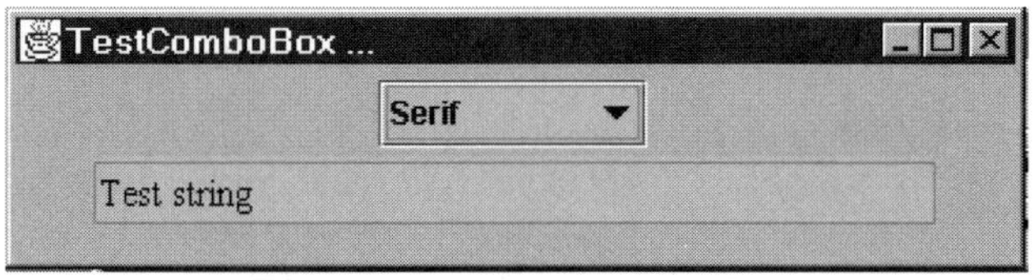

(a)

(b)

(c)

which returns an `Object` reference. This reference is downcast to a `String`, and that font name is used to create a new font with that name and the original font style (`PLAIN`, `BOLD`, etc.) and size. Finally, the `JTextField` is repainted to display the new font. When this program is executed, the results are as shown in Figure 9.13.

GOOD PROGRAMMING PRACTICE

Use `JComboBox`es to make a single selection from a list of mutually exclusive choices.

9.4.5 Checkboxes and Radio Buttons

A **checkbox** is a type of button that toggles between two possible states: on and off. When the checkbox is on, a small check mark appears in it, and when the checkbox is off, the mark is removed. Each time that the mouse clicks on a checkbox, the checkbox toggles to the opposite state. Checkboxes look like small square boxes with check marks in them, but they are in fact fully-fledged buttons with all of the features that we learned about when studying pushbuttons. For example, it is possible to include both text and images on a check box, and it is possible to define keyboard shortcuts for them. Checkboxes are implemented by the **`JCheckBox`** class.

A `JCheckBox` is created with the constructor:

TABLE 9-8 Selected `JCheckBox` and `JRadioButton` Methods

METHOD	DESCRIPTION
`public String addActionListener(ActionListener l)`	Adds the specified listener to receive `ActionEvents` from this `JCheckBox`.
`public Icon getIcon()`	Returns the image from a `JCheckBox`.
`public String getLabel()`	Returns the label from the `JCheckBox`.
`public boolean isSelected()`	Returns the state of the `JCheckBox` (true or false).
`public void setLabel(String s)`	Sets a new label into the `JCheckBox`.
`public void setActionCommand(String s)`	Set the action command string generated by this button when it is pressed.
`public void setDisabledIcon(Icon icon)`	Sets the icon to display when the button is disabled.
`public void setEnabled(boolean b)`	Enables or disables this button.
`public void setHorizontalAlignment(int alignment)`	Sets the horizontal alignment of the text and images. Legal values are `LEFT`, `CENTER`, and `RIGHT`.
`public void setHorizontalTextPosition(int textPosition)`	Sets the position of the text relative to the images. Legal values are `LEFT`, `CENTER`, and `RIGHT`.
`public void setLabel(String s)`	Sets the label of this button to the specified String.
`public void setIcon(Icon icon)`	Sets the default icon for this button.
`public void setMnemonic(char mnemonic)`	Set the keyboard character combination used to activate the button from the keyboard.
`public void setPressedIcon(Icon icon)`	Sets the icon to display when the button is pressed.
`public void setSelected(boolean b)`	Sets a new state into the `JCheckBox`.
`public void setToolTipText(String text)`	Sets the tool tip text to display when the cursor rests over the checkbox or button.
`public void setVerticalAlignment(int alignment)`	Sets the vertical alignment of the text and images. Legal values are `TOP`, `CENTER`, and `BOTTOM`.
`public void setVerticalTextPosition(int textPosition)`	Sets the position of the text relative to the images. Legal values are `TOP`, `CENTER`, and `BOTTOM`.

```
public JCheckBox( String s );
public JCheckBox( String s, boolean state );
public JCheckBox( Icon Image );
public JCheckBox( Icon image, boolean state );
public JCheckBox( String s, Icon image );
public JCheckBox( String s, Icon image, boolean state );
```

In these constructors, `s` is the text label for the checkbox, `image` is the image to display on the checkbox, and `state` is the initial on/off state of the checkbox. If `state` is absent from a constructor, then the new checkbox defaults to off. Some of the methods in class `JCheckBox` are described in Table 9-8.

Class `JCheckBox` implements the `ActionListener` interface, which means that it generates `ActionEvents`. When a `Checkbox` is clicked, the `Checkbox` generates an `ActionEvent` and sends that event to any objects that have been registered with it as listeners.

The program in Figure 9.14 shows how to create `JCheckBoxes`, and to implement choices based on the state of the `JCheckBoxes`. This class expands on the previous program by adding two check boxes to select bold and/or italic font types for display.

```
// This program tests Checkboxes.
import java.awt.*;
import java.awt.event.*;
import javax.swing.*;
public class TestCheckBox extends JPanel {

   // Instance variables
   private JCheckBox cb1, cb2;      // Check boxes
   private JComboBox c1;            // Combo box
   private JTextField t1;           // TextField
   private ActionHandler h1;        // ActionEvent handler

   // Initialization method
   public void init() {

      // Set the layout manager
      setLayout( new FlowLayout() );

      // Create ActionEvent handler
      h1 = new ActionHandler( this );

      // Create the JComboBox for font names
      String[] s = {"Serif","SansSerif","Monospaced",
                    "Dialog"};
      c1 = new JComboBox(s);
      c1.addActionListener( h1 );
      add( c1 );

      // Create the text field with default font
      Font font = new Font( c1.getItemAt(0).toString(),
                            Font.PLAIN, 14);
      t1 = new JTextField("Test string",20);
      t1.setEditable( false );
      t1.setFont( font );
      add( t1 );

      // Create check boxes for bold and italic
      cb1 = new JCheckBox("Bold");
      cb1.addActionListener( h1 );
      cb1.setMnemonic('b');
      add( cb1 );
      cb2 = new JCheckBox("Italic");
      cb2.addActionListener( h1 );
      cb2.setMnemonic('i');
      add( cb2 );
   }

   // Method to update font
   public void updateFont() {
      int valBold, valItalic;

      // Get current font info
      int fontStyle   = t1.getFont().getStyle();
      int fontSize    = t1.getFont().getSize();

      // Get new font name
      String fontName = (String) c1.getSelectedItem();

      // Get new font style
      valBold   = cb1.isSelected() ? Font.BOLD   : Font.PLAIN;
      valItalic = cb2.isSelected() ? Font.ITALIC : Font.PLAIN;
      fontStyle = valBold + valItalic;
```

Figure 9.14. *(cont.)*

```
        // Set new font
        t1.setFont( new Font(fontName, fontStyle, fontSize) );

        // Repaint the JTextField
        t1.repaint();
     }

     // Main method to create frame
     public static void main(String s[]) {

        // Create a frame to hold the application
        JFrame fr = new JFrame("TestCheckBox ...");
        fr.setSize(380,100);

        (rest of main is the same as previous examples ...)
     }
  }

  class ActionHandler implements ActionListener {
     private TestCheckBox tcb;

     // Constructor
     public ActionHandler( TestCheckBox t ) { tcb = t; }

     // Execute when an event occurs
     public void actionPerformed( ActionEvent e ) {

        // State has changed, so call updateFont
        tcb.updateFont();
     }
  }
```

Figure 9.14. A program using `JCheckBoxes` to select bold and italic font styles

When a check box is clicked (or activated by the proper keyboard combination), an `ActionEvent` is created and the `actionPerformed` method in class `ActionHandler` is called. This method in turn calls `updateFont()` to display the new font. Method `updateFont()` gets the status of the Bold and Italic checkboxes using the `isSelected()` method, and creates a new font with that style.

When this program is executed, the results are as shown in Figure 9.15.

GOOD PROGRAMMING PRACTICE

Use `JCheckBoxes` to select the state of items represented by boolean variables, which can only be `true` or `false`.

Radio buttons are a group of checkboxes in which *at most one checkbox can be on at a time.* Radio buttons look like small circles with a dot inside the selected one, but otherwise have the same characteristics as any button. Radio buttons are implemented by the **`JRadioButton`** class.

A `JRadioButton` is created with the constructor:

```
public JRadioButton( String s );
public JRadioButton( String s, boolean state );
public JRadioButton( Icon Image );
public JRadioButton( Icon image, boolean state );
public JRadioButton( String s, Icon image );
public JRadioButton( String s, Icon image, boolean state );
```

Figure 9.15. The results of program `TestCheckBox` as different combinations of the `Bold` and `Italic` check boxes are turned on

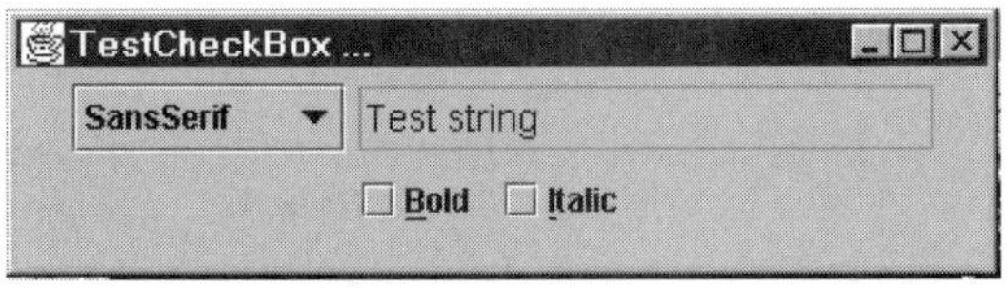

(a)

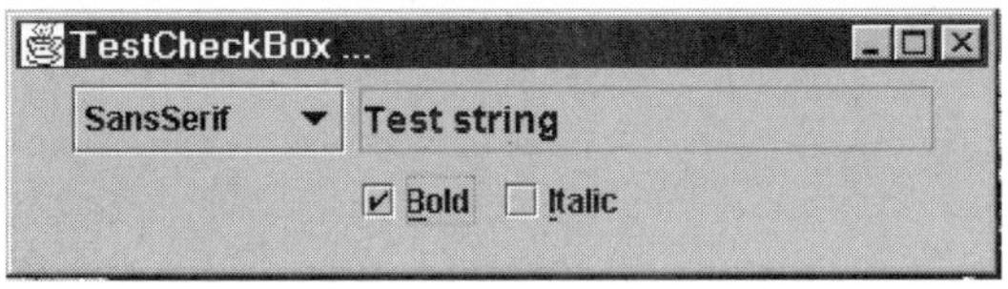

(b)

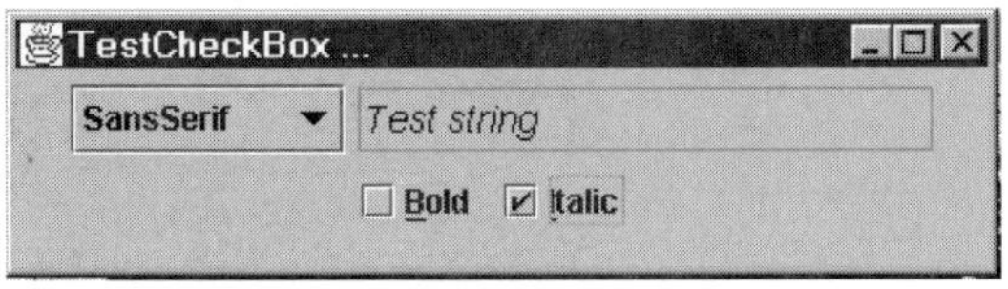

(c)

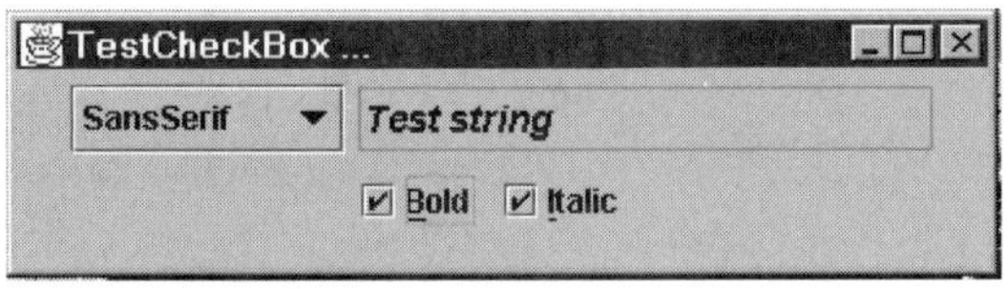

(d)

In these constructors, `s` is the text label for the radio button, `image` is the image to display on the radio button, and `state` is the initial on/off state of the radio button. If `state` is absent from a constructor, then the new radio button defaults to off. The methods in class `JRadioButton` are the same as those in class `JCheckBox`, and are described in Table 9-8.

A group of radio buttons is made mutually exclusive (only one can be on at a time) by placing them in a **`ButtonGroup`**. The `ButtonGroup` ensures that when one of the `JRadioButtons` is turned on, the others are all off.

A `ButtonGroup` is created with the constructor:

```
public ButtonGroup();
```

Some of the methods in class `ButtonGroup` are described in Table 9-9.

TABLE 9-9 Selected `ButtonGroup` Methods

METHOD	DESCRIPTION
`public void add(AbstractButton b)`	Add a button to the `ButtonGroup`.
`public void remove(AbstractButton b)`	Remove a button from the `ButtonGroup`.

The program in Figure 9.16 shows how to create a set of radio buttons using `JRadioButtons` in a `ButtonGroup`. This class modifies the previous program b using four radio buttons instead of two checkboxes to select bold and/or italic font types for display.

```
// This program tests radio buttons.
import java.awt.*;
import java.awt.event.*;
import javax.swing.*;
public class TestRadioButton extends JPanel {

   // Instance variables
   private ButtonGroup bg;               // ButtonGroup
   private JRadioButton b1, b2, b3, b4;  // Check boxes
   private JComboBox c1;                 // Combo box
   private JTextField t1;                // TextField
   private ActionHandler h1;             // ActionEvent handler

   // Initialization method
   public void init() {

      // Set the layout manager
      setLayout( new FlowLayout() );

      // Create ActionEvent handler
      h1 = new ActionHandler( this );

      // Create the JComboBox for font names
      String[] s = {"Serif","SansSerif","Monospaced",
                    "Dialog"};
      c1 = new JComboBox(s);
      c1.addActionListener( h1 );
      add( c1 );

      // Create the text field with default font
      Font font = new Font( c1.getItemAt(0).toString(),
                            Font.PLAIN, 14);
      t1 = new JTextField("Test string",20);
      t1.setEditable( false );
      t1.setFont( font );
      add( t1 );

      // Create radio buttons
      b1 = new JRadioButton("Plain", true );
      b1.addActionListener( h1 );
      add( b1 );
      b2 = new JRadioButton("Bold", false );
      b2.addActionListener( h1 );
      add( b2 );
      b3 = new JRadioButton("Italic", false );
      b3.addActionListener( h1 );
      add( b3 );
      b4 = new JRadioButton("Bold Italic", false );
      b4.addActionListener( h1 );
      add( b4 );

      // Create button group, and add radio buttons
      bg = new ButtonGroup();
      bg.add( b1 );
      bg.add( b2 );
      bg.add( b3 );
      bg.add( b4 );
   }
```

Figure 9.16. *(cont.)*

```
    // Method to update font
    public void updateFont() {
       int valBold, valItalic;

       // Get current font info
       int fontStyle    = t1.getFont().getStyle();
       int fontSize     = t1.getFont().getSize();

       // Get new font name
       String fontName = (String) c1.getSelectedItem();

       // Get new font style
       if ( b1.isSelected() )
          fontStyle = Font.PLAIN;
       else if ( b2.isSelected() )
          fontStyle = Font.BOLD;
       else if ( b3.isSelected() )
          fontStyle = Font.ITALIC;
       else if ( b4.isSelected() )
          fontStyle = Font.BOLD + Font.ITALIC;

       // Set new font
       t1.setFont( new Font(fontName, fontStyle, fontSize) );

       // Repaint the JTextField
       t1.repaint();
    }

    // Main method to create frame
    public static void main(String s[]) {

       // Create a frame to hold the application
       JFrame fr = new JFrame("TestRadioButton ...");
       fr.setSize(380,100);

       (rest of main is the same as previous examples ...)
    }
 }
 class ActionHandler implements ActionListener {
    private TestRadioButton tcb;

    // Constructor
    public ActionHandler( TestRadioButton t ) { tcb = t; }

    // Execute when an event occurs
    public void actionPerformed( ActionEvent e ) {

       // State has changed, so call updateFont
       tcb.updateFont();
    }
 }
```

Figure 9.16. A program using radio buttons to select font styles

When a radio button is clicked, an `ActionEvent` is created and the `action-Performed` method in class `ActionHandler` is called. This method in turn calls `updateFont()` to display the new font. Method `updateFont()` determines which radio button is selected using the `isSelected()` method, and creates a new font with that style. Note that the `ButtonGroup` ensures that only one of the radio buttons will be selected at a time.

When this program is executed, the results are as shown in Figure 9.17 .

Figure 9.17. The results of program `TestRadioButton` as different radio buttons are turned on

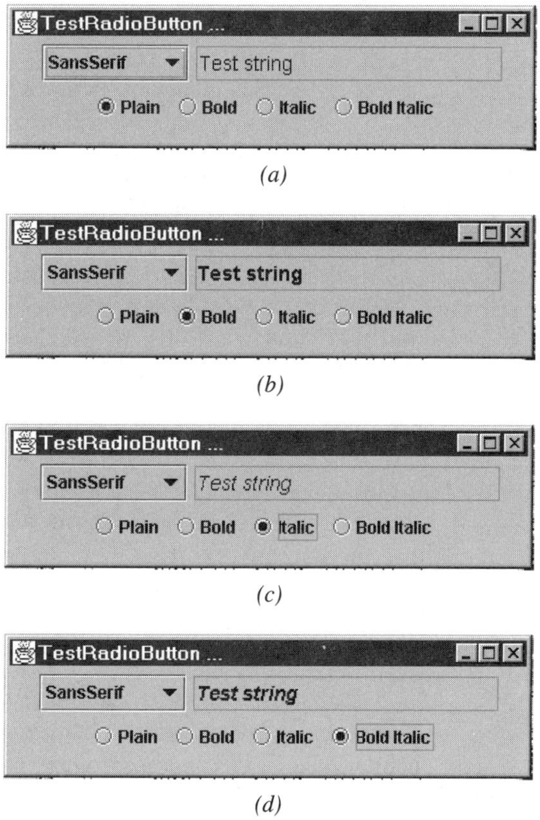

GOOD PROGRAMMING PRACTICE

Use `JRadioButton`s to select the state of a set of items represented by `boolean` variables, only one of which can be `true` at any time.

9.5 LAYOUT MANAGERS

A **layout manager** is a helper class that is designed to automatically arrange GUI components within a container for presentation purposes. The layout managers allow a user to add components to a container without worrying about the specifics of how to place them within the container. They are especially useful for cross platform applications, because the size of a particular component may vary slightly from platform to platform, and the layout manager will automatically adjust the spacing of components to accommodate these differences.

The six standard layout managers in Java are summarized in Table 9-10. Only the `BorderLayout`, `BoxLayout`, `FlowLayout`, and `GridLayout` managers will be

TABLE 9-10 Standard Layout Managers

ELEMENT	DESCRIPTION
`BorderLayout`	A layout manager that lays out elements in a central region and four surrounding borders. This is the default layout manager for a `JFrame`.
`BoxLayout`	A layout manager that lays out elements in a row horizontally or vertically. Unlike `FlowLayout`, the elements in a `BoxLayout` do not wrap around. This is the default layout manager for a `Box`.
`CardLayout`	A layout manager that stacks components like a deck of cards, only the top one of which is visible.
`FlowLayout`	A layout manager that lays out elements left-to-right and top-to-bottom within a container. This is the default layout manager for a `JPanel`.
`GridBagLayout`	A layout manager that lays out elements in a flexible grid, where the size of each element can vary.
`GridLayout`	A layout manager that lays out elements in a rigid grid.

described in this book. Refer to the JDK documentation for information about the `CardLayout` and `GridBagLayout` managers.

9.5.1 `BorderLayout` Layout Manager

The `BorderLayout` layout manager arranges components in five regions, known as *North, South, East, West,* and *Center* (with North being the top of the container). A `BorderLayout` is created with one of the following constructors:

```
public BorderLayout();
public BorderLayout(int horizontalGap, int vertical Gap);
```

The first constructor creates a `BorderLayout` with no pixel gaps between components, while the second constructor creates a `BorderLayout` with the programmer-specified gaps between components.

After a layout object has been created, it is associated with a container by the container's **`setLayout`** method. For example, the statement required to create a new `BorderLayout` object and to associate it with the current container is:

```
setLayout( new BorderLayout() );
```

Objects should be added to a `BorderLayout` using the `add` method qualified by one of the constants `NORTH`, `SOUTH`, `EAST`, `WEST`, or `CENTER`. For example, the following statements produce a GUI containing five buttons, one in each region:

```
setLayout(new BorderLayout());
add(new Button("North"), BorderLayout.NORTH);
add(new Button("South"), BorderLayout.SOUTH);
add(new Button("East"), BorderLayout.EAST);
add(new Button("West"), BorderLayout.WEST);
add(new Button("Center"), BorderLayout.CENTER);
```

These statements will produce the GUI shown in Figure 9.18.

In a `BorderLayout`, the component in the center expands to use up all remaining space in the container. `BorderLayout` is the default layout manager for `JFrame`, and all of our frames have used it. Examine the `main` method in any of the programs in this chapter. Note that the `JPanels` that we have created were added to the `CENTER` section of the `JFrame`, and none of the other sections were used. Since the center section of a `BorderLayout` expands to use all available space, the `JPanels` have completely filled their frames.

Figure 9.18. A typical `BorderLayout`

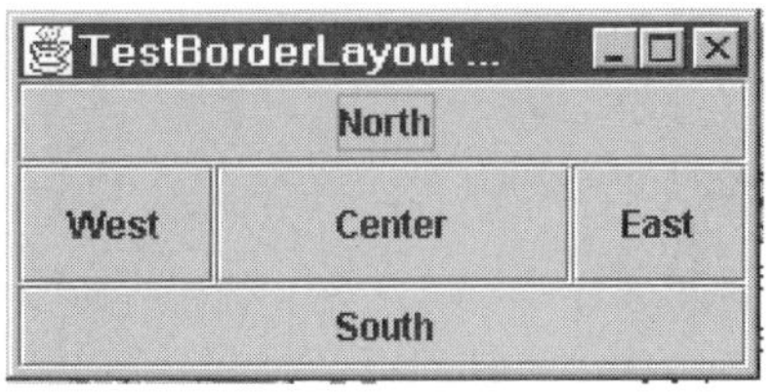

Programs `FirstGUI` in Figure 9.1 and `TestLabel` in Figure 9.4 also illustrate the use of the `BorderLayout` layout manager.

9.5.2 `FlowLayout` Layout Manager

The `FlowLayout` layout manager arranges components in order from left to right and top to bottom across a container. Components are added to a line until there is no more room, and then they are added to the next line. The components that do fit into a line are displayed centered horizontally on the line. A `FlowLayout` is created with one of the following constructors:

```
public FlowLayout();
public FlowLayout(int align);
public FlowLayout(int align, int horizontalGap, int verticalGap);
```

The first constructor creates a `FlowLayout` with no pixel gaps between components and with the components centered on each line, while the second constructor creates a `FlowLayout` with no pixel gaps between components and with the specified alignment. Possible alignments are `LEFT`, `RIGHT`, and `CENTER`. The third constructor allows the programmer to specify both the alignment on each line and the horizontal and vertical gap between components.

Objects should be added to a `FlowLayout` using the `add` method. For example, the following statements produce a GUI containing five buttons, centered on each line, with a five-pixel gap between buttons:

```
setLayout(new FlowLayout(FlowLayout.CENTER,5,0));

add(new JButton("Button 1"));
add(new JButton("Button 2"));
add(new JButton("Long Button 3"));
add(new JButton("B4"));
add(new JButton("Button 5"));
```

These statements will produce the GUI shown in Figure 9.19.

Figure 9.19. A typical `Flow-Layout`

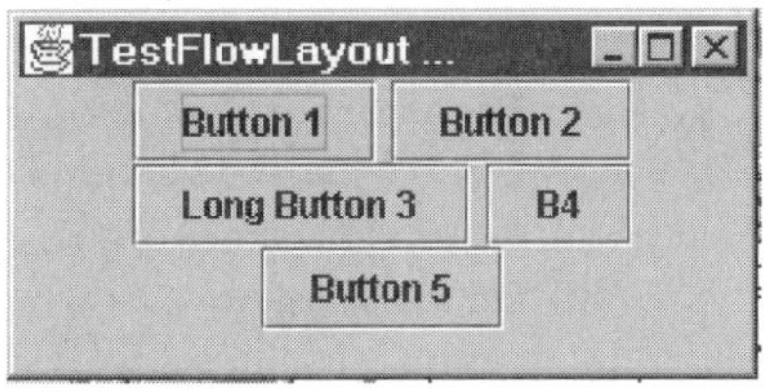

Figure 9.20. A typical `Grid-Layout`

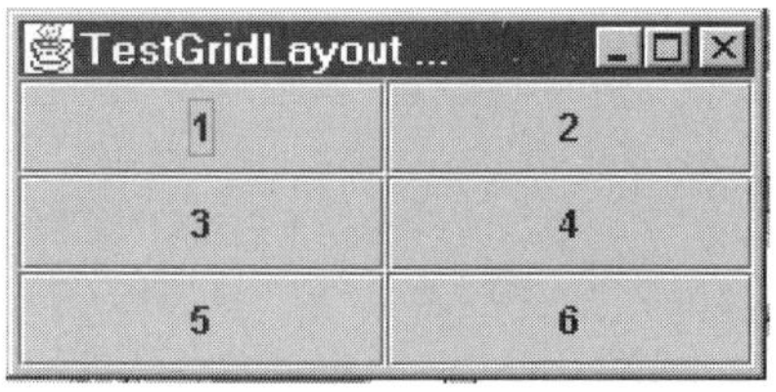

Most of the programs in the chapter have used `FlowLayout` to lay out the components on their panels. `FlowLayout` is the default layout manager for `JPanel` and for applets (class `JApplet`).

9.5.3 `GridLayout` Layout Manager

The `GridLayout` layout manager arranges components in a rigid rectangular grid structure. The container is divided into equal-sized rectangles, and one component is placed in each rectangle. A `GridLayout` is created with one of the following constructors:

```
public GridLayout(int rows, int cols);
public GridLayout(int rows, int cols, int horizGap, int vertGap);
```

The first constructor creates a `GridLayout` with `rows` rows and `cols` columns, and no pixel gaps between components. The second constructor creates a `GridLayout` with `rows` rows and `cols` columns, and with the specified horizontal and vertical gaps between components.

Objects are be added to a `GridLayout` using the `add` method. For example, the following statements produce a GUI containing six buttons in a 3×2 grid.

```
setLayout(new GridLayout(3,2));
add(new JButton("1"));
add(new JButton("2"));
add(new JButton("3"));
add(new JButton("4"));
add(new JButton("5"));
add(new JButton("6"));
```

These statements will produce the GUI shown in Figure 9.20.

9.5.4 `BoxLayout` Layout Manager

The `BoxLayout` layout manager arranges components within a container in a single row or a single column. It is more flexible than the other layout managers that we have discussed, since the spacing and alignment of each element on each row or column can be individually controlled. Containers using `BoxLayout` layout managers can be nested inside each other to produce arbitrarily complex structures that do not change shape when the size of a component or container is changed.

A `BoxLayout` is created with the following constructor:

```
public BoxLayout(Container c, int direction);
```

The first parameter in the constructor specifies the container that the layout manager will control, and the second component specifies the axis along which the components will be laid out (legal values are `BoxLayout.X_AXIS` and `BoxLayout.Y_AXIS`).

Objects are added to a `BoxLayout` using the `add` method. For example, the following statements produce a GUI containing three buttons arranged vertically.

Figure 9.21. A typical vertical `BoxLayout`

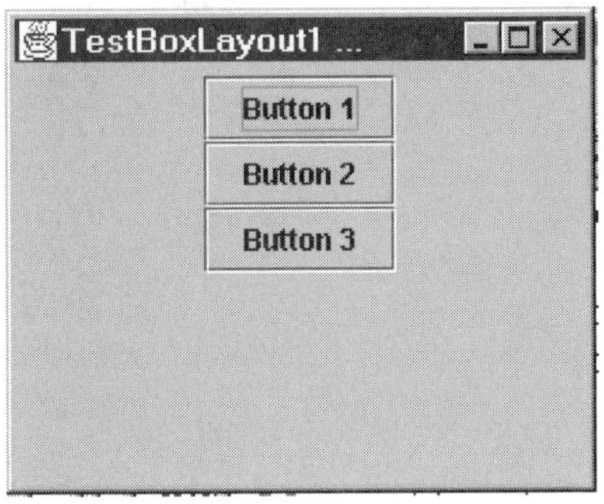


```
// Create a new panel
JPanel p = new JPanel();

// Set the layout manager
p.setLayout(new BoxLayout(p, BoxLayout.Y_AXIS));

// Add buttons
p.add( new JButton("Button 1") );
p.add( new JButton("Button 2") );
p.add( new JButton("Button 3") );

// Add the new panel to the existing container
add( p );
```

These statements will produce the GUI shown in Figure 9.21.

`BoxLayout` honors the preferred size specified for each component, so you can change the size of any component in a `BoxLayout` using a method call of the form

```
comp.setPreferredSize( new Dimension(200,100) );
```

This method call will set the size of component `comp` to be 200 pixels wide by 100 pixels high. Most components such as labels and text fields default to reasonable sizes, so you will not need to set a preferred size for them. However, this method can be very useful for setting the size of graphics components like `JPlot2D`.

The flexibility of the `BoxLayout` manager is enhanced by two additional constraints that can be added to the layout process: **rigid areas** and **glue regions**. Rigid areas are fixed horizontal and/or vertical spacings between components that can be individually specified between any two adjacent components. **Glue regions** are regions that expand or contract to absorb any extra space present when a container changes size. Rigid areas and glue regions are created using the methods in Table 9-11.

To illustrate the use of these components, we will create a new GUI that lays out the same three buttons vertically as before, but places a fixed 20-pixel vertical spacing

TABLE 9-11 Methods to Control Spacing in a `BoxLayout`

METHOD	DESCRIPTION
`Box.createRigidArea(Dimension d)`	Creates a rigid spacing in pixels between two components in a `BoxLayout`. The `Dimension` object specifies the vertical and horizontal spacing between the two components.
`Box.createHorizontalGlue()`	Creates a "virtual component" that uses up all the extra horizontal space in a container.
`Box.createVerticalGlue()`	Creates a "virtual component" that uses up all the extra vertical space in a container.

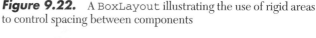

Figure 9.22. A `BoxLayout` illustrating the use of rigid areas to control spacing between components

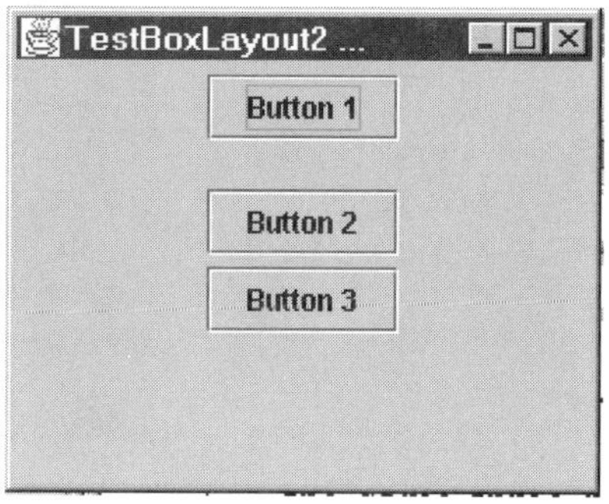

between Button 1 and Button 2, and a fixed 5-pixel spacing between Button 2 and Button 3. The code for this example is

```
// Create a new panel
JPanel p = new JPanel();

// Set the layout manager
p.setLayout(new BoxLayout(p, BoxLayout.Y_AXIS));

// Add buttons
p.add( new JButton("Button 1") );
p.add( Box.createRigidArea(new Dimension(0,20)) );
p.add( new JButton("Button 2") );
p.add( Box.createRigidArea(new Dimension(0,5)) );
p.add( new JButton("Button 3") );

// Add the new panel to the existing container
add( p );
```

and these statements will produce the GUI shown in Figure 9.22.

9.5.5 Combining Layout Managers to Produce a Result

It is often difficult to create exactly the GUI that we want using a standard layout manager, and this fact can be very frustrating. However, it is sometimes possible to *combine* layout managers to achieve a desired appearance. Only one layout manager can be used with a given container at any time, but one container can be placed inside another container, and the two containers can have different layout managers.

To understand how multiple layout managers can be better than a single one, let's reconsider the temperature conversion program of Example 9-1. That program used a `FlowLayout`, which means that the components of the program were laid out horizontally until the end of a line, and then starting over on the next line. Unfortunately, such as design will fail if a user resizes the application, or if the size of the components differs significantly from platform to platform. Figure 9.23 illustrates this problem, showing how the appearance of the application changes as the program frame is resized.

Instead of using a `FlowLayout` with a single container, we can create the same interface using three containers and `BoxLayouts`. The first container (`pHoriz`) will use a horizontal `BoxLayout`, and the other two containers (`pVertL` and `pVertR`) will be placed inside the first one and use vertical `BoxLayouts`. The labels will be placed in container `pVertL` and the text fields will be placed in container `pVertR`. The horizontal space between these two containers will be set by adding a rigid area to the top-level container. The code to build this structure is shown on the next page.

Figure 9.23. (a) The `TempConversion` program laid out with the default width and height. (b) The program after the frame width has been increased. (c) The program after the frame width has been decreased.

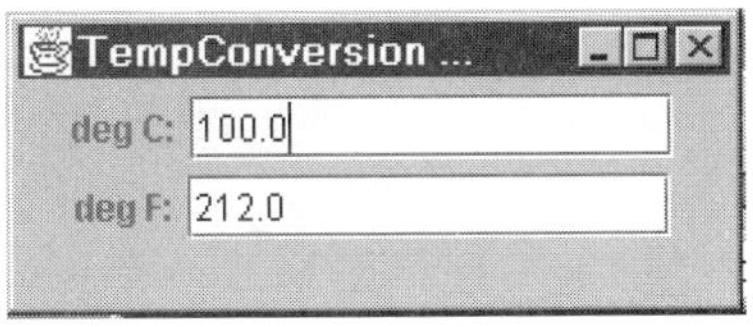

(a)

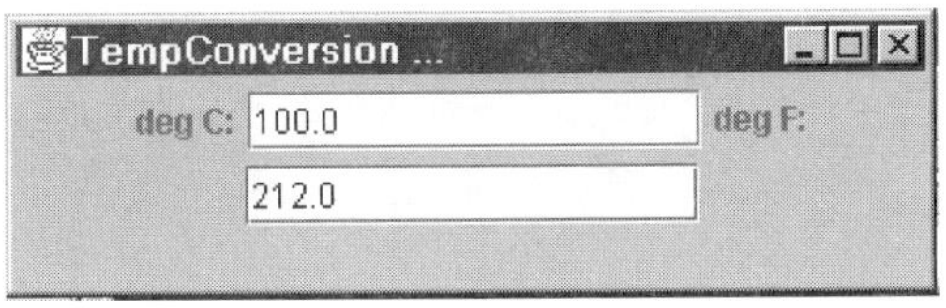

(b)

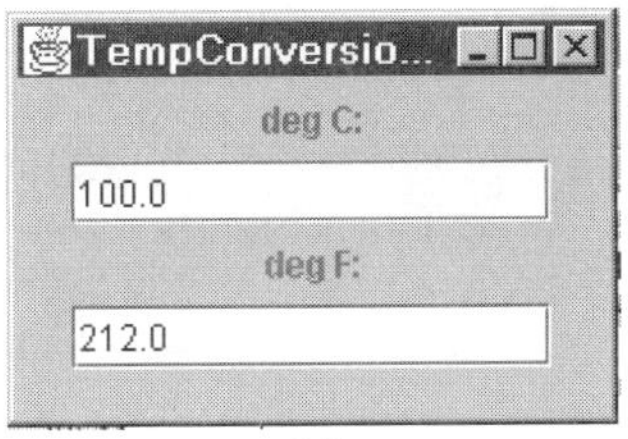

(c)

```
// Create a new high-level panel
JPanel pHoriz = new JPanel();
pHoriz.setLayout(new BoxLayout(pHoriz, BoxLayout.X_AXIS));
add( pHoriz );

// Create two subordinate panels
JPanel pVertL = new JPanel();
JPanel pVertR = new JPanel();
pVertL.setLayout(new BoxLayout(pVertL, BoxLayout.Y_AXIS));
pVertR.setLayout(new BoxLayout(pVertR, BoxLayout.Y_AXIS));

// Add to pHoriz with a horizontal space between panels
pHoriz.add( pVertL );
pHoriz.add( Box.createRigidArea(new Dimension(20,0)) );
pHoriz.add( pVertR );

// Create degrees Celsius field
l1 = new JLabel("deg C:", JLabel.RIGHT);
pVertL.add( l1 );
T1 = new JTextField("0.0",15);
T1.addActionListener( cHnd );
pVertR.add( t1 );

// Create degrees Fahrenheight field
l2 = new JLabel("deg F:", JLabel.RIGHT);
pVertL.add( l2 );
t2 = new JTextField("32.0",15);
t2.addActionListener( fHnd );
pVertR.add( t2 );
```

Figure 9.24. (a) The `TempConversion2` program after the frame width has been increased. (b) The program after the frame width has been decreased. Note that the relative positions of the labels and text fields have been preserved in each case.

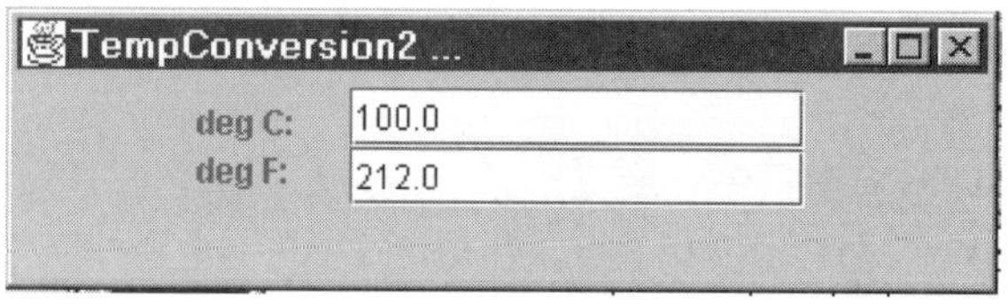

(a)

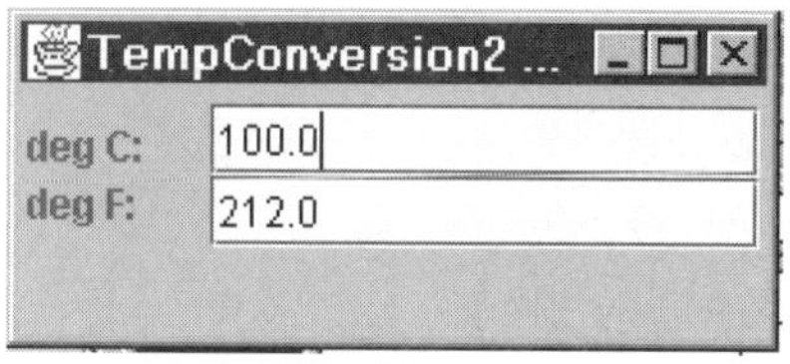

(b)

Figure 9.24 shows the behavior of this program as the frame containing the GUI is resized. This time, the GUI preserves its shape despite changes in frame size.

EXAMPLE 9-2: CREATING A CALCULATOR GUI

Write a program that creates the Graphical User Interface for a calculator.

SOLUTION

A calculator GUI should have a display window for results all across the top, with a rectangular grid of buttons below it. We cannot create such a display with a `BorderLayout` manager, `FlowLayout` manager, a `GridLayout` manager, or a `BoxLayout` manager by itself, but we *can* create it if we can combine a `BorderLayout` manager with a `GridLayout` manager.

The code shown in Figure 9.25 creates two containers, both `JPanels`. The outer container uses the `BorderLayout` manager, and the inner container uses the `GridLayout` manager. The inner container `p2` uses the `GridLayout` manager to lay out a keypad, and then the entire inner container is placed in the center region of the outer container. The results window is placed in the `NORTH` region of the outer container. The resulting GUI is shown in Figure 9.26.

```
// Create a GUI for a calculator.
import java.awt.*;
import java.awt.event.*;
import javax.swing.*;
public class CalculatorGUI extends JPanel {

   // Initialization method
   public void init() {
```

Figure 9.25. *(cont.)*

```
        // Set the layout manager
        setLayout( new BorderLayout() );

        // Add the result field to the panel
        JTextField t1 = new JTextField(10);
        t1.setEditable( false );
        t1.setBackground( Color.white );
        add( t1, BorderLayout.NORTH );

        // Create another Panel for the keypad, and place it
        // in the high-level panel
        JPanel p2 = new JPanel();
        p2.setLayout( new GridLayout(4,5) );
        add( p2, BorderLayout.CENTER );

        // Add keys to the panel
        p2.add( new JButton("7") );
        p2.add( new JButton("8") );
        p2.add( new JButton("9") );
        p2.add( new JButton("/") );
        p2.add( new JButton("sqrt") );
        p2.add( new JButton("4") );
        p2.add( new JButton("5") );
        p2.add( new JButton("6") );
        p2.add( new JButton("*") );
        p2.add( new JButton("%") );
        p2.add( new JButton("1") );
        p2.add( new JButton("2") );
        p2.add( new JButton("3") );
        p2.add( new JButton("-") );
        p2.add( new JButton("1/x") );
        p2.add( new JButton("0") );
        p2.add( new JButton("+/-") );
        p2.add( new JButton(".") );
        p2.add( new JButton("+") );
        p2.add( new JButton("=") );
    }
```

Figure 9.25. Creating a calculator GUI by combining two different containers with different layout managers

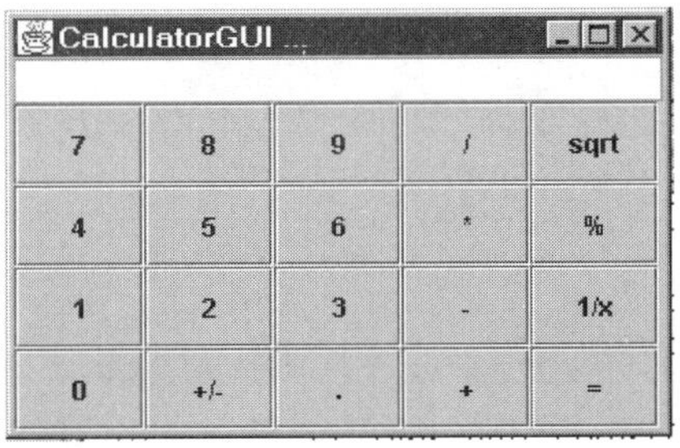

Figure 9.26. A calculator GUI

PRACTICE!

This quiz provides a quick check to see if you have understood the concepts introduced in Sections 9.1 through 9.5. If you have trouble with the quiz, reread the section, ask your instructor, or discuss the material with a fellow student. The answers to this quiz are found in the back of the book.

1. What is a container? Which type of containers(s) are we using in this chapter?
2. What is a component? Which type of components are we using in this chapter?
3. What is a layout manager? What does it do?
4. Why would you wish to use more than one layout manager in a single program?
5. What is an event handler? How are events handled in Java?
6. What listener interface(s) were introduced in this chapter? What does each one do, and which components produce it?

9.6 INTRODUCTION TO APPLETS

An **applet** is a special type of Java program that is designed to work within a World Wide Web browser, such as Netscape Communicator, Microsoft Internet Explorer, or Sun's HotJava. Applets are usually quite small, and are designed to be downloaded, executed, and discarded whenever the browser points to a site containing the Java applet.

Applets are quite restricted compared to Java applications. For security reasons, applets are not allowed access to the computer on which they are executing, so an applet cannot read or write disk files, for example. These restrictions make them less useful than applications for many data analysis purposes.

Applets are most commonly used for creating eye-catching graphics and animations on Web pages.

9.6.1 The `JApplet` Class

An *applet* is any class that extends class `javax.swing.JApplet`. Class **`javax.swing.JApplet`** is a container into which components can be placed. It is very similar to `JFrame`, in that the components must be added to a content pane retrieved by the `getContentPane()` method. In addition, class `JApplet` implements a set of methods that form the interface between the applet and the Web browser. These methods are summarized in Table 9-12.

Every applet has five key methods: `init`, `start`, `stop`, `destroy`, and `paintComponent`. Method `init()` is called by the browser when the applet is first loaded in memory. This method should allocate resources and create the GUI required for the applet. Method `start()` is called when the applet should start running. This call can be used to start animations, etc. when the applet becomes visible. Method `stop()` is called when the applet should stop running, for example when its window is covered or when it is iconized. This call is also made just before an applet is destroyed. Method `destroy()` is the final call to the applet before it is destroyed. The applet should release all resources at the time of this call. Finally, method `paintComponent` is called whenever the applet must be drawn or re-drawn.

Class `JApplet` implements all five of these methods as simple calls that do nothing. A practical applet will override as many of these methods as it needs to perform its function. Simple applets that only respond to user events will only need to override the `init()` and possibly the `paintComponent` methods.

TABLE 9-12 Selected `JApplet` Methods

METHOD	DESCRIPTION
`public boolean isActive();`	Determines if this applet is active.
`public void init();`	Method called by the browser to inform this applet that it has been loaded into the system.
`public void start();`	Method called by the browser to inform this applet that it should start execution. This call will always follow a call to `init()`.
`public void stop();`	Method called by the browser to inform this applet that it should stop execution
`public void destroy();`	Method called by the browser to inform this applet that it is being destroyed, and it should deallocate an resources that it holds. This call will always follow a call to `stop()`.
`public String getAppletInfo();`	Returns information about this applet. This can be a version number, copyright, etc.
`public String getParameter(String name);`	Returns the value of the named parameter in the HTML tag.
`public String[][]getParameterInfo();`	Returns information about the parameters that are understood by this applet.
`public void paintComponent(Graphics g);`	Method to paint the applet.
`public void repaint();`	Repaints the applet.
`public void showStatus(String msg);`	Displays the argument string in the applet's "status window".

9.6.2 Creating and Displaying an Applet

The basic steps required to create and run a Java applet are as follows:

1. Create a container class to hold the GUI components. This class will always be a subclass of `JApplet`.
2. Select a layout manager for the container, if the default layout manager (`FlowLayout`) is not acceptable.
3. Create components and add them to *the content pane* of the `JApplet` container.
4. Create "listener" objects to detect and respond to the events expected by each GUI component, and assign the listeners to appropriate components.
5. Create an HTML text file to specify to the browser which Java applet should be loaded and executed.

These steps are very similar to the ones required to create a Java application, except that we use `JApplet` instead of `JPanel` as the container class, and we create an HTML file instead of a `JFrame` to execute the program. Note that all components must be added to the `JApplet`'s content pane, just as with `JFrame`.

Figure 9.27 shows an applet that creates a simple GUI with a single button and a single label field. The label field contains the number of times that the button has been pressed since the program started. This applet is identical to the `FirstGUI` application in Figure 9.1, except that it has been converted to run as an applet. Note that class `FirstApplet` extends `JApplet`, and this class serves as the container for our GUI components (step 1 above).

```java
// A first Applet. This class creates a label and
// a button. The count in the label is incremented
// each time that the button is pressed.
import java.awt.*;
import java.awt.event.*;
import javax.swing.*;
public class FirstApplet extends JApplet {

   // Instance variables
   private int count = 0;        // Number of pushes
   private JButton pushButton; // Push button
   private JLabel label;         // Label

   // Initialization method
   public void init() {

      // Set the layout manager
      getContentPane().setLayout( new BorderLayout() );

      // Create a label to hold push count
      label = new JLabel("Push Count: 0");
      getContentPane().add( label, BorderLayout.NORTH );
      label.setHorizontalAlignment( label.CENTER );

      // Create a button
      pushButton = new JButton("Test Button");
      pushButton.addActionListener( new ButtonHandle (this) );
      getContentPane().add( pushButton, BorderLayout.SOUTH );
   }

   // Method to update push count
   public void updateLabel() {
      label.setText( "Push Count: " + (++count) );
   }
}

class ButtonHandler implements ActionListener {
   private FirstApplet fa;

   // Constructor
   public ButtonHandler ( FirstApplet fa1 ) {
      fa = fa1;
   }

   // Execute when an event occurs
   public void actionPerformed( ActionEvent e ) {
      fa.updateLabel();
   }
}
```

Figure 9.27. An applet that creates a container, sets a layout manager for the container, and adds components and listeners for the components. This is the basic core structure required to create Java applets.

This applet contains two classes: class `FirstApplet` to create and display the GUI, and class `ButtonHandler` to respond to mouse clicks on the button.

Class `FirstApplet` contains two methods: `init()` and `updateLabel()`. Method `init()` overrides the `init()` in class `JApplet`, and it initializes the GUI. It

```
<html>
<applet code="FirstApplet.class" width=200 height=100>
</applet>
</html>
```

Figure 9.28. HTML file required to execute the `FirstApplet` applet

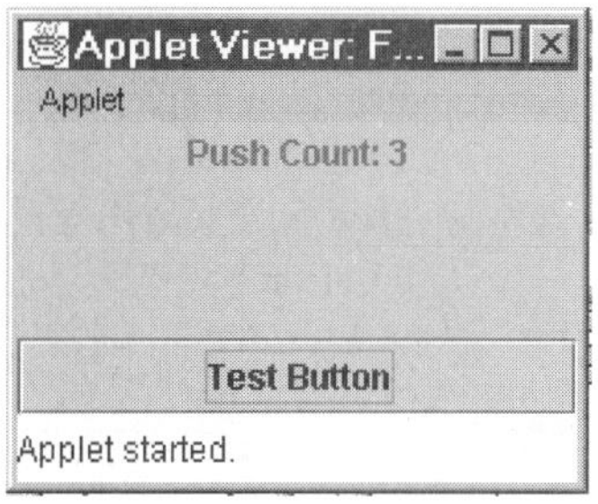

Figure 9.29. The display produced by applet `FirstApplet`

will be called by the browser when the applet is just starting. This class specifies which layout manager to use with the container (line 18), creates the `JLabel` and `JButton` components (lines 21 and 26), and adds them to the container (lines 22 and 28). In addition, it creates a "listener" object of class `ButtonHandler` to listen for and handle events generated by mouse clicks, and assigns that object to monitor mouse clicks on the button (line 27).

Method `updateLabel()` (lines 32–34) is the method that should be called by the event handler every time that a button click occurs. It updates the label with the number of button clicks that have occurred. Class `ButtonHandler` is a "listener" class identical to those we discussed earlier. Its `actionPerformed` method calls method `updateLabel()` whenever a click occurs on the button.

To execute this applet in a Web browser, we must also create an HTML (Hypertext Markup Language) document to tell the browser to load and execute the applet. An example HTML document is shown in Figure 9.28. An HTML document consists of a series of **tags** marking the beginning and ending of various items. For example, the beginning of an HTML document is marked by an `<html>` tag, and the end of the document is marked by the `</html>` tag. Similarly, the beginning of an applet description is marked by an `<applet>` tag, and the end of the applet description is marked by the `</applet>` tag. The series of values after the applet tag are known as *attributes*. The three required attributes specify the name of the class file to execute and the size of the applet in units of pixels.

The applet is executed by loading the corresponding HTML document into a Java-enabled Web browser, such as Netscape Navigator, Microsoft Internet Explorer, or Sun's HotJava. In addition, an applet can be tested with a special program called `appletviewer`, which is supplied in the JDK. The applet can be executed with the following command, and the results are as shown in Figure 9.29.

```
D:\book\java\chap9>appletviewer FirstApplet.html
```

Note that the GUI produced by this applet is identical to the GUI produced by the `FirstGUI` application, except that the applet includes a status line at the bottom of the window (the status line says "Applet started.").

9.6.3 Using Packages Within Applets

Because applets are designed to be transferred through the internet from a server to a browser, there must be a special convention to tell the applet where to find the classes that the applet needs to execute. This convention is illustrated in Figure 9.30.

If an applet uses a class that is *not* built into a package, then that class must be present in the *same directory* as the HTML file used to start the applet (see Figure 9.30*a*). This directory could be a local directory on your computer or a directory on a server on the other side of the world—it doesn't matter as long as the class files and the HTML file are in the same directory. When the HTML file is loaded into the browser, all of the required classes will be transferred over the network to the browser, so that they can be executed on the local computer. Note that applets should be small because they are transferred over the network each time that they are used.

If an applet uses a class that appears in a package that is not a part of standard Java, then *the package must appear in the appropriate subdirectory* of the directory containing the HTML file. For example, suppose that an applet uses the class `Fmt` from package `chapman.io`. When a browser executes this applet, it will look for the `Fmt` class in subdirectory `chapman` under the directory containing the HTML file (see Figure 9.30*b*). This structure is always the same whether the directory containing the HTML file is local to your computer or located on a remote server across the network. Unlike Java applications, applets ignore the `CLASSPATH` environment variable. They only look in subdirectories of the directory containing the HTML file to locate the classes in packages.

9.6.4 Creating an Applet That is Also an Application

It is possible to design a single Java program to run as either an applet or an application. We can create such a dual-purpose program by first creating a working applet,

Figure 9.30. (*a*) The classes required by an applet that *are not* included in a package must appear in the same directory as the HTML file. (*b*) The classes required by an applet that *are* included in a package must appear in an appropriate subdirectory of the directory containing the HTML file. Class `chapman.io.Fmt` is shown in this figure.

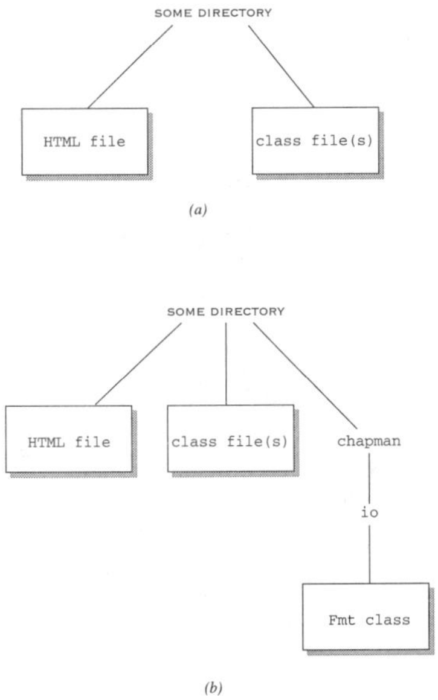

and then adding a main method to the applet. The main method will be designed to create a JFrame, place the applet into the frame, and execute the applet's init() and start() methods. The window listener associated with the frame will be designed to execute the applet's stop() and destroy() methods when the program is closed down.

A typical dual-purpose application/applet is shown in Figure 9.31. This is a version of the temperature conversion application from Example 9-1. When this program is executed as an applet, the main method is ignored and it runs as a conventional applet. When the program is executed as an application, the main method creates a JFrame, places the applet into the frame, and runs the applet's init() and start() methods. Note that the AppletWindowHandler executes the applet's stop() and destroy() methods when the frame is closed.

Figure 9.32 shows the GUI created by this program when it is executed as an application and as an applet. Note that the program behaves identically in either case.

Finally, note that this program uses the Fmt class from package chapman.io. In order for this program to work successfully as an applet, the file Fmt.class must be found in the subdirectory chapman\io below directory containing the file TempConversionApplet.html.

```
/*
   Purpose:
      This GUI-based program converts temperature in
      degrees Fahrenheit to degrees Celsius, and vice versa.

   Record of revisions:
      Date        Programmer         Description of change
      ====        ==========         =====================
     10/14/98    S. J. Chapman      Original code
     12/22/98    S. J. Chapman      Modified for applet
*/
import java.awt.*;
import java.awt.event.*;
import javax.swing.*;
import chapman.io.*;
public class TempConversionApplet extends JApplet {

   // Instance variables
   private JLabel l1, l2;          // Labels
   private JTextField t1, t2;      // Text Fields
   private DegCHandler cHnd;       // ActionEvent handler
   private DegFHandler fHnd;       // ActionEvent handler

   // Initialization method
   public void init() {

      // Set the layout manager
      getContentPane().setLayout( new FlowLayout() );

      // Create ActionEvent handlers
      cHnd = new DegCHandler( this );
      fHnd = new DegFHandler( this );


      // Create degrees Celsius field
      l1 = new JLabel("deg C:", JLabel.RIGHT);
```

Figure 9.31. *(cont.)*

```
         getContentPane().add( l1 );
         t1 = new JTextField("0.0",15);
         t1.addActionListener( cHnd );
         getContentPane().add( t1 );

         // Create degrees Celsius field
         l2 = new JLabel("deg F:", JLabel.RIGHT);
         getContentPane().add( l2 );
         t2 = new JTextField("32.0",15);
         t2.addActionListener( fHnd );
         getContentPane().add( t2 );
      }

      // Method to convert deg F to deg C
      // and display result
      public void toC( double degF ) {
         double degC = (5. / 9.) * (degF - 32);
         t1.setText( Fmt.sprintf("%5.1f",degC) );
         t2.setText( Fmt.sprintf("%5.1f",degF) );
      }

      // Method to convert deg C to deg F
      // and display result
      public void toF( double degC ) {
         double degF = (9. / 5.) * degC + 32;
         t1.setText( Fmt.sprintf("%5.1f",degC) );
         t2.setText( Fmt.sprintf("%5.1f",degF) );
      }

      // Main method to create frame
      public static void main(String s[]) {

         // Create a frame to hold the application
         JFrame fr = new JFrame("TempConversionApplet ...");
         fr.setSize(250,100);

         // Create and initialize a TempConversionApplet object
         TempConversionApplet tf = new TempConversionApplet();
         tf.init();
         tf.start();

         // Create a Window Listener to handle "close" events
         AppletWindowHandler l = new AppletWindowHandler(tf);
         fr.addWindowListener(l);

         // Add the object to the center of the frame
         fr.getContentPane().add(tf, BorderLayout.CENTER);

         // Display the frame
         fr.setVisible( true );
      }
   }

   class DegCHandler implements ActionListener {
      private TempConversionApplet tc;

      // Constructor
      public DegCHandler( TempConversionApplet t ) { tc = t; }
```

Figure 9.31. *(cont.)*

```
   // Execute when an event occurs
   public void actionPerformed( ActionEvent e ) {
      String input = e.getActionCommand();
      double degC = new Double(input).doubleValue();
      tc.toF( degC );
   }
}

class DegFHandler implements ActionListener {
   private TempConversionApplet tc;

   // Constructor
   public DegFHandler( TempConversionApplet t ) { tc = t; }

   // Execute when an event occurs
   public void actionPerformed( ActionEvent e ) {
      String input = e.getActionCommand();
      double degF = new Double(input).doubleValue();
      tc.toC( degF );
   }
}

public class AppletWindowHandler extends WindowAdapter {
   JApplet ap;

   // Constructor
   public AppletWindowHandler ( JApplet a ) { ap = a; }

   // This method implements a listener that detects
   // the "window closing event", shuts down the applet,
   // and stops the program.
   public void windowClosing(WindowEvent e) {
      ap.stop();
      ap.destroy();
      System.exit(0);
   };
}
```

Figure 9.31. A dual-purpose application/applet version of the temperature conversion application from Example 9-1

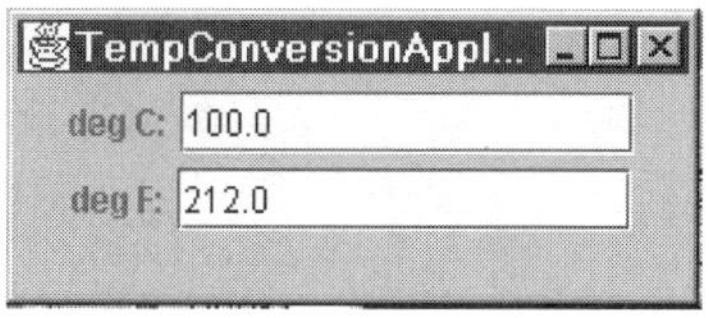

(a)

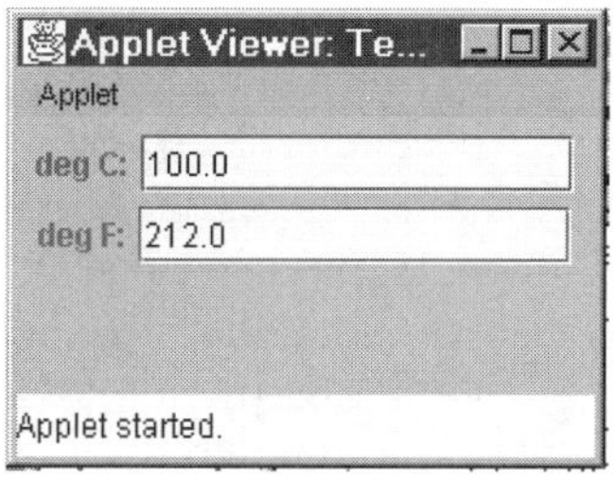

(b)

Figure 9.32. The GUI produced when `TempConversionApplet` is run as an application. (*b*) The GUI produced when `TempConversionApplet` is run as an applet.

GOOD PROGRAMMING PRACTICE

Whenever possible, design your programs to be dual purpose applications / applets.

EXAMPLE 9-3: PLOTTING DATA

Create a program that plots sin x, cos x, or both depending on the values of two checkboxes. The curves should be plotted for the range $0 \le x \le 2\pi$. Use `BoxLayout` so that the locations of the components will be preserved when the application changes size. Also, the program should be capable of running as either an application or an applet.

SOLUTION

We can plot the data using `chapman.graphics.JPlot2D`, which is a Swing component just like any other component. This program will require one `JPlot2D` object and two checkboxes, as well as an `ActionListener` to monitor the state of the checkboxes.

1. **State the problem**. Create a GUI-based program that plots sin x, cos x, or both depending on the values of two checkboxes. Ensure that the layout of the components does not change as the window is resized. Design the program to function as either an application or an applet.
2. **Define the inputs and outputs**. The inputs to this program are the status of the two checkboxes "Plot sine" and "Plot cosine". The output from the program is a plot of the sine and/or cosine, depending on the status of the checkboxes.
3. **Decompose the program into classes and their associated methods**. This program requires two classes to function properly, if we make the principal class also function as an `ActionListener`. The principal class (called `PlotSinCos`) must contain the following methods: (1) a method `init()` to generate the GUI display, (2) a method `display()` to display the desired curve(s), (3) a method `actionListener` to respond to mouse clicks on the checkboxes, and (4) a `main` method to start up the application when it is used in that mode. The second required class is a `WindowListener` class to shut down the program when it is running as an application and the user clicks on the "Close Window" box. Finally, an HTML file will be needed to execute the program as an applet.
4. **Design the algorithm that you intend to implement for each method**. The `init()` method of the `PlotSinCos` class must create the GUI. It must lay out the graphical elements. We would like to arrange the GUI so that the plot appears on top of the display, and the two checkboxes appear side-by-side below it. One way to achieve this design is with nested panels using `BoxLayouts`, as shown in Figure 9.33. The top-level panel `pVert` will use a vertical `BoxLayout`, and the `JPlot2D` object and a horizontal panel `pHoriz` will be added to it. Then, the two checkboxes can be added to the horizontal panel.

Figure 9.33. The structure of containers and layout managers required to create the GUI for the `PlotSinCos` program

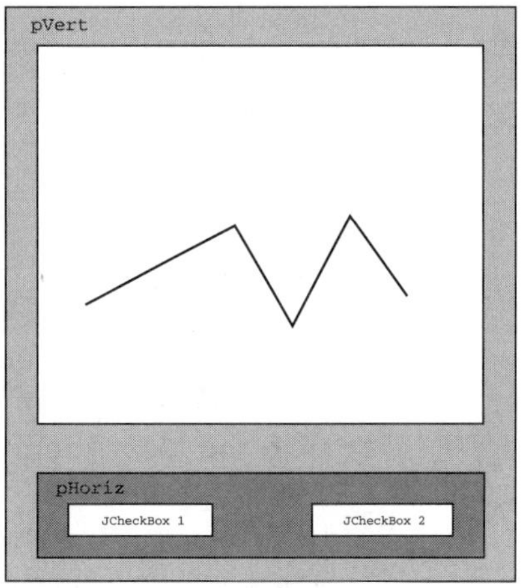

The code required to create this GUI is

```
// Create a new high-level panel
JPanel pVert = new JPanel();
pVert.setLayout(new BoxLayout(pVert, BoxLayout.Y_AXIS));
getContentPane().add( pVert );

// Add a blank plot
pl = new JPlot2D();
pl.setPreferredSize(new Dimension(400,400));
pVert.add( pl );

// Create a subordinate panel for the bottom
JPanel pHoriz = new JPanel();
pHoriz.setLayout(new BoxLayout(pHoriz, BoxLayout.X_AXIS));
pVert.add( pHoriz );

// Create the "Add sine" checkbox
b1 = new JCheckBox("Add sine");
b1.addActionListener( this );
pHoriz.add( b1 );
pHoriz.add( Box.createRigidArea(new Dimension(40,0)) );

// Create the "Add cosine" checkbox
b2 = new JCheckBox("Add cosine");
b2.addActionListener( this );
pHoriz.add( b2 );
```

Method `init()` must also create the data to plot when a plot is requested. The code required for this step is the following:

```
for ( int i = 0; i < x.length; i++ ) {
   x[i]  = (i+1) * 2 * Math.PI / 40;
   y1[i] = Math.sin(x[i]);
   y2[i] = Math.cos(x[i]);
}
```

Method `display()` must display the requested curves. To do this, it must first remove any existing curves on the plot with the `JPlot2D` method `removeAll()`, and then add the requested curves back into the plot. The method can check the state of each checkbox using the `isSelected()` method. This code is shown in the complete program below.

Method `actionListener()` is very simple. It must listen for any change in either checkbox, and call method `display()` when one occurs. Method `display()` does all the hard work.

Finally, method `main()` and the `WindowListener` class are essentially identical to the ones shown in the `TempConversionApplet` earlier. The required HTML file is identical to the ones shown earlier in the chapter, except that it executes program `PlotSinCos.class` with a size of 400 pixels wide by 430 pixels high. It is not shown.

5. **Turn the algorithm into Java statements**. The resulting Java program is shown in Figure 9.34.
6. **Test the resulting Java program**. To test this program, we will execute it both as an application and as an applet, and observe the results. Figure 9.35 shows the appearance of the program as an application and as an applet. Execute the program youself to veryify that it functions properly. Note that *for the applet to work properly, package* `chapman.graphics` *must be in subdirectory* `chapman\graphics` *below the directory containing the applet's HTML file.* Also, re-size the application and see if the GUI components preserve their relative locations.

```
/*
   Purpose:
      This program plots sin x and/or cos x for 0 <= x <= PI
      depending on the state of two checkboxes.

   Record of revisions:
      Date         Programmer      Description of change
      ====         ==========      =====================
    12/16/98     S. J. Chapman    Original code
*/
import java.awt.*;
import java.awt.event.*;
import javax.swing.*;
import chapman.graphics.JPlot2D;
public class PlotSinCos extends JApplet
                     implements ActionListener {

   // Instance variables
   private JCheckBox b1, b2;  // Check boxes
   private JPlot2D pl;        // Plot
   double[] x, y1, y2;        // Data to plot


   // Initialization method
   public void init() {

      // Create a new high-level panel
      JPanel pVert = new JPanel();
      pVert.setLayout(new BoxLayout(pVert, BoxLayout.Y_AXIS));
      getContentPane().add( pVert );
```

Figure 9.34. *(cont.)*

```
      // Add a blank plot
      pl = new JPlot2D();
      pl.setPreferredSize(new Dimension(400,400));
      pVert.add( pl );

      // Create a subordinate panel for the bottom
      JPanel pHoriz = new JPanel();
      pHoriz.setLayout(new BoxLayout(pHoriz, BoxLayout.X_AXIS));
      pVert.add( pHoriz );

      // Create the "Add sine" checkbox
      b1 = new JCheckBox("Add sine");
      b1.addActionListener( this );
      pHoriz.add( b1 );
      pHoriz.add( Box.createRigidArea(new Dimension (40,0)) );

      // Create the "Add cosine" checkbox
      b2 = new JCheckBox("Add cosine");
      b2.addActionListener( this );
      pHoriz.add( b2 );

      // Define arrays to hold the two curves to plot
      x = new double[41];
      y1 = new double[41];
      y2 = new double[41];

      // Calculate a sine and a cosine wave
      for ( int i = 0; i < x.length; i++ ) {
         x[i] = (i+1) * 2 * Math.PI / 40;
         y1[i] = Math.sin(x[i]);
         y2[i] = Math.cos(x[i]);
      }
   }

   // Method to display sine and cosine plots
   public void display() {

      // Remove old curves
      pl.removeAll();

      // Add sine curve
      if ( b1.isSelected() ) {
         pl.addCurve(x, y1);
         pl.setLineColor( Color.blue );
         pl.setLineWidth( 2.0f );
         pl.setLineStyle( JPlot2D.LINESTYLE_SOLID );
      }

      // Add cosine curve
      if ( b2.isSelected() ) {
         pl.addCurve(x, y2);
         pl.setLineColor( Color.red );
         pl.setLineWidth( 2.0f );
         pl.setLineStyle( JPlot2D.LINESTYLE_LONGDASH );
      }

      // Turn on grid
      pl.setGridState( JPlot2D.GRID_ON );

      // Repaint plot
      pl.repaint();
   }
```

Figure 9.34. *(cont.)*

```
   // Execute when an event occurs
   public void actionPerformed( ActionEvent e ) {
      // Update display
      display();
   }

   // Main method to create frame
   public static void main(String s[]) {

      // Create a frame to hold the application
      JFrame fr = new JFrame("PlotSinCos ...");
      fr.setSize(400,430);

         (method not shown to save space)
      }
   }
```

Figure 9.34. The `PlotSinCos` application / applet

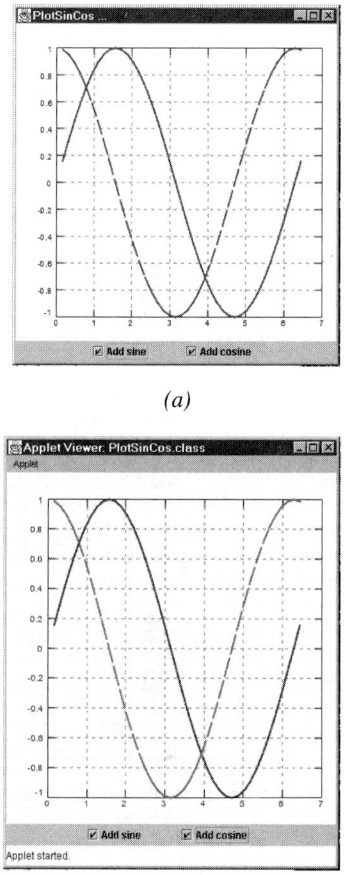

(a)

(b)

Figure 9.35. (*a*) Program `PlotSinCos` running as an application. (*b*) Program `PlotSinCos` running as an applet.

PRACTICE!

This quiz provides a quick check to see if you have understood the concepts introduced in Section 9.6. If you have trouble with the quiz, reread the section, ask your instructor, or discuss the material with a fellow student. The answers to this quiz are found in the back of the book.

1. What is an applet? How does it differ from an application?
2. What are the five key methods in an applet? What do they do?
3. How can you write a single program that can run as either an applet or an application?

SUMMARY

- The principal elements required to create a Java GUI are components, a container to hold them, a layout manager, and event handlers.
- The `JLabel` class creates a GUI component that displays read-only text.
- The `JButton` class creates a GUI component that implements push buttons. This class generates an `ActionEvent` containing the button label when a button is clicked.
- The `JTextField` class creates a GUI component that allows a user to display and edit text. This class generates an `ActionEvent` containing the field's text when the Enter key is pressed.
- The `JPasswordField` class is identical to the `JTextField` class, except that asterisks are displayed in the field instead of the typed text.
- The `JComboBox` class creates a drop-down list of choices, and allow the user to select one of the by clicking with the mouse. It may optionally be set to allow the user to type text directly into the combo box. This class generates an `ActionEvent` when a selection is made.
- The `JCheckBox` class creates a checkbox, which is a type of button that is either on or off. This class generates an `ActionEvent` when a state change occurs.
- The `JRadioButton` class creates a radio button, which a type of checkbox designed to be grouped into sets. A `ButtonGroup` object is used to group together all the radio buttons that form a set. Only *one* button within a set of radio buttons may be on at any given time.
- A layout manager is a helper class that is designed to automatically arrange GUI components within a container for presentation purposes. There are six standard layout managers: `BorderLayout`, `CardLayout`, `FlowLayout`, `BoxLayout`, `GridLayout`, and `GridBagLayout`.
- An applet is a special type of Java program that is designed to work within a World Wide Web browser.
- Applets are usually quite small, and are designed to be downloaded, executed, and discarded whenever the browser points to a site containing the Java applet.
- For security reasons, applets are not allowed access to the computer on which they are executing.
- An applet is created by extending class `javax.swing.JApplet`.
- An applet is executed within a browser by loading an HTML page containing a `tag` referring to the applet.

- If an applet uses a class that appears in a package, then the package must appear in a subdirectory of the directory containing the HTML file.
- We can create a dual-purpose application/applet by first creating a working applet, and then adding a `main` method to the applet.

APPLICATIONS: SUMMARY OF GOOD PROGRAMMING PRACTICES

The following guidelines introduced in this chapter will help you to develop good programs:

1. To handle button events, use a class that implements the `ActionListener` interface, and contains an `actionPerformed` method. Register an object from this class with each button, and code the `actionPerformed` method to do whatever is required when the button is pressed.
2. One `ActionListener` object can monitor many buttons, using the result of the `getActionCommand` method to determine the button that created a particular event.
3. Use `JTextFields` to accept single lines of input data from a user, or to display single lines of read-only data to the user.
4. Use `JPasswordFields` to accept input data from a user that you do not wish to have echoed to the screen, such as passwords.
5. Use `JComboBoxes` to make a single selection from a list of mutually-exclusive choices.
6. Use `JCheckBoxes` to select the state of items represented by boolean variables, which can only be `true` or `false`.
7. Use `JRadioButtons` to select the state of a set of items represented by boolean variables, only one of which can be `true` at any time.
8. Whenever possible, design your programs to be dual purpose applications / applets.

KEY TERMS

`ActionEvent` class
`ActionListener` interface
`actionPerformed` method
applet
`applet` tag
`BorderLayout` class
`BoxLayout` class
`ButtonGroup` class
`CardLayout` class
callback method
checkbox
component
container
`destroy()` method
event handler
`FlowLayout` class
`GridBagLayout` class
`GridLayout` class
`java.awt.event` package
`init()` method
`JApplet` class
`JButton` class
`JCheckBox` class
`JComboBox` class
`JFrame` class
`JLabel` class
`JPanel` class
`JPasswordField` class
`JRadioButton` class
`JTextField` class
keyboard shortcut
layout manager
radio button
`setLayout` method
`start()` method
`stop()` method
tool tips

Problems

1. Explain the steps required to create a GUI in Java.
2. Convert the calculator GUI of Figure 9.26 into a fully functional calculator. (Caution: this is a challenging problem.)
3. Create a GUI that uses a `JComboBox` to select the background color displayed by the GUI.

4. Write a class that displays a circle of random size and color, and calculates and displays the radius, diameter, area, and circumference of the circle. Use a canvas to draw the circle, and use read-only `JTextFields` to display the information about the circle. Include a button that can be clicked to cause the program to generate a new randomly-selected circle. (*Note:* In determining the size of the circle, assume that there are 72 pixels per inch.)
5. Write a GUI program that plots the equation $y(x) = ax^2 + bx + c$. The program should use class `JPlot2D` for the plot, and should have GUI elements to read the values of *a, b, c,* and the minimum and maximum *x* to plot.
6. Re-write the previous problem as an applet.
7. Modify the temperature conversion GUI of Example 9-1 to add a "thermometer". The thermometer should be a canvas with a drawing of a thermometer shape and a fluid level corresponding to the current temperature in degrees Celsius. The range of the thermometer should be 0° – 100° C.
8. Create four GUIs that place five buttons into a `JPanel` using the `BorderLayout`, `FlowLayout`, `GridLayout`, and `BoxLayout` layout managers. What do the resulting GUIs look like?
9. Create a GUI that displays a user-selected image. The GUI should contain a combo box (drop-down list) to select the desired image, and a label to display the image. The Chapter 9 files available at the book's Web site include GIF files containing pictures of a dog, a cat, a cow, a pig, and a rabbit. The user should be able to select one of these pictures in the drop-down list, and the appropriate picture should be displayed.
10. Modify the GUI created in Exercise 9.9 to use a set of five radio buttons to select the image to display. The radio buttons should be lined up along the left-hand side of the GUI, with the image on the right-hand side of the GUI. What sort of layout manager(s) are required to create this GUI?
11. Write an applet that draws ten randomly-sized shapes in randomly-selected colors. The type of shape to draw (square, circle, ellipse, *etc.*) should be selectable through a `JComboBox`, and the display should be redrawn whenever the user presses a "Go" button on the GUI.
12. Convert the applet of Exercise 9.11 into a dual application / applet.
13. **Least Squares Fit** Write a GUI-based application that reads a series of (x,y) values from a disk file, performs a linear least squares fit on the values, and displays both the points and the least-squares fit line using class `JPlot2D`. The least-squares fit algorithm is described in Exercise 6.16, and the method developed there can be used with this application.

    The GUI elements in the program should include a `JTextField` for the input file name, a `JButton` to read the file, two read-only `JTextFields` for the slope and intercept of the fitted line, and class `JPlot2D` to display the input points and the fitted results. Use class `chapman.io.FileIn` to actually read the data.

    How many containers and which layout managers are required to create this GUI?

Appendix A: ASCII Character Set

The ASCII character set is a subset of the Unicode character set. It contains the first 127 characters of the Unicode character set, which are the ones most commonly used in Java programs. The full details of the Unicode character set can be found by consulting the World Wide Web site http://unicode.org.

The table shown below includes the first 127 characters, with the first two digits of the character number defined by the row, and the third digit defined by the column. For example, the letter `'R'` is on row 8 and column 2, so it is character 82 in the ASCII (and Unicode) character set.

	0	**1**	**2**	**3**	**4**	**5**	**6**	**7**	**8**	**9**
0	nul	soh	stx	etx	eot	enq	ack	bel	bs	ht
1	nl	vt	ff	cr	so	si	dle	dc1	dc2	dc3
2	dc4	nak	syn	etb	can	em	sub	esc	fs	gs
3	rs	us	sp	!	"	#	$	%	&	'
4	(	)	*	+	,	-	.	/	0	1
5	2	3	4	5	6	7	8	9	:	;
6	<	=	>	?	@	A	B	C	D	E
7	F	G	H	I	J	K	L	M	N	O
8	P	Q	R	S	T	U	V	W	X	Y
9	Z	[	\	]	^	_	`	a	b	c
10	d	e	f	g	h	I	j	k	l	m
11	n	o	p	q	r	s	t	u	v	w
12	x	y	z	{	|	}	~	del		

Appendix B: Operator Precedence Chart

The Java operators are shown in decreasing order of precedence from top to bottom, with the operators in each section having equal precedence.

OPERATOR	TYPE	ASSOCIATIVITY
`()`	parentheses	left to right
`[]`	array subscript	
`.`	member selection	
`++`	unary preincrement	right to left
`++`	unary postincrement	
`--`	unary predecrement	
`--`	unary postdecrement	
`+`	unary plus	
`-`	unary minus	
`!`	unary logical negation	
`~`	unary bitwise complement	
`(type)`	unary cast	
`*`	multiplication	left to right
`/`	division	
`%`	modulus	
`+`	addition	left to right
`-`	subtraction	
`<<`	bitwise left shift	left to right
`>>`	bitwise right shift with sign extension	
`>>>`	bitwise right shift with zero extension	
`<`	relational less than	left to right
`<=`	relational less than or equal to	
`>`	relational greater than	
`>=`	relational greater than or equal to	
`instanceOf`	type comparison	
`==`	relational is equal to	left to right
`!=`	relational is not equal to	
`&`	bitwise AND	left to right
`^`	bitwise exclusive OR boolean logical exclusive OR	left to right
`|`	bitwise inclusive OR boolean logical inclusive OR	left to right
`&&`	logical AND	left to right

<table>
<tr><td>||</td><td>logical OR</td><td>left to right</td></tr>
<tr><td>?:</td><td>ternary conditional</td><td>right to left</td></tr>
<tr><td>=</td><td>assignment</td><td>right to left</td></tr>
<tr><td>+=</td><td>addition assignment</td><td></td></tr>
<tr><td>-=</td><td>subtraction assignment</td><td></td></tr>
<tr><td>*=</td><td>multiplication assignment</td><td></td></tr>
<tr><td>/=</td><td>division assignment</td><td></td></tr>
<tr><td>%=</td><td>modulus assignment</td><td></td></tr>
<tr><td>&=</td><td>bitwise AND assignment</td><td></td></tr>
<tr><td>^=</td><td>bitwise exclusive OR assignment</td><td></td></tr>
<tr><td>|=</td><td>bitwise inclusive OR assignment</td><td></td></tr>
<tr><td><<=</td><td>bitwise left shift assignment</td><td></td></tr>
<tr><td>>>=</td><td>bitwise right shift with sign extension assignment</td><td></td></tr>
<tr><td>>>>=</td><td>bitwise right shift with zero extension assignment</td><td></td></tr>
</table>

Appendix C: Answers to Practice Boxes

CHAPTER 2

Practice Box, page 22

1. Valid `double` constant
2. Invalid-commas not allowed
3. Valid `double` constant
4. Valid `char` constant

   Invalid-to create a `char` constant containing a single quote, use the backslash escape character: `'\''`
5. Valid `double` constant
6. Valid `String` constant
7. Valid `boolean` constant
8. Same value
9. Same value
10. Different value
11. Valid name—would be a variable or a method name
12. Valid name—would be a class name
13. Invalid—name may not begin with a number
14. Valid name—would be a constant (or final variable)
15. Valid
16. Invalid—`MAX_COUNT` is a `short`, and 100000 is an `int`. An explicit cast is required to convert `int` to `short`. In addition, 100000 is too large a number to be represented in a `short`.
17. Invalid-can't assign a `String` to a `char`.
18. These statements are illegal. They try to assign a new value to the constant (final variable) `k`.

Practice Box, page 29

1. The order of evaluation is:
   a. Expressions in parentheses, working from the innermost parentheses out
   b. Multiplications, divisions, and mod, working from left to right
   c. Additions and subtractions, working from left to right
2. *(a)* Legal-result is 12 *(b)* Legal-result is 42 *(c)* Legal-result is 2 *(d)* Legal-result is 2.25 *(e)* Legal-result is 2.3333333 *(f)* Legal-result is 1
3. *(a)* 7 *(b)* -21 *(c)* 7 *(d)* 9

4. These statements are legal: `x` = 16; `y` = 3; `result` = 17.5.
5. These statements are illegal. The expression evaluates to a `double` 17.5, but the variable result is an `int`. This assignment is a narrowing conversion, which is illegal unless an explicit cast is used.

Practice Box, pages 39-40

1. `rEq = r1 + r2 + r3 + r4;`
2. `rEq = 1 / (1/r1 + 1/r2 + 1/r3 + 1/r4);`
3. `t = 2 * Math.PI * Math.sqrt( 1 / g );`
4. `v = vm * Math.exp( -alpha*t ) * Math.cos( omega*t );`
5. $d = \frac{1}{2}at^2 + v_0 t + d_0$
6. $f = \frac{1}{2\pi\sqrt{LC}}$
7. $E = \frac{1}{2}Li^2$
8. *(a)* Illegal-mismatched parentheses. *(b)* Illegal—explicit cast needed to convert `double` to `int`. *(c)* Illegal—explicit cast needed to convert `double` to `int`. [Note: This one is tricky. Because the cast operator (`int`) is evaluated before division, `a` is converted to an `int`. Since `a / b` is an `int` divided by a `double`, the result is a `double`, and it is illegal to assign the `double` value to `k`.] *(d)* Legal—`b` = 3.666667 *(e)* This is legal, but the calculation includes a floating-point division by zero; the result is `infinite`.
9. The results are: `a` = 2.0, `b` = 3.0, `c` = 4.666666666666667, `i` = 5, `j` = 0, `k` = 2.

CHAPTER 3

Practice Box, pages 83-84

1. *(a)* `false` *(b)* Illegal—can't use the not (`!`) operator with a `double` value *(c)* `true` *(d)* `true` *(e)* `true` *(f)* `false`

2.
```
if ( x >= 0 ) {
     sqrtX = Math.sqrt(x);
}
else {
System.out.println('Error: x < 0');
     sqrtX = 0;
}
```
3.
```
if ( Math.abs(denominator) < 1.0E-30 )
System.out.println('Divide by 0 error.');
else {
     fun = numerator / denominator;
System.out.println('fun = '+fun);
}
```

4.
```
if ( distance <= 100. )
    cost = 0.50 * distance;
else if ( distance <= 300. )
    cost = 50. + 0.30 * (distance - 100);
else
    cost = 110. + 0.20 * (distance - 300);
```
5. These statement will compile correctly, but they will not do what the programmer intended. Since there is no "`else`" in front of the second `if` statement, the second `if` statement will be executed regardless of the result of the first `if` statement. Thus if `volts` = 130, both "`WARNING: High voltage on line.`" and "`Line voltage is within tolerances.`" will be printed out.
6. Since `i < j`, the expression `j / i` will be executed, and the result will be `k = 1.6666666666666667`.
7. These statements are incorrect—a colon is required after the keyword `default`.
8. These statement will compile correctly, but they will not do what the programmer intended. If the `temperature` is 150, these statements will print out "`Human body temperature exceeded.`" instead of "`Boiling point of water exceeded.`", because the `if` structure executes the first `true` condition and skips the rest. To get proper behavior, the order of these tests should be reversed.

CHAPTER 4

Practice Box, pages 111-112

1. 4 times
2. This is an infinite loop. The values of `j` are 7, 6, 5, 4, 3, 2, 1, 0, -1, ... Since the loop terminates when `j` > 10, the loop will never terminate.
3. 1 time
4. 9 times
5. 7 times
6. 9 times
7. infinite loop
8. `ires` = 10, and the loop executes 10 times
9. `ires` = 55, and the loop executes 10 times
10. `ires` = 15, and the loop executes 5 times
11. `ires` = 15, and the loop executes 5 times
12. `ires` = 15, and the loop executes 5 times
13. `ires` = 18, and the loop executes 6 times
14. `ires` = 3, and the loop executes 3 times
15. `ires` = 25; the outer loop executes 5 times and the inner loop executes 25 times
16. `ires` = 15; the outer loop executes 5 times and the inner loop executes 15 times
17. `ires` = 2; the outer loop executes 1 time and the inner loop executes 3 times

18. `ires` = 10; the outer loop executes 5 times and the inner loop executes 15 times
19. Invalid. Variable `i` is used to control both loops.
20. Invalid. These statements will compile and execute, but they will produce an infinite loop. The semicolon after the `while` statement terminates the `while` loop without changing the value of `x`, so `x` will never be less than or equal to 0.
21. Invalid. The `i--` modifies the value of the loop variable, producing an infinite loop.

CHAPTER 5

Practice Box, page 143

1. An array is a special object containing *(1)* a group of contiguous memory locations that all have the same name and same type, and *(2)* a separate instance variable containing an integer constant equal to the number of elements in the array. An element of an array is addressed by the array name followed by an integer subscript in square brackets ([]). The components of the array are the elements of the array plus the constant containing the length of the array.
2. A reference is a "handle" or "pointer" to an object that permits Java to locate the object in memory when it is needed.
3. An array object is created with the `new` operator. For example:

   ```
   double[] x = new double[5];
   ```

   creates a new five-element `double` array.
4. An array may be initialized by assignment statements, or by the use of an initializer when the array is created. For example,

   ```
   int a[] = {1, 2, 3, 4, 5};
   ```

   creates a new five-element `int` array, and initializes the values of the array elements to 1, 2, 3, 4, and 5.
5. A 100-element array would be addressed with the subscripts 0 to 99. Any other subscripts would produce an `ArrayIndexOutOfBoundsException`.
6. Valid. These statements create a new 10-element `double` array.
7. Invalid. A `double` reference cannot refer to an `int` array.
8. Invalid. An initializer can only be used in an array declaration, not in an assignment statement.
9. Valid. These statements will print out the second through fifth elements in the array. They will *not* print the first element, since the subscript for the first element is 0, and the loop begins at 1.
10. Valid. These statements will print out the first through fifth elements in the array *in reverse order.*

Practice Box, page 167

1. Invalid. These statements will compile successfully, but the will produce an `StringIndexOutOfBoundsException` at runtime, since `s1` is not eight characters long.

2. Valid. The statement "`s3 = s1.substring(1,3);`" selects the characters at indices 1 and 2 of `s1`, which is the string "`bc`". The following statement concatenates the characters "`123`" to it, so the final result is "`bc123`".
3. Valid. The `equals` test will be `false`, because the two `Strings` are not identical. However, the `equalsIgnoreCase` test will be `true`.
4. Valid. The result is `false`, since the two references point to physically different objects, even though the contents of the objects are identical.

CHAPTER 6

Practice Box, page 182

1. Incorrect. The `int` array is the first parameter in `method1`, but the second parameter in the call to method1.
2. Incorrect—`method2` is declared `void` but returns a value.
3. Correct. The `main` method calls `method3` with array `x`, and `method3` sums the values in the elements of array `x`, and divides that result by the number of elements in the array. The `main` method then prints out this result, which will be -0.5.

Practice Box, page 201

1. The duration of a variable is the time during which it exists. The types of duration in Java are automatic duration and static duration.
2. The scope of a variable is the portion of the program from which the variable can be addressed. The types of scope are class scope and block scope. Variables with block scope can be defined for any block size, such as a method body, a `for` loop, etc.
3. Variables defined within a Java method have automatic duration and block scope. The block in which they are visible is the method body, which is delineated by the open and closing braces `{}`.
4. A recursive method is a method that either directly or indirectly calls itself.
5. Method overloading is the process of defining several methods with the same name but different sets of parameters (based on the number, types, and order of the parameters). When an overloaded method is called, the Java compiler selects the proper method by examining the number, type, and order of the calling arguments.
6. This program is incorrect. Variable `i` is redefined within the `while` loop, which willl cause the loop to behave improperly.
7. This program is incorrect. The two overloaded methods `m1` have the same signature, and so cannot be distinguished from each other.

CHAPTER 7

Practice Box, page 229

1. The major components of a class are fields, constructors, methods, and finalizers. Fields define the instance variables that will be created when an object

is instantiated from a class. Constructors are special methods that specify how to initialize the instance variables in an object when it is created. Methods implement the behaviors of a class. A finalizer is a special method that is called just before an object is destroyed to release any resources allocated to the object.

2. The types of member access modifiers are `public`, `private`, `protected`, and package. Private access is normally used for instance variables, and public access is normally used for methods.
3. A variable with class scope is visible anywhere within the class in which it is defined, while a variable with block scope is only visible within the block in which it is defined.
4. If a method contains a local variable with the same name as an instance variable in the method's class, the instance variable will be "hidden", and so will not be directly accessible from the method. However, the method can still access the instance variable using the `this` reference.
5. To use classes in packages other than `java.lang`, you must include an `import` statement for each package at the beginning of the source file. Note that the `import` statements must appear *before* the class definition.
6. To create a user-defined package, include a `package` statement in the source file of each class to go into the package, and compile each class with the "`-d`" option to specify the location of the package directory structure. Include an `import` statement in each class using the package, and be sure to set the `CLASSPATH` environment variable so that the package can be found by the Java compiler.
7. The `CLASSPATH` environment variable tells the Java compiler and the Java run-time system where to look for packages being imported.
8. A variable or method declared with `public` access may be accessed from any class anywhere within a program. A variable or method declared with `private` access may only be accessed from within the class in which it is defined. A variable or method declared with package access may be accessed from within the class in which it is defined, or from any class within the same package. A variable or method declared with `protected` access may be accessed from within the class in which it is defined, from any class within the same package, or from any subclass of the class in which it is defined.

Practice Box, page 238

1. The garbage collector is a low-priority thread that searches for and destroys objects that are no longer needed. It runs automatically in the background while a Java program is executing. A Java object is eligible for garbage collection when no reference to the object exists, because the object can no longer be used once there are no longer any references to it.
2. Static variables are variables that are *shared* by all objects created from the class in which the variables are defined. These variables are automatically created as soon as a class is loaded into memory, and they remain in existence until the progam stops executing. Static variables are useful for keeping track of global information such as the number of objects instantiated from a class, or the number of those objects still surviving at any given time. They are also useful for defining single copies of final variables that will be shared among all objects of the class.

3. Static methods are commonly used to perform calculations that are independent of any instance data that might be defined in a class. The methods in class `java.lang.Math` (`sin`, `cos`, `sqrt`, etc.) are good examples of `static` methods.

CHAPTER 8

Practice Box, page 270

1. A `Container` is a graphical object that can hold `Components` or other `Containers`. The type of container used in this chapter is a `JFrame`.
2. A `Component` is a visual object containing text or graphics, which can respond to keyboard or mouse inputs. The type of component used in this chapter is a `JCanvas`.
3. The basic steps required to display graphics in Java are:
   a. Create the component or components to display.
   b. Create a frame to hold the component, and place the component into the frame.
   c. Create a "listener" object to detect and respond to mouse clicks, and assign the listener to the frame.
4. Java employs a coordinate system whose origin (0,0) is in the upper left hand corner of the screen, with positive x values to the right and positive y values down. The units of the coordinate system are pixels, with 72 pixels to an inch.
5. Method `getSize()` belongs to class `java.awt.Component`. Since this class is a superclass of any component or container, all components and containers include this method. The method is used to return the width and height of a particular component or container in pixels. This information can be used by the component to re-scale itself whenever the size of the window in which it is drawn changes.
6. The style of lines and borders is controlled by class `BasicStroke`.
7. Text is displayed on a graphics device using the `Graphics2D` method `drawstring`. The font in which the text is displayed is specified by creating a new `Font` object, and using the `Graphics2D` method `setFont` to specify the use of that font. Information about a font can be recovered with the `FontMetrics` class.

CHAPTER 9

Practice Box, page 312

1. A `Container` is a graphical object that can hold `Components` or other `Containers`. We are using `JPanels` and `JFrames` in this chapter.
2. A `Component` is a visual object containing text or graphics, which can respond to keyboard or mouse inputs. The types of components used in this chapter are: `JButton`, `JCheckbox`, `JComboBox`, `JLabel`, `JPasswordField`, `JRadioButton`, and `JTextField`.

3. A layout manager controls the location at which components will be placed within a container.
4. Cascaded layout managers permit a program to construct layouts that are more complex than can be accomplished with a single layout manager.
5. An event handler is a special method within a listener class that is called whenever a specific type of event occurs in a GUI component. The listener must first be registered with the GUI component.
6. The listener interface introduced in this chapter is the `ActionListener` interface. The `ActionListener` interface handles action events, which can be produced by all the components that we studied in this chapter, except for `JLabels`.

Index